The Amazing
SECRETS
of
NATURE

The Amazing SECRETS *of* NATURE

Reader's Digest

NEW YORK • SYDNEY • MONTREAL • HONG KONG

THE AMAZING SECRETS OF NATURE

was produced by The Reader's Digest Association

First English language edition 2004

Original edition first published by Sélection du Reader's Digest, SA,
212, Boulevard Saint-Germain, 75007 Paris, 2001

CONSULTANTS
Philippe Dubois, ecological engineer and ornithologist
Jean-Claude Bousquet, doctor of science

WRITERS
Vincent Albouy, entomologist
Jean-Claude Bousquet, doctor of science
Jean-Marie Dautria, professor of geology
Frédéric Denhez, scientific journalist
Philippe Dubois, ecological engineer and ornithologist
Serge Lallemand, director of research at CNRS, specialist in tectonics and subduction
Jacques Malavieille, doctor of science, director of research at CNRS
Philippe Masson, professor at Paris-Sud University
Marc Morell, hydrological engineer
Bernard Pellequer, doctor of astrophysics, director of Géospace
Patrick Piro, journalist, specialist in environmental issues

EDITOR **Sandy Shepherd**
ART DIRECTOR **Peter Bosman**
DESIGN **Mathias Durvie, Colman Cohen**
ILLUSTRATIONS **Christine Adam, Régis Mac, Jean Soutif, Jean-Louis Verdier**
PICTURE RESEARCH **Marie-France Naslednikof, Marie-Christine Petit,
Véronique Masini, Danielle Burnichon**
TRANSLATOR **Dr Jane Robertson**
PROOFREADER **Anne Wevell**
INDEXER **Mary Lennox**

ISBN 962-258-306-7
Book Code : US 3564/G

Printed in China

Contents

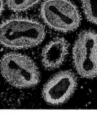

5. CONQUERING THE WORLD

6. THE PRICE OF SURVIVAL

7. NATURE'S THEATRE

8. WATER, WATER EVERYWHERE

14. THE HEALING POWER OF NATURE

15. EAT OR BE EATEN

16. NATURE'S TOWERING RAGE

17. DEALS, ALLIANCES AND SOLIDARITY

18. THE CHAMPIONS OF THE NATURAL WORLD

Chapter
1

BIRTH

The beginning of all things

The hatching of a Nile crocodile, Botswana.

Heaven's nurseries

Enormous hydrogen clouds in space are like hatcheries which, with the help of gravitational forces, create thousands of balls of hydrogen gas, which we know as stars.

In 1995, the Hubble Space Telescope examined a huge mass of gas and dust particles called the Eagle nebula, in the constellation of Serpens, 7 000 light-years from Earth. In the centre of this mass were uneven patches, looking like darker clouds against a lighter background.

Between the bright area, composed of hydrogen, and the remaining dark cloud, called a molecular cloud, were three columns of very dense gas more than one light-year in length.

◄ *The Eagle nebula, which lies in the Serpens constellation, takes its name from the shape of its huge clouds of ionised hydrogen which shelter baby stars.*

A multitude of embryos

These areas of compressed gas are like the nurseries of the heavens, where groups of stars are born. Some are huge (more than eight times the size of our Sun), and others are much smaller.

The young stars emit strong ultraviolet radiation. This radiation, like a sort of solar wind, sweeps away all the dust that makes the surrounding space cloudy. In the area seen in the photograph (left), it has dispersed most of the surrounding matter and left only three gigantic columns. It was here that the Hubble telescope revealed the existence of little globules just a few astronomical units in size, the embryos of future stars.

An astronomical unit represents the distance between the Earth and the Sun, some 149 million kilometres. The size of these globules thus corresponds to little solar systems in the making.

▼ *The Hunting Dogs galaxy, in the Great Bear. Baby stars come into being in the arms of the galaxies.*

▲ *In the galaxies, hydrogen (shown in pink) indicates the presence of gaseous clouds, the nurseries of space.*

Chain reactions

As the pillars of gas evaporate, these star 'eggs', made essentially of hydrogen, become visible. Under gravitational force, like the rest of the Universe, they gradually change. The gas which is at first tenuous becomes progressively more dense. As the density and the temperature increase, matter becomes more compact. These changes are the result of gravitational collapse which can be halted only by the advent of nuclear force.

Indeed, when the temperature reaches between 15 and 20 million degrees Celsius, hydrogen fuses to create helium and it is at this stage that thermonuclear reactions set in. The force of gravity has given rise to another force opposed to it and the star's life is assured by the ongoing struggle between these two forces. If one grows weaker, the other grows stronger.

The drop-outs

But not all these embryo stars will automatically shine. They may stay as little dark spheres. When this happens, it means that there is not enough matter to produce a sufficiently strong gravitational force for thermonuclear reactions to take place.

These stillborn stars, also called brown dwarfs, are of the utmost interest to astronomers. They might be an important part of what is called dark matter, which is highly sought after by those who investigate the Universe because of the information it contains. The question of the expansion of the Universe depends on how much of this invisible matter exists. If there is only a little, the Universe will expand indefinitely. If there is a lot, the Universe will probably contract.

A death is predicted

At the heart of a star, a cauldron is boiling away transforming hundreds of millions of tons of hydrogen into helium every second and giving off a colossal amount of energy.

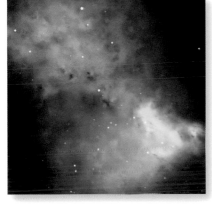

But at the same time as this energy feeds the life of the star, it will be the cause of its death. The smaller the mass of the star, the slower the process of destruction, but the inevitable end awaits. If the star is a large one, death may come in the form of a supernova explosion, certain elements of which may perhaps generate the birth of other stars in other molecular clouds. If the star is larger still, it may collapse until not even light can escape from it, and it forms a 'black hole'.

Alternatively, the dying star may swell to form a 'red giant' and throw off its outer layers of gas which will form a planetary nebula around it. Finally, all that will remain will be the dead body of the star, a 'white dwarf', thousands of which litter our Universe. But that is another story.

▲ *A star is born, only to die one day. The future of our Sun will be like that of Dumbell, this planetary nebula.*

THE BIRTH OF THE SOLAR SYSTEM

The life of our Sun, like that of any other star, is determined by the balance of two antagonistic forces – gravitational and nuclear. Thus, 4.7 billion years ago in the outskirts of the Milky Way galaxy, a cloud of interstellar matter condensed. At the heart of the system, the density was so great that the temperature reached between 15 and 20 million degrees Celsius. The thermonuclear fusion of hydrogen into helium gave off an enormous amount of energy, which acted against the process of collapse, and our Sun was created. As a result of the rotation of this central star, the peripheral gas flattened out and took the shape of a disc, in which particles of matter gathered to form planetesimals, minute planetary bodies which accreted to form planetoids (larger planetary bodies), which in turn formed the planets. Over a period of 30 million years our Solar System was born.

A continent breaks up, an ocean is born

Twenty million years ago, Arabia began to break away from the African continent. The two land masses, which since then have continued to separate by a further centimetre a year, left a space for the Red Sea to form.

▲ *The Sinai Peninsula, with the Gulf of Suez to the west (separating it from the African continent), the Gulf of Aqaba to the east (which separates it from the Arabian Peninsula), and the Red Sea to the south.*

In what is now East Africa, everywhere was stable 20 million years ago. Season followed season. Rain and storms followed the stillness of a blue sky. And then things began to change dramatically. As the earth shook with great tremors, gigantic, sheer cliffs rose up out of the ground.

The surface split open and created a long ravine, with lakes forming along its edges. Volcanoes swelled up, spitting ash and spewing lava. The evening skies were lit with red. Something was going on underground.

From the depths of the earth, a solid spring wells up

A slow but fearsome process was indeed under way. Something resembling a 'solid spring' of liquid rock of a very high temperature was welling up under the land. From the depths of the mantle, the hottest rock rose at the rate of a few centimetres a year.

The scale of the area affected was gigantic. It began from what is today the Gulf of Suez and extended over almost 2 000 kilometres, as far as what is now Djibouti. 'Wedges' of hot mantle broke through the base of the continental crust and cracked it, while on the surface the distended crust tore open and a multitude of volcanoes erupted. In fact, as the solid mantle rose slowly upwards, part of it (five percent) melted, producing liquid basalt which leaked through the fractures and spread over the surface of the crust.

An ocean evolves

About five million years ago, the Arabian Peninsula began to drift away from Africa. Water flowed into the gap, and the Red Sea was born. As the drift continued, the crust stretched so thin that the molten rock of the mantle itself seeped through it, coming up under the sea. Today the Red Sea is no longer just a sea, but a young ocean. And in the middle it has a ridge through which lava exudes, continuously laying down a new ocean floor. No one can say for how long this split will continue to widen, or how big the ocean born from it will eventually be.

▲ **The birth of the Red Sea**
1. *Africa*
2. *Red Sea*
3. *Asia*
4. *Continental crust*
5. *Mantle*
6. *Central region where the hottest mantle rocks rise to the surface*

THE OPEN BOOK OF ZABARGAD

The rocks of Zabargad, a Red Sea island, are a virtual archive of the area's geology. They reveal gneiss from the Afro-Arabian continent, cooled down 400 million years ago before the split. The first cracks in the crust allowed peridotite (the greenish rock shown right) in the mantle to rise. It is made largely of olivine, which contains precious stones sought after since ancient times. Conglomerates are also found, showing the sediment that accumulated in the cracks caused by the separation. And finally, on the upper layers of the ocean floor, they contain basalt, cooled lava.

From the bowels of the earth

The birth of a volcano is a spectacular event. Today the warning signs can be detected, measured and interpreted.

The steeple of a village church rises in Mexico, not above houses but from the field of lava surrounding Paricutín, a volcano. The story has become legend. On February 20, 1943, it all happened very fast. A farmer was shocked to see the earth he was ploughing first shake, then crack, and puffs of steam come out of his furrows. In no time at all, his village was destroyed by an earthquake. Nothing was left but ruins, and his field had become an immense hole spewing ash, steam and incandescent rocks. By the next day, it had been filled in by a volcanic cone eight metres high. Paricutín had been born.

A year later, the cone was 450 metres high. The first eruptions had started and the volcanic activity was to continue for nine years. Several streams of lava escaped from this new volcano, one of which literally buried the village.

It is quite exceptional to witness the birth of a volcano. The new-born always appears in the company of older neighbours, which may be active or dormant, in zones known for their volcanic activity. These sites are determined by the geometry of the plates that make up the earth's crust and the underground forces governing their movements.

Live from Surtsey

When it first erupted in 1963, the volcanic island of Surtsey, in Iceland, looked like a long ripple with a gaping fissure running along its length. The sea flowed into the fissure, mixed with the magma and formed gigantic plumes of water vapour shaped like cypress trees or cocks' tails, which in places rose eight kilometres into the air. When the cone was high enough to stop the seawater getting in, the activity abated and mainly took the form of eruptions. Several streams of lava flowed down the slopes of the young volcano and into the sea, throwing up great curtains of steam. The active period of Surtsey was to last almost three years and spread to the north-east.

Warning signs

Every eruption is preceded by a multitude of warning signs. Some indications are clearly perceptible, like the rising and bulging of the ground, the appearance of smoke and gases, and a rapid succession of slight tremors, which often precede by several weeks the arrival of the magma at the surface. Other signs can be discerned only by using appropriate instruments to determine a rise in soil temperatures, localised changes in the gravitational and magnetic fields and a change in the composition of smoke and gas emissions (for instance, an increase in sulphur and carbon dioxide).

Today, automatic measuring instruments are permanently set up on the slopes of volcano-laboratories, like those in Hawaii and Réunion, and on volcanoes that may become active again. Measurements are taken continuously and immediately sent by satellite to centralised laboratories. By means of remote infrared detection, for example, a thermal image of the whole of a volcano is always available.

◀ *The ruins of the church of San Juan emerging from a field of lava are the only remains of a village, buried as Paricutín volcano was born.*

▼ *Live birth. The emergence of the volcano Surtsey in 1963, amid a plume of steam and pieces of vitrified lava.*

▼ *Surtsey 30 years on: an island of little over 2 km² in size, whose scoria cones are surrounded by an unbroken ring of basaltic lava flow.*

When travel is a means of survival

Plants must travel in order that the species may continue. But only the seeds need to be mobile. Air, water or animals are their means of transport.

A coconut palm bending towards the sea on a soft sandy beach makes a very attractive subject for a holiday snapshot. But can you tell if it was taken in the Antilles, Africa, the Seychelles, Indonesia or Polynesia?

It may be difficult to say, because this tree, the archetypal symbol of a relaxing holiday in the sun, is found on all beaches in tropical regions. So just how has this plant been able to travel all around the world?

Sailors

For the coconut palm, the answer is seawater. Currents are a practical means of dispersal for shoreline vegetation. The thick fibrous shell of the coconut makes an excellent float, highly resistant to the onslaught of the waves.

The nuts drift for several days, months or even years. Many are lost on the way. But some finally land on a welcoming beach and set root there.

The alder, a tree which grows on water-logged land, entrusts its seeds to the river.

The seeds float because of their high oil content and can still germinate after a year in the water. But rivers have a disadvantage: they always flow in the same direction. So the seeds of the alder must also be dispersed by the wind in order to conquer new territory upstream.

Aeronauts

The wind, which blows everywhere, is the chief means of seed dispersal. Plants have invented a great variety of methods to take advantage of it. One of the most efficient and economical is the parachute. The parachute of the dandelion, made of a sprig of fine hairs above the seed, is a model of its kind. When the seed is ripe and dry, the sprig unfolds and at the first breath of wind the parachute takes off. The seed, being heavier, lands first and so germinates more easily.

This dispersal method is very common, and more or less complex according to the species. Sometimes the parachute consists merely of a tuft of hairs, like cotton – which we collect to make cloth.

▼ *Washed up on a sandy beach, this coconut is beginning to germinate, using the stores of energy it already has on board.*

▲ *The prickly fruit of the burdock often sticks to the coats of animals which brush against it and thereby unwittingly provide a taxi service.*

Another means is the propeller. In plants which use this structure, the heavy seed has a broad appendage on top, which is slightly curved. When it falls, it spins so that it is slowed down. If the wind blows, it can be carried a long way. It is because the seeds of maple trees have propellers that they can colonise new territories and have a wide distribution throughout the world. Other seeds are able to use flat wings, which catch the wind easily enough for them to be carried a long way off.

Hitch-hikers

Plants are anchored in the ground, but the mammals that live alongside them move about easily, sometimes covering a great distance. So why not make use of this passing transport? The technique is simple: hang onto their coats and then drop off again some time later, maybe onto a fertile spot.

Take the case of the burdock, found everywhere on fallow land in Europe and North America, and a vegetable grown for its root in Japan. The fruit containing the seeds is a fat capsule bristling with hooks that cling onto the wool or fur of mammals as they pass by. The capsule is open at the top, so the seeds are dispersed gradually. In addition, irritating hairs among the seeds cause itching, which means that the animal scratches itself and the seeds fall to the ground at random intervals.

Many kinds of seed – like those of the walnut, hazelnut and grasses – are dispersed by animals that carry them away to eat. Of the seeds taken by rodents or ants, some are dropped along the way, and those that are stored sometimes remain forgotten in their hiding place, their owners unable to find the way back, or having themselves become food.

Ingested passengers

The smooth surface of feathers makes it difficult to hitch a surreptitious lift on them. However, birds offer an unparalleled opportunity for seed dispersal because they fly over great distances. So plants have had to make an effort and pay for their services. Countless numbers of berries and fleshy fruits which occur everywhere, from the

▲ *Fleshy berries (like those of this climbing bramble) provide a feast for many animals, particularly birds. The seeds of the berries remain undigested and are dispersed by way of the droppings, sometimes over a wide area.*

◀▲ *Dandelion. The parachute (left) takes off in the slightest breeze. But it is sometimes an insect (above) which carries the seed away.*

◀ *Cotton. The fibres, used by people to make cloth, are produced for efficient seed dispersal by wind.*

◀ *It is not only seeds that can travel. The pollen of this flower (Nanocnide japonica) is ejected into the air.*

arctic tundra to equatorial forests, are a kind of payment to birds and other animals for carrying the seeds and later releasing them again in their droppings in the course of eating the fruit.

In the South American rainforest, the quetzal eats the fruit of wild avocados. Not only does the bird eat the oily flesh, which is a very rich food, but it also eats the large central seed. The next day it excretes the seed which drops to the ground and can then grow.

The avocado tree and the quetzal are dependent on each other for their survival. Without the quetzals, the avocados would not be able to disperse their seeds. Without avocado seeds scattered by earlier generations of quetzals, these birds would have no food.

Budding engineers

The ecballium, or squirting cucumber, has developed a type of pressurised syringe for sending out its seeds. This plant, belonging to the gourd family, with a large yellow flower and found on fallow land in the Mediterranean basin, has fleshy, hairy fruit the size and shape of a large gherkin. As the fruit ripens, it fills with liquid. When it reaches maturity, the seeds are floating in a medium which is under considerable pressure. If the fruit receives a very slight knock, it falls off, and the liquid and seeds spurt out of the end like the cork from a bottle of champagne. The fruit remains unbroken, but empty.

If a plant relies solely on its own resources, it can spread over only a few metres, but this is often enough for herbaceous plants. The techniques they use are very varied and quite comparable with the most ingenious machines invented by human beings.

The catapult method for seeds, for instance, is based on distortion of the plant tissue, which contracts as it dries out and then is released suddenly when the seeds become loose. Plants of the *Impatiens* genus, some species of which are commonly grown in gardens, owe their name to their ability to eject their seeds with a cracking sound: the mechanism goes off as soon as the ripe fruits are touched.

Time travellers

The corn-cockle produces fat, round seeds which fall to the ground at the base of the plant. No provision has been made for their dispersal. So what remedy is there for this lack of mobility? Time travel.

Once they have fallen to the ground, these seeds become dormant and wait until there is less competition from other species before they germinate, perhaps after a flood or after ploughing. After several years or even several decades, the seed will germinate in a very different environment from the one that existed when it fell to the ground.

Even the most arid sand may contain dormant seeds, which is why a desert blooms in the days following a storm. Its plants grow, flower, bear fruit and die within the course of a few weeks. The seeds produced will wait patiently for the next rains.

▲ *The dry fruit of the balsam (*Impatiens balsamina) *releases its tension at the slightest disturbance and expels its seeds with a little cracking sound.*

IT'S ALL IN THE SEED

A plant produces seeds in order to reproduce, for the survival of the species. All seeds, even those of the orchid, which are as fine as dust particles, contain an embryo capable of creating a whole new plant. In the cross-section of the castor oil seed on the right, you can see within the integument (1), the embryos of the two future leaves, side by side (2), and the future stem and root (3). A seed also contains a food supply to help the young plant develop (4). This store of food is much greater in an acorn than in an orchid, but the principle remains the same.

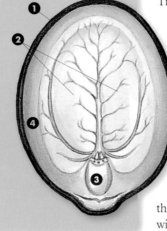

Born perfectly formed

Mammals, a few reptiles and amphibians, certain invertebrates and a number of sharks all give birth to live young. After a long gestation period, these young are born with a good chance of survival.

When a giraffe-to-be sends a hormonal message to its mother telling her it is time she gave birth, little does it know what lies ahead. When she does deliver, the birth takes place two metres above the ground. The baby giraffe falls with a thud, and after about 20 minutes it stands up. After an hour it is already suckling. Its birth weight is between 45 and 70 kilograms and it is almost 1.9 metres tall.

A speedy birth to avoid a death

Giving birth is an extremely complex process. Once contractions of the uterus have begun, the time factor is crucial. The foetus's blood supply is cut off from the mother's. Its lungs begin to work and must very soon be filled with air. Its digestive system prepares to receive the nourishment that until now came into its bloodstream by way of the umbilical cord. This is why the young of land animals are born head first: in order to breathe. On the other hand, newly born dolphins enter this world tail first, as the tail helps them out. In any case, with a tolerance for apnoea, cetaceans are at less risk of asphyxiation than mammals.

Almost independent

Like most other ungulate mammals, giraffes give birth to their young in full view of their predators. For their own safety, therefore, baby giraffes have to be fairly independent right from birth. After a gestation period of 13 to 16 months, they enter the world as miniature adults, their weight already more than 10 percent of their mother's. Well

developed and very soon mobile, they are not completely helpless in the face of predators. It is the same in the case of marine mammals and hares.

The elephant calf, after the longest gestation period in the animal world – more than 22 months – falls out onto the ground. It soon stands up, in spite of weighing 100 kilograms, and goes to its mother which suckles the calf for a good six years, in some cases. It is cared for by a group of female elephants and is never left alone. The size of its elders deters predators.

Animals which have holes or burrows, on the other hand, and are born under cover, like rabbits, come into the world naked, almost blind, weak and very small. A newly born bear cub weighs about 500 grams, nearly a thousand times less than its mother does.

The luxury of being viviparous

Many sharks also give birth to live young. Female lemon sharks, for instance, always find their way to the same lagoons to have their young. Unlike mammals, whose babies are only semi-independent, young sharks very soon begin to hunt for food.

In fact, viviparity is relatively common among what are called the higher animals, those that have an elaborate physiology and bonding mechanism, and are capable of adapting to sudden changes in their environment. Most of them experience more or less stable living conditions and have to face few dangers during their lifetime. So they are able to devote a lot of energy to a few offspring that will have the best chances of survival when they are born. This applies even to the minute daphnia, or water flea, which is an arthropod.

▶ *Unlike newborn mammals, baby sharks are immediately left to their own devices.*

▲ *For giraffes, birth is a somewhat brutal experience, but recovery has to be as rapid as possible. If the baby giraffe manages to keep close to its mother, the risk of being eaten by a predator will be only about 10%.*

The Birth of the Universe

Imagine all matter as one single mass, all light concentrated into one source. Imagine a world with more than the three dimensions our surroundings are limited to. Before the Big Bang perhaps the Universe was like this, but perhaps matter and light did not even exist. The Universe as we understand it was born from a single initial point, began expanding, and slowly took shape. Over the unimaginable period of 15 billion years it gave rise to vast fields of galaxies, populated with hundreds of billions of stars and planets. Since the beginning of the 20th century, scientists have been piecing together the history of these 15 billion years. But it is a huge task. And from the first observations of Edwin Hubble to the most recent space missions, we have only just begun to understand our Universe.

REVEALING ANOMALIES

1924. The American astronomer, Edwin Hubble, discovered a surprising feature. Analysing the light coming from stars, he noticed the colour of the light shifting towards the red end of the light spectrum. This indicated that the wavelengths of the light were getting longer, as if the light were being emitted from an object moving away from the point of observation. Hubble realised that there was a connection between this discrepancy and the speed at which the galaxies were moving away from the Sun. In fact, it is not really that the galaxies are moving farther away, but that the distances between them are becoming greater. In other words, the Universe is expanding.

1965. The Americans Penzias and Wilson discovered microwave radio signals in space that were continuous: wherever they were picked up they were equally strong. They could not be coming from a single object, because they were the same everywhere. It was a discovery of great importance. It demonstrated the existence of a cosmic microwave background, left over from the time when light first came into existence in the early Universe. Today the fossilised remains of this radiation can be detected (as depicted in the colour-coded image opposite taken by the Cobe satellite in 1993, showing the portion of sky observable from the Earth).

AND THEN THERE WAS LIGHT

At the beginning of time there was an instant of great heat and density called the Big Bang, see opposite (1). Particles in constant interaction with each other formed a medium containing quarks, photons, neutrinos and electrons (2). During the first three minutes following the Big Bang, these particles combined permanently, and from the quarks, protons and neutrons were born (3).

When the temperature began to fall, these protons and neutrons combined to form the first nuclei (4). Atoms of the lightest elements, particularly hydrogen and helium, were synthesised.

Thus were laid the first building blocks of the Universe. The density of this whole medium was so great that gigantic numbers of photons were blocked by electrons and unable to disperse, so the Universe was opaque and completely without light.

Three hundred thousand years later, matter became electrically neutral, its density decreased and the photons were able to move about and begin to propagate: light appeared. These same photons, spread thinly throughout expanding space, today make up the cosmic microwave background, also known as 'fossil' radiation.

IN SEARCH OF LOST MATTER

After the Big Bang, expansion continued for a billion years. With the separation of matter from energy, the Universe gradually built itself up from hydrogen (90 percent of matter) and helium (10 percent) **(5)**. Over time, slightly less dense regions appeared. These gave birth to the precursors of the galaxies **(6)**. Under the influence of gravitational forces, the construction of matter was completed. This phenomenon put a brake on the expansion of the Universe.

But to have brought about the Universe as it is today **(7)**, there must have been a lot more matter than we can account for. We cannot detect it, so we call it dark matter. But how much of it exists? And what is it made of? There are numerous theories, including those that suggest the existence of structures called antiparticles, neutrinos and neutralinos, among other things.

Perhaps the dark matter hides structures with the amusing acronym of WIMPs (Weakly Interacting Massive Particles), or maybe MACHOs (Massive Astronomical Compact Halo Objects), or even 'aborted' or dead stars. The answer may emerge from a combined study of the infinitesimally small and the immensely great.

BIG BANG, BIG CRUNCH; BIG BANG, BIG CRUNCH...

Perhaps the Universe has a rhythm like a gigantic pulse. Or perhaps it is still expanding, and always will be. It all depends on the density of matter. Too great a density would mean too much gravity. And too much gravity would mean a reversal of the expansion: galaxies attracted by gravitational forces would collide. This is the theory of the Big Crunch. But if the density had less than a certain critical value, the Universe would continue its expansion until all the galaxies lost contact with each other and the stars died out. Certain scientists hypothesise that the Universe pulsates, alternately undergoing expansion (Big Bang) and contraction (Big Crunch). Did our Big Bang follow a Big Crunch? The answer to this question is still a long way off.

THE PORTION OF SKY VISIBLE FROM THE EARTH

The delicate art of birth

Fish, amphibians, reptiles and birds must break out of their shells to enter the world. It is not always an easy task, and is one which sometimes requires the parents' help.

▶ *Birth of a Nile crocodile. The female, alerted by the cries of its egg-bound offspring, frequently helps the baby to break the shell by rolling it between her tongue and the roof of her mouth.*

The female Nile crocodile seems edgy. She is pacing up and down around a sort of mound in the sand. The time is near. In a few minutes her young will start to come into the world, but not, of course, all at once: the rite of passage into life will take several hours yet.

Soft eggs easily torn open

After mating, the female crocodile had dug a hole with her hind feet. She then laid her eggs, between 15 and 80 of them, in this nest. In certain species the female adds plant matter, which decomposes, contributing to the warmth necessary for incubation, which lasts from two to three months. The extra heat is needed because the mother does not sit on the eggs. Instead, she guards her nest jealously, and does not hesitate to attack any curious intruder.

Suddenly, the pacing mother goes up to one of the eggs, hearing little squeaks and grunts from inside the shell. She busily scratches aside the fine layer of sand to free the egg. If the youngster does not manage to get out of the shell on its own, its mother will pick up the egg gently with

her teeth and roll it between her tongue and the roof of her mouth until the shell tears apart, which it will do because it is fairly soft. She will repeat this performance with the whole clutch of eggs, though not all the eggs will hatch.

Then, with the same delicate touch, she gathers the tiny crocodiles one by one with her teeth, collecting them in a pouch in her mouth, and transports them to the nearby water for their first bath and a meal.

A floating rosary

In the spring, the common toad lays its eggs in a pond. These are not single eggs, but thousands joined in a long string which floats on the water, sometimes getting caught up in the pond weed. Each egg is made up of an embryo wrapped in a sticky jelly, which binds them together like a string of beads. If they are not eaten by a hungry fish, the beads will gradually lose their jelly, which the embryos feed on in order to develop into a mass of tadpoles. But many of them will fall victim to the appetites of other pond-dwellers and not reach adulthood. On a country walk in spring, you might come across several puddles of water with bunches of jelly eggs on them, each with a tiny dark dot in the middle marking the embryo of a future toad.

▲ *The common toad. Each year, this amphibian lays thousands of eggs in ponds. But few are destined to hatch: most become fish food.*

CENTRAL HEATING FOR CROCODILES

This species of South American cayman lays its eggs in the heart of the tropical forest. But the sun rarely reaches in as far as the undergrowth, so these creatures need another heat source to incubate their eggs. The female makes her nest near a termite mound, which produces sufficient heat for the eggs to hatch. However, the termites soon incorporate the nest into their mound, so that the female has to knock down part of it before she can break the eggs to free the babies.

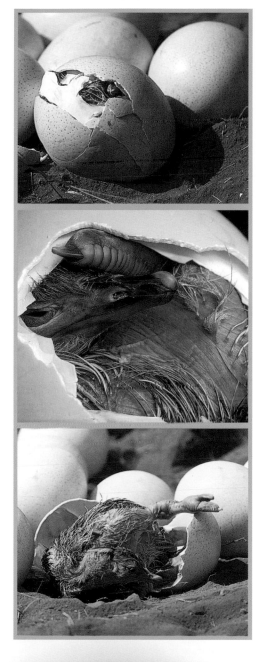

◀ *The birth of an ostrich. After an incubation period of six weeks, the baby tries to force its way out of the shell. Before hatching, it utters little cries which alert the parent birds. It has a hard struggle to free itself from its calcareous straitjacket. However, the story of its hatching almost always ends happily.*

Tools for the job

Everyone knows that birds lay eggs, which develop a hard, calcareous shell. But how can a chick, such a fragile little thing curled into a tight ball, break out of this straitjacket into its brand new world?

Like crocodiles, to which they are relatively close on an evolutionary level, chicks utter little cries before they are born, so the parents (usually the female) hear them and fuss over the eggs. But most species do not go as far as to break open the shells. For chicks are provided with a very useful tool for getting out of the egg: an egg tooth. In fact it is really a triangular projection at the end of the upper jaw, which disappears soon after birth. Of course, if you are a baby ostrich you need a really hefty peck to break the eggshell, which is particularly thick in this species.

The babies of the Australian giant cuttlefish have a strange way of breaking the eggshell. After no fewer than four months spent inside the gelatinous egg, the young cuttlefish frees itself using a sort of brush at the end of its tail which works like a file and gradually wears away the shell. Like a chick's egg tooth this little brush also disappears soon after birth.

Baby sharks in capsules

The dogfish is a small oviparous shark with spiny dorsal fins, and is named after its habit of travelling and hunting in packs. It is well known off shores around the world. The female dogfish lays her eggs encased in a soft, rectangular envelope with long curling filaments attached to the ends. These allow the egg to attach itself to seaweed while the embryo develops inside its protective casing. As soon as the baby dogfish is born – it extricates itself with its teeth – the capsule floats away and often ends up on the tide line on beaches. The cases are frequently found by people walking along the beach, looking like large, dark-coloured plant pods, with the remains of the spiral filaments forming four points at the ends.

▼ *A dogfish egg case attached to seaweed. Rich in nutrients, it allows the embryo within to develop over a period of several months. This long gestation period is characteristic of oviparous sharks and is unique among fish.*

▼ *A newly hatched ostrich chick. Wet and exhausted, the bird needs several minutes to recover.*

BRINGING UP CHILDREN IN THE ANIMAL KINGDOM

Crèches and baby-sitters

Bringing up the young is not exactly restful. It is a busy time, fraught with many dangers. So it seems a good idea for animals to find clever ways to make this early stage of life go as smoothly as possible.

It can be heard a long time before it is seen – a large colony of pink flamingoes containing several thousand adult birds. By July, they have finished breeding and the young have left the nest. But where do they go? Only the adults are visible, their pink uniforms standing out against the white and blue of the salt marshes.

It is not that breeding has been unsuccessful. Over on the other side of the marsh there is a dense group of flamingoes with brown plumage. These are the young ones. Hundreds of them are paddling in the shallow water under the watchful eye of a few adults.

The pink crèche

The pink flamingo, like certain other birds, has a 'crèche' system. As soon as the young are old enough to move about, they assemble under the leadership of a few adult birds. But this does not mean that their parents have abandoned them. They keep coming to feed their chicks for as long as they are still dependent, and the extraordinary thing is they recognise their own offspring just by their cheeping. Each

one must therefore have its own call. When you think that in certain crèches of the dwarf flamingo, which lives in tropical Africa, the number of young can reach 30 000, you can imagine what an incredibly good 'ear' the flamingoes must have.

As the young flamingoes grow up, the bonds between them and their parents gradually weaken

A FISHY ROLE-REVERSAL

Numerous species of apogon (a type of fish which lives mainly in tropical waters) practise buccal incubation, which means that the female lays her eggs in the male's mouth. The male keeps them safely for the whole incubation period, generally about a fortnight, although the time varies according to the species. Some observers say that sometimes the young fish take shelter in their father's mouth when danger threatens. Scientists, however, have their doubts about such behaviour.

as they become able to find their own food and leave the crèche. But throughout the whole period of communal life, the young have been well protected against predators. It is much more difficult to seize a young bird in such a crowd than it is to catch one that has wandered away from its baby-sitters.

Other species of bird act the same way, such as the penguins of Antarctica and terns, especially the Sandwich tern. In these birds, too, the adults find their offspring by the cries they make. In the case of the shelduck, several families band together in the same way as flamingoes do, and two adults, for instance, can be seen followed by some 40 chicks – a suspiciously large family!

The custom of grouping the young in a crèche is also found in certain bats and even among the larger sea mammals, notably humpback whales.

Baby-sitting
Bird families have developed a really effective mutual assistance scheme. Among the Florida shrub jays and the European bee-eaters, the young born the previous year often help their parents to feed the new nestlings, sometimes providing up to 30 percent of their food. Some scientists think this behaviour trains them for their future role as parents.

Among common moorhens, even the young from the first clutch of eggs laid in spring sometimes help the parents with the second brood, which hatches a few weeks later.

Protectors of all kinds
Sociable species which live in sometimes highly complex societies have numerous ways of helping and protecting each other when bringing up their young. Among lions, for example, several females may live together, and a lioness may easily suckle a cub other than her own.

Among several species of bat, the females live apart from the males. If a female goes off to hunt and leaves her offspring behind, it can happen that when she returns she suckles the first baby she encounters. Mostly, however, in spite of the large numbers, she can locate her offspring using ultrasound even amid the clamour and darkness of a cave.

In the same way, among wolves the young may be entrusted to a 'protector', male or female, when the mother goes off to hunt.

In a herd of elephants, there are no males, except young ones. When they grow up, males live alone. The herd, under the leadership of an old female, contains only sisters, daughters and granddaughters. When there is a birth, the females all stand round the baby and trumpet. Later on, they take as much trouble over the calf as its own mother does, helping the elephant calf to cope with obstacles and tending to it with great care.

▲ *A line of young blue tits clamours to be fed by the adult bird. These well-known little woodland birds are found in Europe, North Africa and the Middle East.*

◀▲ *Elephants and lions are two entirely different species of mammal, but among both the females unite to bring up the young and will look after offspring other than their own.*

▼ *When it is still very young, a baby whale swims right up against its mother as it travels. This is a way of protecting itself against a possible attack by an orca.*

◀ *For a few weeks, this young flamingo will be fed by its parents. Then it will go to the 'crèche' where it will be among chicks its own age.*

BIGGER THAN ITS PARENTS

The young of certain birds, like penguins and the northern gannet, are fed by their parents to the point where, just before they leave the nest, they are heavier than the adults. The reason for this is that once they leave the nest, the youngsters will have to find their own food, and for the first week they will not be able to feed themselves properly (this is especially true of gannets). So their extra weight is a means of survival, giving them energy reserves they can live off for a while.

It's time to eat

Among certain animals, the young are fed by their parents until they are able to feed themselves. Other newly born animals, on the other hand, are independent right from birth. Among mammals, suckling is another stage between birth and the emancipation of the young.

It is a May day on the Californian coast. The great reserve bordering the Pacific resounds with thousands of cries. Adult birds of numerous species, and their young, are busy with an activity of prime importance to the chicks: feeding.

In one area, hairy young brown pelicans stretch out their open beaks towards an adult bird that is regurgitating the catch it has stored in its pouch, under the lower jaw. The chicks, a few days old, their bodies covered with a thin fuzz, remain huddled in the nest on the ground for most of the day.

▲ *This adult grey pelican fills up with fish in order to feed its offspring still in the nest.*

In another area, a family of avocets, small waders with long upturned beaks, stride calmly along in the shallow water of a salt pan. The downy chicks leave the nest at birth and paddle briskly along behind their parents, finding their own food. At first things are not easy, and their beaks usually dip fruitlessly into the water. The parents often help by calling to show them places where food is plentiful.

Nidicolous and nidifugous

Pelicans are nidicolous, like many other species of birds. This means the chicks are born blind and naked and grow up in the nest where the parents feed them until they can see and fly, and when their plumage has grown.

Avocets, on the other hand, are nidifugous. As soon as they hatch, the chicks are physically developed enough to leave the nest and feed themselves.

Among the nidicolous birds, the development of the embryo is not complete at birth, whereas in nidifugous species development is complete at birth, or virtually so.

Suckling: a matter for mammals

The distinguishing feature of mammals is that the females suckle their young. Most of the time, the udders or teats with which they suckle are chiefly made of adipose tissue. However, as soon as there is an embryo in the uterus, the sex hormones, oestrogen and progesterone, cause mammary glands and channels for excretion to be formed in the teats. The glands are constructed like a bunch of grapes. There are several of them, each made of cells that manufacture milk which flows into a series of ducts and from there into the channels that end in the nipples.

The mammary glands begin to work only after the young are born. The amount of oestrogen and progesterone in the mother's body drops after birth and there is an abrupt increase in the amount of prolactin, which results in the secretion of the precious milk. Suckling further stimulates the production of prolactin, as well as the synthesis of another hormone, oxytocin, which has the function of helping the mammary muscles to contract, which is necessary to expel the milk.

▲ *After they are weaned, wolf cubs receive all their food from the adults. These two have brought the cubs whole pieces of meat that have been softened a little so the youngsters will be able to chew the meat themselves.*

▼ *Young ferrets are born blind and naked. They are suckled for several weeks before they start to discover the world.*

A FLY THAT SUCKLES ITS YOUNG

The ill-reputed tsetse fly gives birth to one baby at a time. But this only child is cosseted. The female carries it in a sort of pouch and feeds it with a liquid that she secretes, which comes out of her abdomen. The larva breathes through two tubes which exit near the cloaca of the mother and periodically moults in the pouch. When it is big enough, it comes out to become a pupa – the immobile stage – from which it will emerge as an adult fly.

▼▶ The embryo of the Australian echidna, or spiny anteater (below), does not finish growing in its mother's body. After hatching from an egg, which the mother rolls from her cloaca to an incubation groove on her abdomen, the hatchling (right) attaches itself to the fur where it licks milk which seeps from a mammary gland, and continues to grow.

When the female eventually stops lactating, the manufacture of milk also ceases, and the udder or teat returns to its original state.

The young of marine mammals have no lips and therefore cannot suckle. But the mother has very strong muscles round the teats, and as they contract, they send a strong spurt of milk straight into the baby's mouth. This avoids the nutritious liquid being diluted with seawater.

The variable composition of milk

Highly nutritious, milk is made of water, proteins (including the easily digestible casein), sugars and lipids, or fats. The proportion of the ingredients – the sugars generally vary in inverse proportion to the lipids and proteins – depends on the living conditions and the needs of the baby. The milk of animals living in dry environments, like desert or savanna, has a much higher proportion of water in it than the milk of mammals which live in the meadows of

western Europe. The young of marine mammals and animals in cold regions receive milk that is particularly rich in lipids.

Fats are, in fact, the best fuel for their interior heating system. In the sea and in the polar regions, young animals must grow very quickly so that they do not succumb to the harsh temperatures of their environment. Their energy requirements are enormous, and can be met only by a diet very high in fats.

But milk is not everything. Once weaned, baby mammals have to feed themselves. Herbivores do this unaided, because it does not require great skill. They just have to bend their heads and graze. But for predators, it is a more complicated business learning to obtain food. First the parents feed their young by regurgitating part of what they have caught, chewed and half digested. Then the hunting lessons begin.

Pigeon milk

The chicks of doves and pigeons could not survive without what is called 'pigeon milk'. This substance is a whitish, liquid paste rather like cottage cheese, produced in the adult bird's crop by the hormone prolactin. The cells of the crop walls, which become engorged with this secretion that is really very like the milk of mammals, are shed and eaten by the young, which peck at and remove this food from deep inside the beak of the adult bird.

Eaten alive

Other species have other customs. Ichneumon flies lay their eggs inside the caterpillars of butterflies. The larvae grow by eating their host from the inside, while keeping it alive. They pass through all the moulting stages inside, until the last, when they turn into a pupa at the same time as the caterpillar turns into a chrysalis (which will not survive) and emerge as adults. Some ichneumon flies are parasites on flies of the same kind, which in turn are parasites of caterpillars.

AGEING AND DEATH

Everything changes, everything comes to an end

*A Jeffrey pine
(Yosemite Valley, California)*

Changing skins

Moulting, not to be confused with metamorphosis, is a periodic or regular phenomenon which affects the hair, fur, feathers, skin or shell of animals.

It is the end of summer in Greenland. The wheatear has finished raising its young. It must now undertake a long migration that will carry it as far as tropical Africa, where it will spend the winter. But there is no question of its leaving yet, with its flight feathers worn out by sitting on the nest and by the perpetual daylight of the far north. The 'postnuptial' moult happens just when it is needed.

Flight feathers and festive feathers

Not all migratory birds necessarily moult before their long journey. Some species, like the Sabine's gull, which breeds in the Arctic and then migrates south below the equator, prefer to count on their stores of fat and moult later, once they have reached their winter quarters off the coasts of Africa and South America.

Of all the groups of animals, birds have the most varied forms of moulting. With constant use the feathers become worn, and therefore have to be replaced regularly.

A bird moults for the first time at a young age when its down changes to proper feathers. But once it reaches adulthood, it may moult once or twice a year (sometimes three times, or even four, as in the grouse). During the moult, many species of birds do not renew the whole of their plumage, but only part of it, which means that they can continue to fly.

However, in some ducks and geese, and also in cranes and certain rails, all the flight feathers fall off at the time of the post-breeding moult, so that for a certain period the birds cannot fly and are very vulnerable to predators. To protect themselves, they gather on inaccessible stretches of water or impenetrable marshland.

In other ducks, like the mallard, pintail and shoveler, and also the goldeneye, the red-breasted merganser and the magnificent mandarin duck, winter is the time for the males to assume their best wedding finery. For each male is already seeking his mate, and a richly coloured plumage obviously helps to win her over. The females, on the other hand, take on dull colours so that later they can sit on the nest unnoticed, in relative safety. At the end of winter, the males strut about to show off their plumage. When this period is over, they go through another moult and their plumage becomes as dull as that of the females. It is their turn to become invisible.

A new invisibility cloak

Moulting is also the chance to change colour completely. Certain species living in the mountains or the tundra assume a spotlessly white costume in autumn so that by winter they can merge into the snowy background. In the spring, their coats or feathers change back to the same colour as the rocks and lichen, which allows them to be as invisible to predators as possible.

This is the case, for instance, with the ptarmigan, a species related to the partridge, and with the ermine, a small carnivorous mammal with a long slim body and fur that is brown in summer and white in winter, except for the tip of its tail which remains dark. The same thing happens to the Arctic fox. Thus predators, too, play along with the seasons and change the colour of their coats by moulting, in order to become in their turn invisible to their prey.

▲ *The male wheatear sheds all its feathers in autumn. Moulting enables it to change its breeding plumage of grey-blue, white and black (top) for a russet winter costume (above).*

◀ *The Arctic fox has two coats. In the spring, it changes its totally white winter clothing for a darker, grey-brown outfit. In this way it always melts into its surroundings.*

THE ROLE OF LIGHT

Some species moult many times during their lives, others only a few. It is not known precisely what initiates this complex phenomenon. However, it is thought that the increase in hours of daylight as summer approaches plays a vital role in the process. The extra light stimulates the pituitary gland, which is responsible for the production of certain hormones. These hormones activate glands, among them the thyroid, which play a part in tissue growth and the pigmentation of plumage.

No room in the shell

Many invertebrates moult essentially in order to reach the adult stage. Arthropods, for instance, which include spiders, insects, crustaceans and millipedes, possess what is called an exoskeleton, a sort of hard carapace which protects them. Growth takes place throughout life, and is made much easier by a series of moults or sloughs during which the animal casts off its old shell and grows another one, but a size larger. The woodlouse, for example, moults in two stages. First it sheds its shell from the back of the body, and then sheds it from the front a few hours later.

In certain millipedes, moulting no longer takes place once they become adults. Instead, with each moult, a new segment of the body appears, and once the maximum number of rings is reached, the animal stops growing and becomes sexually mature.

▼ This snake (bottom) may look as though it is being devoured by another snake, but in fact it is simply shedding its old skin (below).

▲ To grow larger, spiders must regularly shed their carapace because it becomes too tight. This mygale has just moulted. It has shed its old skin (the upper shape), revealing a new one which is still soft. Over the next few hours the soft skin will harden.

A skin which comes off like a glove

In snakes, moulting – which usually occurs once a year – is not a gradual process as it is in birds. Periodically, these reptiles slough off the whole upper layer of their skin in one go, not because it gets too tight but because it gets worn out. A short while before the moult takes place, snakes become lethargic and their colours fade, they lose their appetite, and some spend a lot of time in water, perhaps to soften their skin.

When the time comes to moult, the snakes rub up against a rock or any rough surface in order to break open the old, outer layer, starting at the head. As soon as the head is free, the snake emerges with its new skin and the rest of the old skin comes off in one single piece, like a glove turned inside out. Once the old skin has come off – it is called the slough – the snake regains its healthy colour.

Two bodies in one lifetime

The larva and the adult of the same insect species are sometimes so different that it is difficult to believe they are the same creature. The adult is a metamorphosed larva, its anatomy totally transformed.

For a long time, it was believed that bad meat made maggots. The little white worms were thought to be completely unrelated to flies. In the same way, it is difficult to imagine that a hairy caterpillar can change into a magnificent butterfly. Yet all insects undergo this astonishing process of metamorphosis.

FROM A TADPOLE TO A FROG

The metamorphosis of frog larvae, better known as tadpoles, seems to sum up the conquest of dry land by fishes 350 million years ago. The tadpole is a swimmer and a purely aquatic animal. It breathes through internal gills and orients itself in its environment by means of the same organ as the fish, the lateral line. As the tadpole grows bigger, it develops first one pair of legs and then another. And at last, it metamorphoses into a miniature frog which breathes through its lungs and skin. The larvae of salamanders and newts also breathe by means of external gills. But they leave the water to metamorphose when they are two or three months old. Certain larvae, however, do not metamorphose. Those of the axolotl, a Mexican salamander, grow and reproduce without ever reaching the adult stage. This phenomenon is of hormonal origin and is called neoteny.

▲ *To reach the adult stage, the frog, like an insect, undergoes a metamorphosis. It is born as a tadpole and lives in the water, before growing lungs and legs, which enable it to breathe air and live partially on land.*

Growing up in a box

Insects, wonderful articulated machines, are imprisoned in their bodies. They are protected by a sort of box with thick walls, made of a protein called chitin, which is also found in our hair. In order to grow, the insect must periodically destroy this coat of armour and put on another suit that fits its new size better. It has to moult. The larva (or young insect) does not change much in the intervals between its frequent moults, and it is only during the last stage that it really becomes an adult. This process is metamorphosis, and is more or less radical according to the species.

Change can be gradual...

The larvae of grasshoppers and crickets live in the same places as the adults and feed themselves in the same way. They moult four or five times before their metamorphosis. Their development is very gradual: first the eyes and the antennae, then the beginnings of wings, which grow larger with each moult. Young bedbugs develop similarly, their wings only appearing at the fourth stage. During their five moults, they come to resemble adults more and more closely.

...or sudden

Among mayflies and dragonflies the larvae are very different from the adults, which makes their transformation all the more spectacular. They do not look at all like the adults and live in a different habitat – the insects live in the air and the larvae in water. The larvae are renowned for their voraciousness. Their modified lower lip, shaped like a horrible mask, can unfurl to catch small fish. After about 10 moults, and a little less than a year in the water, the young insects stop eating, and

begin to change. But from the outside, there is no clue to what is going on inside their rigid box.

The larva finally climbs out of the water and holds onto a leaf. The carapace cracks open along the back and the adult crawls out, head first (see left). Then, folded in two, it drags its thorax out of its prison, and then its legs and wings. Head down, upside down in relation to its old body, it waits until its legs can give it the push necessary to turn round. Then it is able to place itself the right way up on its old body, in order to extract its abdomen. After an hour, the effects of increased circulation enable the wings to unfold and the abdomen to tighten. The dragonfly can fly away to its life as an adult, which is shorter than its life as a larva has been. It leaves behind its old body, known as the slough, to be carried away by the wind.

The nymph, an extra stage

Beetles, flies, mosquitoes, bees, fleas, ants and butterflies all go through an intermediate stage of moulting known as the nymph phase. This amazingly complex stage occurs between the final larval stage and the moult that produces the adult insect. The nymph stage is in fact a resting phase. The larva curls up in a shell, where its body undergoes an almost complete renewal, already begun during the preceding larval stage.

In all these insects, metamorphosis is complete, in that the larvae are totally unlike the adults. The last change is the most important one. With each moult, all the larvae have done is change their size, nothing more. It is only the last moult that brings about a real transformation.

And what a transformation! Take the case of the antlion. The adult looks like a slow, night-flying dragonfly. But the larva has a large abdomen connected to the rest of the body by a tiny thorax, at the other end of which is a head equipped with impressively large jaws. Crouched at the bottom of a funnel-shaped hole, the antlion larva waits patiently for a stray ant to slide down the funnel, whereupon it seizes the hapless ant, carves it up and eats it.

▲ *The nymph is a moulting stage characteristic of various insects, including bees. This is the nymph of a future queen bee 12 days after the hatching of the egg.*

▲ *The adult emperor dragonfly is able to fly for a whole day without settling.*

▲ *The metamorphosis of an emperor dragonfly. Now ready to emerge, the larva leaves its old body. This dragonfly is one of the largest insects in Europe (about 10 cm long) and a redoubtable predator in freshwater streams.*

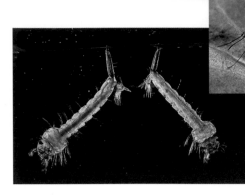

stage. They soon metamorphose, and emerge as an adult on the surface of the water. The caddis fly larva has become a new insect, related to the butterfly, with extra large antennae adapted to nocturnal life.

◀▲ The larvae of the mosquito Culex pipiens *breathe on the surface of the water through the anus, by way of a siphon. Above, the adult mosquito.*

Maggots, legless larvae

We have dealt with larvae with legs and larvae with false legs. In maggots, however, we have larvae without legs. Only the members of the Diptera family, which includes flies, mosquitoes, craneflies and midges, have legless larvae. Some could be said to be relatively attractive (mosquito larvae), others not (flies). But after four or eight moults (in the mosquito and the fly respectively), they are all enclosed in their final skin, called the pupa, which is equally unattractive. Metamorphosis will take place inside this skin. In order to get out of the pupa, the adult will make it burst open by inflating a pouch on its head. As a constant reminder of this explosive exit, it will for ever have a slight dent at the base of its antennae.

The phenomenon of metamorphosis is not peculiar to insects: it also takes place in the larvae of marine invertebrates, the tadpoles of amphibians and even young fish. But only insects undergo such a radical, and sometimes even total, transformation at the end of a lengthy, repetitive process.

▲ At the larval stage (top), caddis flies, which are nocturnal moths, live in the water, inside silk cases with small pebbles and shells adhering to them. The adult caddis fly (inset) bears no resemblance at all to its larva.

A silken suit: the chrysalis

The larvae of butterflies and moths are caterpillars with five pairs of false legs on the abdomen and three pairs of true legs on the thorax. After three or four moults, each caterpillar wraps itself in a covering called a chrysalis. It is a shell which may be made of silk, like that of the silkworm or bombyx. The larvae of caddis flies also produce silk, but only to line their cocoon. Some look like caterpillars, whereas others have legs. Most of them roll themselves up in silken sheaths to which scraps of plant matter and small stones adhere. Protected and weighted down by this coating, caddis fly larvae can hope to escape the attention of trout and avoid being swept away by the current in the fast-flowing streams where they live.

After five to seven moults, the larvae close up the sheath and enter the nymph

THE STRANGE METAMORPHOSIS OF THE FLAT FISH

The larvae of flat fish (like plaice, sole and turbot) look like 'normal' fish, with a symmetrical body and an eye on each side. They later undergo a metamorphosis. Their bodies widen, their undersides grow bigger, and one eye moves from one side across to the other (right, a turbot aged one month). Then the larvae stop swimming and rest at the bottom of the water, on the side now without eyes. The process ends with the loss of pigmentation on this side and the fish becoming flat.

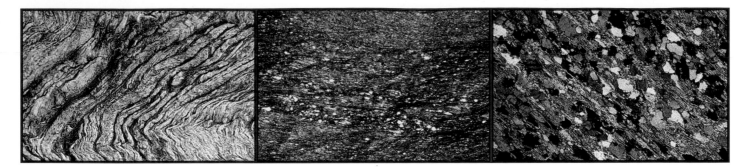

METAMORPHISM

From limestone to marble

Where two plates of the earth's crust converge, one is forced under the other and gets buried deep in the semi-molten rock below. There the buried rock recrystallises and reappears much later, 'metamorphosed'.

Today, marble is quarried in the open, although once it was a part of the rocks that came into being several kilometres under the ground. The development of marble, like that of shale and mica, is quite particular, as it is the result of the metamorphosis of old rocks (limestone in the case of marble, clay in the case of shale and mica) drawn downwards. They are called metamorphic rocks. If they are now near the surface, it is because the kilometres of material that initially lay on top of them have been worn away by erosion.

Dragged downwards

The lithosphere, the outermost crust of our planet, is made of plates of rigid and fragmented rocks, which are always moving. Sometimes these plates move apart and sometimes they converge.

At a convergent boundary, one of the plates usually slides underneath the other. And while a mountain chain is formed above the surface, the rocks of the plate underneath sink progressively lower into the earth's crust to a depth of several tens of kilometres. This is a process called subduction, and is quite disruptive. It often results in earthquakes and volcanic eruptions as the tensions caused within the crust find release elsewhere on the surface.

The newly buried rocks are subjected not only to a considerable increase in pressure (about 300 atmospheres a kilometre), but also to a dramatic increase in temperature (about 30°C a kilometre).

In addition, because these rocks sink during clashes between plates, the immense forces to which they are subjected cause them to cleave in thin layers perpendicular to the forces. This flaky characteristic of metamorphic rock is termed schistosity (after which schist gets its name).

Deformation and recrystallisation

Increases in temperature and pressure usually go hand in hand with the deformation of rock. They also cause the complete recrystallisation of the rocks, which do not change their chemical composition, but only their mineralogical composition and the size of their crystals. The new minerals develop a characteristic alignment according to the direction of the pressure.

▲ From left to right: metamorphic slate; a thin layer of non-metamorphic clay (the small light flecks are quartz crystals); the same rock turned into mica by the process of metamorphism. The recrystallised minerals (the grey and white is quartz, the yellow-orange is mica) have increased in size and are aligned according to their schistosity.

▶ 1. Two continental plates are drawn together by the subduction mechanism. 2. The collision sets in motion an upwelling of land in the form of a mountain range, and sedimentary rock which originally covered the ocean floor is buried. 3. After the collision, erosion of the landscape gradually exposes the metamorphic rocks.

THE FORMATION OF METAMORPHIC ROCKS

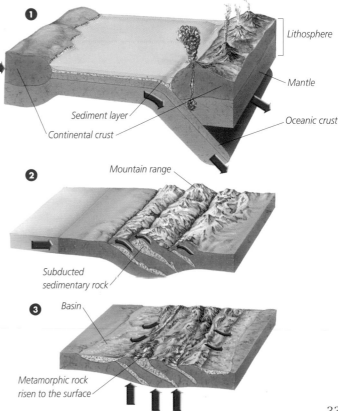

❶ Lithosphere

Mantle

Sediment layer

Continental crust

Oceanic crust

❷ Mountain range

Subducted sedimentary rock

❸ Basin

Metamorphic rock risen to the surface

The secret of longevity

Unlike animals, plants grow throughout their lifetime by simply increasing in size, as there is regular renewal of plant tissue.

▲ *The trunk of an old tree is in essence the same as a fragile stem emerging from the soil. It results from the accumulation of wood produced over centuries.*

In a little village in Normandy called Allouville-Bellefosse, there is an oak tree so ancient that it has had to be propped up to withstand bad weather. The hollow trunk is more than four metres wide at the base and some 15 metres in circumference.

In 1696, a chapel dedicated to Our Lady of Peace was created inside the tree, with an oratory above. The tree's crown, which has been struck several times by lightning, has a wooden roof and a cross on the top, like a church steeple. It probably grew from a seed at the time of Charlemagne, which would make it about 1 200 years old. However, each year the huge branches that remain continue to put out leaves and produce an abundant crop of acorns.

The certainty of rebirth

The oak of Allouville has certainly lost a few branches in the course of its long life. But even if it had been cut down at the base of the trunk, this would not have killed it. It would just have made new shoots. For the oak, like all plants, holds a trump card: it always has a supply of tissue capable of producing a complete, new plant, from roots to leaves. This tissue, called meristem, is concentrated in the buds and the ends of the roots. For as long as this exists, the plant will be able to reconstitute itself. Cultivators have exploited this asset for thousands of years, by the use of cuttings, grafting and cloning.

Animals develop in a totally different way, however, because of the differentiation of their embryonic tissues, which evolve to become distinct organs. This process is very soon irrevocable. If one of the vital organs is destroyed, death is inevitable, for the organ cannot be regenerated. But plants have no highly specialised organs. Their main functions, like respiration, water absorption and photosynthesis, are separate.

Plant Meccano

Because of meristem, the growth of plants is flexible. It facilitates the construction of basic units, each of which consists of a section of stem, one or several leaves and a bud. The stem and leaf or leaves represent the present generation; the bud

◀ *The banana palm grows very fast and has a short life. Unlike the oak, which lives for hundreds of years, it does not accumulate wood. Its 'trunk' is made of closely interwoven stems and leaves.*

TO GROW OR TO FLOWER

Plants are constantly developing, their architecture governed by whether the buds turn into leaves or into flowers.

1 to 5. Buds
6 and 7. New units of stem, leaf and bud
8. Flower

Buds 1 and 2 and the terminal bud (5) have made the plant grow by generating new units of stem, leaf and bud (6 and 7). Bud 4 has remained latent, in reserve. Bud 3 has produced a flower (8) so that reproduction from seeds may be possible.

Young plant

The same plant a few weeks later

is the hope for the future, as it possesses a section of meristem.

The way an oak, a banana tree and a dandelion grow is identical. Only their structure is different. In the oak tree, the stem is long in relation to the leaf, and an old tree has a majestic crown. In the dandelion, the stem is very short and the leaves grow so close together that they almost touch and form a rosette on the ground. The banana palm is halfway between these two forms. All plants may be similarly broken down like a Meccano set into their basic elements, whose shape and appearance give the plant its final form.

Forever young, but not immortal

In a tree, the only living parts are the branches with this year's leaves, the buds and a thin layer of cells between the bark and the wood. Meristem ensures that these tissues always remain young because they are renewed on a regular basis. The rest of the trunk and branches are no more than dead wood. This means that a tree can go rotten inside its trunk and still flourish.

Meristem permits an extraordinarily long life. The oldest vertebrates are terrestrial tortoises, which live for less than 200 years. The oldest trees, slow-growing conifers of North America and Australia, live 4 000 years or more. The oldest on record is a Colorado pine 4 765 years old. This ancient was born a short time before the pyramids of Giza were built! However, all trees die in the end, of old age, eaten away from the inside, blown down in a storm, or felled by their own weight. And even meristem finally perishes.

Vegetative reproduction does not allow a plant to adapt to changes in its environment. The climate may change, the soil may become impoverished: plants that are unable to either adapt (by sexual reproduction) or to move (by seed dispersal), will die.

▲ *A bud can produce a stem and a leaf: this is called vegetative reproduction. It may also produce a flower, which will produce seeds: this is known as sexual reproduction.*

▶ *Even the most flourishing pine trees will die in the end, although some species are renowned for their longevity, like the bristlecone pines of North America, thought to be the longest-lived trees on earth.*

▲ *The dandelion, a meadow plant, has a very short stem, and its leaves overlap. But its growth still obeys the same laws as those of the oak or the banana palm.*

VEGETATIVE AND SEXUAL REPRODUCTION

Plants grow continuously, but in the end they reach an 'adult' size. Then there is a mechanism which permits them to evolve: sexual reproduction. The meristem of certain buds produces not another new stem, leaf and bud element, but a flower, which in turn produces pollen (male) and/or ovules (female). When the pollen and the ovule from two individual plants meet, a third individual is produced in the form of a seed, different genetically from the first two. And the seeds sometimes travel a long way. However, as the bud has not created a new bud capable of continuing vegetative growth, the meristem in this bud disappears. If all the buds on a plant form flowers (as annual plants do, for instance), the plant dies once seeds have been produced. Vegetative and sexual reproduction are thus two survival strategies: one guarantees the survival of the individual, the other the continuation of the species.

When death nourishes life

In nature, all life leads to death but death in turn sustains life. When a herbivore is eaten by a carnivore, it enters into a recycling process that will allow its own kind to live and to give life. It is by this process of death and regeneration that the cycle of matter continues, and with it all ecosystems. With the help first of scavengers and then of decomposing agents (bacteria and fungi), the bodies of animals are slowly transformed and broken down into simpler and simpler elements that can eventually be assimilated by plants. The circle is then complete, with plants that constitute the basic food of herbivores feeding in their turn on the remains of what has absorbed them.

▼ 1. THE QUARRY OF PREDATORS

Cheetahs and their young devour their prey, a warthog (below). But they will not finish it all. Other carnivores arrive – hyenas – to pick at the leftovers. It looks as though there may be a bitter struggle. In the daytime, hyenas are usually scavengers and come to claim their share of the animal after it has been killed by the predators. In a group, spotted hyenas, for instance, are capable of chasing off lions in order to appropriate their kill. But at night, hyenas also hunt in a group and then become as predatory as lions, panthers or cheetahs. The jackal, being smaller, waits its turn. Like the hyena, it is a hunter as well as an opportunistic scavenger. However, with this warthog kill it has to be content with the third sitting, which means the cheap cuts that the hyenas leave behind.

▶ 3. A FEAST FOR INSECTS

All that now remains are a few shreds of flesh on some of the bones, some hair and skin, all quite indigestible. The small carnivores of the grassland, winged and wingless, make the most of the scraps. Then the body is left to the insects. Flies always get there first, but not to eat. Instead, they lay eggs. The worm-like larvae that hatch (right) will feed off the body before changing into adults. Then it is the turn of the beetles, such as bacon and carrion beetles (which are perhaps the best destroyers in nature), burying beetles and histerids, which pick the carcass clean. Dipterids and other beetles will finish off the work. But the bones will only be really white and the surrounding ground become really clean after the mites have visited, because they make use of all forms of organic matter. What they leave will finally be food for bacteria which will change it into macromolecules and then molecules, nutriments which can be easily assimilated by plants.

4. THE JOURNEY OF THE LEFTOVERS

Numerous insects, worms, mammals and birds carry off bits of the carcass and eat them elsewhere. The female dung beetle assembles a few bits of their waste matter and rolls it into a ball that she takes to her hole. There she either eats it or lays her eggs in it, in which case it is the larvae that eat it. Burying beetles tear off bits of flesh, bury them in the ground, then make them into little balls that they put in a small chamber. The meat will later provide a feast for the larvae that hatch from eggs laid in the galleries leading to these chambers. Finally, the waste left by decomposers, and anything they have not eaten, finds its way into the digestive tract of minute insects, such as worms and mites.

▲ 2. THE DANCE OF THE VULTURES

A zebra's body has been lying in the grass for several days, but there is still activity round the carcass. A chorus of grunts emanates from a group of vultures, their beaks and necks covered in blood as they pull at the remains of the meat, jumping up and down and flapping their wings to scare off rivals. First on the scene are the white-backed vultures and the Rüppel's griffons. With their long beaks and long necks they tackle the viscera and scraps of muscle remaining on the carcass. Then larger birds arrive, like the lappet-faced vulture and the white-headed vulture, whose massive beaks allow them to attack the skin and cut through the tendons. Finally, the Egyptian vulture and the hooded vulture, distinctly smaller than the others, are content to pick the bones. A little further away stand the marabous, large black birds which look like funereal storks and feed on the waste left by the vultures.

AFTER THE DEATH OF A TREE

In temperate forests, a dead tree disappears in 10 to 15 years. The surface of the trunk in contact with the damp ground is rapidly colonised by thousands of bacteria and fungi which penetrate the heart of the trunk and digest the cellulose. Then come insects like springtails, cockroaches and earwigs. Together with millipedes, earthworms and woodlice they consume the damaged tissue. As they make their way through the trunk, they meet other insects, such as stag and chafer beetles, which lay their eggs there.

Their larvae will feed on the wood until they metamorphose. The trunk gradually becomes more and more fragile. The decomposers that have emerged from the soil gnaw away at it from underneath creating an arch that eventually reaches through to the upper side of the trunk, and causes the dead tree to fall apart. It is now nothing more than bits of bark and a heap of particles. This compost, in turn digested by mites and bacteria, restores to the woodland soil the minerals that are indispensable to plant life.

Mountains, immortal giants

Nothing withstands erosion, not even the highest mountains. The ancient rocks of the continents once belonged to chains of mountains whose tops were gradually smoothed down. Sometimes a rejuvenating process takes years off their age.

Who would guess, when travelling through Britanny, in France, or the Appalachian region in the east of the United States, that 300 million years ago their rocks belonged to the same mountain chain? At this time, before the Atlantic Ocean was formed, earth's land formed one mass known as Pangaea. The mountains were part of what is called the Hercynian chain, with a totally different relief which perhaps looked in some places like the Himalayas do today.

A Herculean task

As soon as this chain of mountains began to form, the forces of construction and destruction entered into battle. The convergence of continental plates *(see pp.192-193)* continued to produce folding and overlapping, with the result that the crust forced below the mountains grew thicker. But continental crust is lighter than the underlying mantle, so it was forced upward. Unlike a cork released under water, though, it did not rise immediately. The upthrust was slow and ended only when the crust regained its original thickness, surface erosion having worn away the surplus rock.

It took nearly 60 million years before the relief of this Hercynian mountain chain was transformed into an extensive surface that was almost flat, known as a peneplain, or 'almost plain'. It was covered by deposits from the beginning of the Mesozoic Era, about 240 million years ago. However, in Brittany, and particularly in the Appalachians, where summits rise to more than 1 000 metres, there are few traces of the peneplain. A recent (geologically speaking) and moderate upthrust created a new cycle of relief and erosion and the appearance of the surface has changed yet again.

Erosion can wear away an extensive thickness of crust and deposit. Many peaks in the Pyrenees and the Alps are made of ancient rock that also once belonged to the Hercynian chain and which have since been uncovered and uplifted, such as the granite intrusion that is now the Mont Blanc massif.

▲ *Part of the ancient Hercynian mountain chain, the Appalachians were eroded before being rejuvenated by a more recent upheaval.*

▼ *The Monts d'Arrée in Brittany belonged to the same Hercynian chain. Their folded rocks were smoothed out at the end of the Paleozoic Era, a geological time about 245 million years ago.*

THE LEVELLING PROCESS

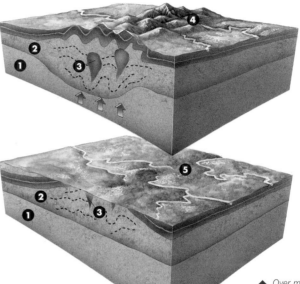

▲ *Over millions of years, erosion can transform a mountain (4) into a peneplain (5). The upward thrust from the mantle (1) makes the continental crust (2) rise together with its granite rocks (3), while weathering erodes and smooths the features on the surface.*

A death is predicted

In five billion years, our Sun will begin to destroy itself, and will take with it the surrounding planets of the Solar System. This will be the end of the Earth.

When the Sun is nearly nine billion years old, it will have reached such a size that it will become unstable. The formidable machine that for so long gave the Earth the benefit of its energy will be burning its last stores of hydrogen. The heat at its heart will reach 100 million degrees Celsius. Its radiation will become more and more intense. Its light will become redder and our Sun, our nearest star, will change into a red giant. It is a stage that the inhabitants of the Earth will not be present to witness, because the planets of the Solar System that are nearest to the Sun will already have turned into fiery furnaces.

Return to the future

Before this time, on Earth, one disaster will have followed another. The temperature will have gradually risen, the polar icecaps melted, desert regions expanded and windstorms, tornadoes and cyclones torn through our skies.

With glaciers gone and lakes dried up, first vegetation and then all living things, unable to thrive, will have died out, species by species. Only the most primitive lifeforms, those that are least dependent on their neighbours and able to withstand the heat, radiation and lack of water, will survive for a while. But humans will not be among them.

As for the dying star, it will continue its fatal dilation and will swell so much that the Earth will be swallowed up in its outer layers. Heated to 4 000°C, our planet will then be definitively uninhabitable and its surface will revert to the state of fusion and liquidity it experienced at its birth, 10 billion years earlier…

Let the sky not fall on our heads!

But the apocalyptic scenario that is predicted may never happen. Between now and the next five billion years, many other changes may take place. For example, an asteroid may strike the Earth or a comet may collide with it. Even if they did not cause the Earth to blow up completely, such events would still have a lasting effect on its equilibrium and set in motion disasters like a succession of volcanic eruptions, a rise in sea levels, and pollution of the atmosphere. All these cataclysms might herald the end of the Earth in the long term.

Would they also mean the end of humankind? Humans will perhaps be condemned to an exodus in a fleet of spaceships to search for a new planet. This is a favourite theme of many science fiction writers, but today its possibilities are also being investigated by space programmes.

The inhabitants of the Earth are gaining some understanding of the Solar System and its future. But they have also acquired the knowledge and means of mass destruction. The future of our planet therefore probably lies more with people than with the fury of the Sun.

▼ *In five billion years, an excessively large Sun will gradually burn up the Solar System. The soil on Earth will become lifeless. Human beings will be dead, or will have travelled far away.*

Explosions in space

All stars are destined to die one day. The length of their life and the manner of their death depend essentially on their size at birth.

▲ The remains of the supernova detected in 1987 in the Large Magellanic Cloud.

On February 23, 1987, at the Cerro Tololo observatory in Chile, a young Canadian astronomer took a photo of a nearby galaxy. When he developed it he noticed an unusual star in the photograph. He discussed it with his colleagues – and word soon flew around the world. After comparisons were made with older photographs of the region, there was no doubt: it was a supernova. The star was dying in a gigantic explosion of energy. It was a significant discovery because the supernova is relatively near, 170 000 light-years away. Today it is still being monitored by the Earth's telescopes. The fact that its matter is still expanding is an indication of the violence of this phenomenon.

A cauldron at the heart of a star

How can a star which, to us, is just a little point of light in the sky, put on such a display of fireworks? In fact, the mechanism that leads to this explosive death has been at work ever since the birth of the star.

It all begins with a dense cloud of gas in which globules are formed, essentially consisting of hydrogen. Under the influence of gravitational force, what is called gravitational collapse occurs, when matter is drawn towards a centre of attraction. At the same time, the temperature and pressure

▼ A white dwarf. This planetary nebula, in the constellation Gemini, has been nicknamed 'clown's head' by astronomers. The white dwarf is the bright dot visible in the centre.

increase to a point at which the fusion of the simplest element, hydrogen, takes place. A force opposing that of gravity then appears: thermonuclear force. While the one force acts to reduce the diameter of the star, the other increases it. The stability of the star depends on their equilibrium. At this stage, the heart of the star is nothing but a cauldron burning a specific fuel, hydrogen, and the product of this combustion is helium.

Hot inside, cold outside

The helium produced in this way slowly fills the heart of the star. After some time the fusion reactions start to diminish and gravitational force gains the upper hand, bringing about the contraction of the heart. The star begins to destabilise. Pressure and temperature increase further, and the surplus energy is sent to the surface of the star, where the outer layers are expanding. The diameter of the heart decreases while the diameter of the star increases. Paradoxically, while the temperature at the heart increases, that on the surface decreases. The light turns redder. The star has become a red giant.

A sudden flash

At the heart of this star, the temperature rises from 20 to 100 million degrees, hot enough for the helium to fuse, which produces carbon. The star changes once more. It goes on to a stage where the heart is producing carbon from helium while on the periphery hydrogen is burning and producing helium. And when a very large quantity of helium fuses, there is a sudden helium flash. It is the ultimate stage of fusion in small stars like the Sun. In larger ones, where the mass is more than six times that of the Sun, the process of dying violently will continue.

White dwarf, supernova...

Small and large stars therefore do not share the same fate. When their mass at birth is less than six times that of the Sun, the process ends because there is not enough energy for the carbon to fuse. After a very bright stage, when the star is called a red

CHRONOLOGY OF A SMALL STAR: THE SUN

Stages	Time
1. Birth	0
2. End of the stable period of the star	10 billion years
3. Beginning of red giant stage (life no longer sustainable on Earth)	10,2 billion years
4. Helium flash	10,5 billion years
5. Temporary survival	10,6 billion years
6. Red supergiant stage and death of the star (white dwarf)	11 billion years
7. End of cooling of white dwarf (black dwarf)	100 billion years

supergiant, the external layers are cast off. They form what is called a planetary nebula round what is left of the star, a white dwarf. Its density is as much as a million tonnes per cubic centimetre. Several billion years later, the white dwarf will become a black dwarf. Large stars, whose initial mass exceeds six times that of the Sun, have an even more brutal end. The fusion reactions synthesise all the elements, even iron. The star explodes in an orgy of energy in the form of a supernova. The central heart is subjected to the law of gravitational force on a massive scale. It collapses until the very constituents of the nucleus come into contact with one another. This is a neutron star, with a density of one billion tonnes a cubic centimetre. Its rotational speed, in the order of milliseconds, transforms the star into a sort of superpowerful lighthouse, whose waves of energy can be seen at regular intervals. These pulsating stars are known by their contracted name, pulsars.

...and black hole

In very large stars, the collapse cannot be halted, because the atomic nuclei sometimes are unable to withstand greater masses. The remaining heart is then so dense that nothing can escape from it, not even light. If a beam of light could be emitted from such an object, it would return, inexorably attracted to its emitter. It is because of this extraordinary gravitational field that this object is called a black hole.

▲ *The Crab nebula, 500 light-years away from the Earth. In the year 1054, the Chinese detected this 'new star'. The scattered matter indicates the violence of the supernova explosion.*

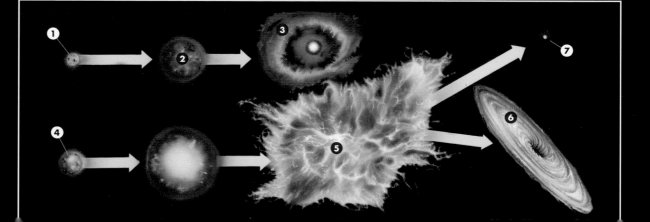

▶ *The smallest stars (1) change into red supergiants (2) then into planetary nebulae surrounding a white dwarf (3). The largest (4) reach the stage of supernova (5) then give rise to black holes (6) or shrink and become neutron stars (7).*

'Old age cannot be cured'

Why do we grow old? Ageing is a phenomenon that is largely unexplained, but is thought to be part of the evolution of species. Its mechanics are being actively investigated as people hope to improve the length and quality of their life.

▲ *Bacteria (here Haemophilus influenzae) reproduce by division. This mode of reproduction allows the transmission of an unchanged genetic inheritance and ensures a theoretical immortality for this type of organism.*

▲ *An old Mongolian woman. It is thought that the biological limit of longevity in humans is about 150 years. The current state of scientific knowledge does not allow this limit to be extended.*

A rat does not live longer than three years. An elephant can hope to live for 70 years, and a whale, 200. The Frenchwoman Jeanne Calment, to date the oldest person ever known to have lived, died in 1997 after 122 summers. These four cases illustrate the same phenomenon of ageing, the various types of deterioration occurring in certain organisms as they grow old. But can one talk of ageing in bacteria, which reproduce exactly simply by division? If such an organism deteriorated, it would do so continuously, from one division to another. And it would already have died a long time ago. So bacteria are potentially immortal, and forever young, whatever their age.

An inevitable process, but still a mystery

On the other hand, organisms which reproduce by what are called germ cells (the spermatozoa and ova in humans) can undergo the ageing process because an entirely new individual is born with each act of procreation. This is the case with all animal life, almost without exception. The phenomenon of ageing therefore seems closely linked to mechanisms for the reproduction of the species. You could even say that it occurs only after sexual maturity. And as it is a phenomenon which systematically affects all organisms, it is natural for a genetic origin to be attributed to it.

There are three theories at present which seek to explain ageing. The first, suggested by the English biologist Sir Peter Medawar, goes like this. If deterioration caused by ageing affected individuals before they reached a reproductive age, the continuation of their lineage and therefore of their genes, would be more uncertain than it would be for individuals in good health. And they would die out. On the other hand, if it affected them after the age of sexual maturity, undesirable changes due to ageing would not be eliminated by natural selection and the genes causing them would be transmitted from one generation to the next.

The American evolutionary biologist George Williams goes further. He thinks

◀ *The giant tortoises of the Galapagos Islands are almost invulnerable inside their shells. They, often live for more than 100 years.*

that if a certain characteristic is found universally, it is because natural selection actively intervenes to preserve it. Therefore he goes so far as to suggest that the genes responsible for ageing have positive effects during the youth of the individual, for instance in improving fertility.

Another English biologist, Tom Kirkwood, has another theory: he suggests that in the allocation of its energy resources, an organism gives priority to the reproductive process to the detriment of the maintenance functions of the organism, and that this causes ageing.

These three main hypotheses, which in fact complement each other, form at present the only basis for explanations of the phenomenon of ageing. Clearly, the fundamental cause of ageing is still a mystery.

The bigger it is, the longer it lives

The study of longevity in animals sheds useful light on our understanding of ageing. Why do elephants live longer than mice? In the first place, because they have fewer predators. When competition for life is tough, there are few old survivors in the world. And mice have more rapacious mouths to fear than large mammals do. Bats, which sleep hidden away in caves and fly about at night, manage to

▼ *Like most animals, this pair of blue herons ensures its succession by means of sexual reproduction. This 'decision' by nature may have led to the emergence of ageing.*

▲ *An orang-utan from Sumatra. The heavy cheeks of this primate are a secondary sexual characteristic in the male, and grow larger with age.*

The isolation of a handful of genes directly responsible for longevity with an aim to manipulate them has now been abandoned. The hypothesis is too simple. The greatest age an individual can hope to reach is the result of a complex set of interactions in which genetics as well as the circumstances of the individual's life play a part.

In spite of everything, the hope of miracle cures keeps illusions alive – and makes for good business. Antioxidants, which help to fight free radicals, molecules that are potentially very damaging to living tissue, have been praised to the skies for their role in staving off age. People have been encouraged to consume foods and pills rich in antioxidants, and even to put them on their skin, but there are no results to show for certain that they are effective in warding off the effects of ageing. Reduction of kilojoule intake has been another suggestion. In rats it increases their lifespan, but not in the fruit fly. The injection of DHEA, a molecule synthesised by the suprarenal glands, has shown some positive effects in slowing down ageing, but only in certain cases. Neither does shortening the telomeres, the ends of chains of DNA, during cell division appear to be the key factor in preventing ageing.

So the era of youth elixirs may be well and truly past, at least in the minds of scientists. But if the research being done in the different branches of gerontology is anything to go by, the next few years will certainly see fruitful new developments in understanding the mechanisms of ageing.

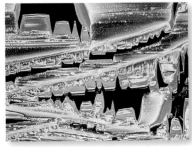

▲ *A crystal of DHEA. This molecule is thought to have, in some conditions, anti-ageing properties. But in fact there is no miracle drug against the ravages of time.*

live for as long as 15 years. Tortoises, almost invulnerable inside their shell, live to be more than 100. For all that, even in a protected environment no mouse will live to be 100. Natural selection has therefore not retained a strong longevity in species that in nature have no chance of living for a long time, probably because it would have been of no advantage for their survival as a species.

In any case, there are logical reasons to explain why larger species (not individuals) live longer than smaller ones. The bigger the animal, the longer it takes for it to grow and reach maturity. So it is natural that an elephant grows old later than a mouse does, especially as its offspring also take more time to gain their independence. The single elephant calf that is born to its mother represents the hope that it will carry on the line, and needs its parents for longer than the baby mouse, which risks an early death but whose lineage is assured by the large number of young born at the same time.

In search of an elixir of youth?

But besides considerations about the evolution of the species, ageing is the object of very concrete studies. And although failing to solve the mystery of immortality, lines of research as old as the world are still being pursued, such as fighting off the appearance of the signs of old age, and delaying death.

PROGERIA, THE TERRIBLE DISEASE THAT MAKES CHILDREN OLD

Progeria affects children and causes certain of their biological functions to deteriorate very fast so that they die before they are 12 years old looking like old people. However their cognitive functions are not affected, so it is not a question of ageing in the real sense. The disease is not transmissible, in the sense that the children who suffer from it die before they are capable of reproduction. It appears after spontaneous mutations and in 1990 affected only 17 children in the world. Research into progeria has not at present furthered our knowledge about the phenomenon of ageing.

SEX AND REPRODUCTION

Perpetuating life

Orchids from Madagascar.

Sex and no sex

Jellyfish, sea anemones and corals have one feature in common: they reproduce by division and also by means of sex cells, taking a different form for each.

▲ *Two animals of the cnidarian group in polyp form, the sea anemone (above) and corals (below).*

Take a jellyfish, turn it upside down and put it on a rock. With its dome underneath and its tentacles in the air, it now looks like a sea anemone. There is nothing strange about this, as jellyfish and anemones, like corals, sea fans and hydroids, belong to the same zoological group, the cnidarians. They all have a dome-shaped body, a mouth, and tentacles that sting.

A cnidarian which swims is said to be a medusa, a jellyfish form. A cnidarian which does not move is a polyp. But nothing is fixed about these creatures. Many pass from one state to the other, according to their reproductive needs.

Medusa and polyp, both the same animal

In primitive cnidarians like the obelia, the medusa or jellyfish form and the polyp form are equally important. The medusa serves only to produce

1. *Larva*
2. *Young polyp*
3. *Budding (asexual reproduction)*
4. *Polyp producing medusae (sexual reproduction)*
5. *Young medusa*
6. *Adult medusa*

THE REPRODUCTIVE CYCLE OF A JELLYFISH
This *Aurelia* takes two different forms in its reproductive cycle. The medusa (a swimmer) produces larvae which change into polyps (immobile) which produce medusae.

and disseminate the sex cells. Fertilised by the sperm, the ova produce larvae called planulae which soon settle on the seabed. Then the larvae become polyps which, in turn, after numerous divisions, become new jellyfish.

The so-called true jellyfish (such as *Aurelia*, *Rhizostoma* and *Pelagia*) and box jellyfish reproduce in a similar way, but their polyp phase is brief and not visible and the jellyfish phase is predominant, which is why they are known as 'true jellyfish'.

In almost all other cnidarians, only the polyp phase exists. So freshwater hydra and fire corals, and more particularly anemones, sea fans and other corals, never take the jellyfish form. Their larvae swim but nevertheless stick on the substrate to produce polyps which, by budding, can create colonies and produce sex cells again.

The cnidarians, one of the most diversified families, have not chosen between sexual reproduction and reproduction by division. In a way, they use one to disseminate themselves and the other to settle on new aquatic territories, such as rocks, shells and the hulls of boats.

▲ *The* Physalia, *or Portuguese man-of-war, is a false jellyfish. It is made of a colony of polyps hanging from a float. Each polyp has a specific function – to protect, digest or reproduce.*

▲ *Male elephant seals can reach the truly impressive weight of four tonnes, or more than three times the weight of the female. No doubt this is a necessary size to defend a large harem.*

SEXUAL DIMORPHISM

Opposites attract

Within each species, males and females are different and each possesses its own morphological characteristics. Some of these are easily visible, often for reasons related to reproduction: to attract, show strength and perpetuate the species.

In the humid jungle somewhere between the Central African Republic and Rwanda, a troop of mountain gorillas makes its way through dense, bushy undergrowth. At the front, young ones jump about all over the place. The females follow, some with very young offspring clinging to them.

A little apart from the others, an individual markedly larger than the others ambles along, taking his time. It is the patriarch of the troop. Besides his impressive size, which stands out compared with that of the females, he has a sort of crest on top of his head, a more marked arch over the eyebrows and a velvety grey back

contrasting with the black colour of the rest of his coat. These are physical characteristics that distinguish the males from the females, and constitute sexual dimorphism.

Big and strong

This dimorphism is not frivolous but has a very clear meaning, first of all on a visual level. A predator, or simply another male, will immediately recognise a male by his height, colouring and certain special traits. Similarly, it will also identify a female, which is often slighter in build and more neutral in colour.

47

special feathers on the neck, the tip of the tail or the back. All this is simply for the sake of pleasing the female, as in general it is only during the breeding season that males display this colourful plumage. Ducks are decked out in multicoloured feathers and numerous passerine songbirds are, too. With the addition of their song, they are well equipped to charm a mate.

◀ The male praying mantis runs a great risk of being devoured after mating. When the female is hungry she considers the males of her own species to be prey like any other.

In the case of the gorilla, the size and breadth of the male will warn off a possible antagonist, be it predator or rival.

In the African savanna, lions are distinguished from the lionesses by a more powerful build, greater height and more distinctive features. In East Africa, the lions have a long brown mane. The lions of West Africa, on the other hand, have a more modest mane. Potential prey or another lion would use these morphological particularities, at least in part, to distinguish a lion from a lioness.

In marine mammals, sexual dimorphism can be quite exaggerated in favour of the male. This is the case, for example, with sea lions or elephant seals, where the male may be three times as large as the female. This feature is commonly found among polygynous species (in which one male lives with several females), where the male is the master of a veritable harem. It is best to be big and tall in order to intimidate potential rivals.

*▲ In the king bird of paradise (*Cicinnurus regius*), from New Guinea, the male's plumage is white and bright red, which is in marked contrast to the olive brown of the unlucky female.*

Fine feathers for handsome charm

In birds, sexual dimorphism is more a matter of decoration than of size. The males of certain species are certainly larger than the females, but this is not the general rule. What characterises the dimorphism in a large number of bird species is the colour of the plumage sported by the male, which is usually far more brightly coloured than that of the females. Or else the male may have

Enchanting costumes and devastating trills

Other species wear these flashy clothes almost all the time, notably the members of the Galliformes family, which include the ruffed grouse, ptarmigans and the ring-necked pheasant, as well as peacocks. In peacocks, the long tail feathers that usually trail along the ground are used to attract the attention of the female. The male lifts them and spreads them out, making a superb blue-green fan to which no peahen in the neighbourhood can remain indifferent. Similarly, birds of paradise compete for the best decorated feathers. The favoured male is the one with the brightest colours and the most stylish display.

In certain species, on the other hand, there is no marked sexual dimorphism, and a male is visually indistinguishable from a female. This is the case, for instance, with nightingales, warblers and thrushes. But it is the song of the male in spring that shows the difference... and what a difference it is!

Frumps and vamps

Females in many bird species have a dull plumage, often brownish and mousy. This may not be an attractive outfit, but it has one important advantage. By merging with the background, the female can escape the attention of predators when she is sitting on her eggs.

In other species, notably small waders like the phalarope or the dotterel, not only are the

▼ Among red-necked phalaropes, birds of the sandpiper and snipe family, the female (bottom) has the most colourful plumage. This is because it is the male (below) that sits on the eggs and also raises the young, and so must be invisible.

females larger, but they also wear a richly coloured breeding plumage. Better still, the females can mate with several males, and once they have laid their eggs in a nest built by …yes, the male… it is the male that is responsible for sitting on the eggs and bringing up the young, while the female mates with another partner.

A large lady and a little fellow

In many insect species, sexual dimorphism is exaggerated in favour of the female in that the female is frequently larger than the male. This characteristic is very often necessary because their reproductive load is heavy – the females carry a

great number of eggs – and demands much more energy. In termites, for instance, the queen is enormous because she bears millions of eggs.

There is also a considerable difference in size in the praying mantis and spiders of the *Nephila* genus. This means that the male runs the risk of being eaten by the female immediately after mating has taken place. In the mantis, it seems as though this nuptial sacrifice in reality allows for the continuation of the species. Having devoured the male and therefore replenished her stores of energy, the female can then rapidly lay her eggs.

There is another explanation that does not contradict the first. In many insects the males are not involved in the protection of the family or group and are rarely in competition with each other. So there is no need for the males to be a greater size.

TOO LITTLE TO LIVE ALONE

In some animals, sexual dimorphism is so marked that the animals could be mistaken for being two species. The female of the marine worm of the genus *Bonellia*, which has a proboscis that helps it to hunt, generally measures about 10 cm in length, whereas the male rarely exceeds 1 mm. In the *Bonellia viridis* species, the female (right) may be as much as 1.5m long whereas the male is a thousand times smaller. The male is at first a larva and changes into a male only once it attaches itself to the body of the female as a parasite.

▼ *The European red deer stag is larger than the doe. His head bears antlers which he uses as a weapon against other males during the rutting period.*

Male and female in one

Individuals of the same species sometimes have difficulty meeting each other. Hermaphroditism may be considered as one of the means nature has found to ensure the survival of certain species.

When two slugs mate, there is first an abundant secretion of mucus, then each of them puts out its penis. These twist together and exchange their sperm. Whereupon each penis retracts and takes the partner's semen into its own female sex organ. For slugs are hermaphrodites equipped with only one sex organ, consisting of a penis and a semi-vagina.

It still takes two

Sexual reproduction, efficient though it may be in ensuring the genetic diversity of the following generation, nevertheless obliges males and females to meet each other.

But meeting up can be very difficult sometimes, especially in a marine environment, where the distances between members of a similar species are often considerable. Hermaphroditism is therefore a very pragmatic approach to reproduction. By possessing both male and female sex organs hermaphrodite animals enable sexual reproduction to take place at each encounter of two individuals. The survival of the species is ensured. This is the case with many invertebrates, particularly snails and slugs. In fact, hermaphrodites are nearly always marine or aquatic animals.

Theoretically, these animals are able to fertilise their own female sex cells themselves. By reproducing itself, a hermaphrodite would ensure singlehandedly the production of a new generation. But its offspring would be genetically identical to it which, if species survival depends on diversity of the gene pool, is not desirable. So the genital anatomy of these animals, either the form of the organs or the distance between them, generally prevents autoimpregnation.

▲ *Slugs are hermaphrodites. When they mate, each extends its penis and the two animals exchange their sperm.*

When solitude is not a problem

However, certain animals do impregnate themselves. For instance, the taenia, or tapeworm, a parasite that lives in the intestines of cattle and human beings, does so. It is formed of about a hundred rings, each with genitalia that are both male and female. If the worm is completely alone, it just folds over itself to reproduce. The rings that come into contact exchange their sperm and the deed is done.

Fish of the genus *Serranus*, to which sea basses and sea perches belong, are also sometimes able to impregnate themselves. For example the painted comber (*Serranus scriba*), which is found in the eastern part of the Atlantic as well as in the Mediterranean and Black seas, may fertilise its own eggs at the beginning of summer. However, this practice is rare and does not seem to have threatened the survival of the species.

Sex changes when necessary

Successive hermaphroditism is more subtle. It means a change of sex according to need. The objective is to avoid producing equal numbers of males and females. As gender is genetically

A DAPHNIA IS FEMALE WHEN ALL IS WELL

Daphnia are minute aquatic invertebrates commonly occurring in ponds and lakes. This primitive crustacean has a reproduction cycle remarkably suited to its living conditions. When these are normal, daphnia are all female. Their eggs develop without having been fertilised, a process known as parthenogenesis, and hatch as females only. If the environmental conditions become bad, a few eggs produce male daphnia which fertilise the females. Each female then lays one or two very strong spare eggs, for storage, which fall to the bottom of the pond or lake and wait until conditions have improved before they hatch.

▼ *A display by groupers in the Mediterranean. The male (silver-coloured) circles the female to force her up to the surface with him so that he can impregnate her there.*

▲ *The taenia, or tapeworm, an intestinal parasite, is one of the few hermaphrodite animals with the ability to self-fertilise.*

determined, the ratio of the two sexes is always in the region of 50 percent, whatever the species. However, what can be measured on the scale of a species, can rarely be measured on the scale of a population. An isolated group of animals of the same species and the same sex may find themselves in a blind alley as far as reproduction is concerned. So what are they to do?

To get round this fate, nature invented transsexuality, in which one sex will change into another. Many fish practise this inversion, notably those from the families Serranidae (sea basses and groupers), Labridae (wrasses) and Sparidae (porgies). The brown grouper from the Mediterranean is thus born both male and female. This undeclared state lasts for about five years, until the fish measures about 40 centimetres long. Then it becomes a female. Between the age of 14 and 17 years, when it has doubled in size, the grouper can change into a male. Whether it changes depends on the density of the local population of the species. When the brown grouper is overfished, the oldest individuals remain female in order to ensure the survival of the species.

She becomes he when she becomes attached

Such demographic constraints do not concern the slipper limpet. This mollusc, dangerously overabundant in the English Channel, reproduces by piling up and forming colonies. At the beginning it is a larva, which attaches itself to a rock. As it changes into an adult, it becomes a female slipper limpet. The larva that attaches itself to her becomes a male and impregnates her. The same thing happens to the next larva, the first female making all the larvae that pile up on top of her into males, one by one. The first male, no longer serving any purpose, becomes a slipper limpet of indeterminate sex.

Reproduction therefore depends on the youngest and smallest individuals, always male, on the top of these piles. They impregnate the same female by means of a long penis. Between the two are slipper limpets of indeterminate gender. For these molluscs, masculinity depends on being attached to another limpet. If a slipper limpet falls off the pile, it changes into a female.

▲ *Slipper limpets. These molluscs, which are hermaphrodite, collect in a heap in order to breed, and end up actually making barriers across currents. Sediment piles up behind their structures and causes the environment to silt up.*

Life at any cost

In a forest in New Guinea, the cries of the birds of paradise grow ever louder. These colourful birds seem extremely agitated and are behaving in a rather strange way. But there is no cause for alarm: they are just guys looking for gals. Anything goes when it comes to ways to attract a mate. This is a very serious matter for animals. It is nothing less than a question of ensuring the survival of the species. And in order to have the chance to reproduce, there first has to be a conquest. Courtship displays, fights, songs, finery, offerings and various kinds of demonstrations are among the many different strategies employed to charm a potential partner. The unsuccessful ones have reason to worry: at best, they will be left standing; at worst, their lineage will die out.

▼ **THE DANCE OF THE JAPANESE CRANES**
Among Japanese cranes, couples are elegantly dressed when they get together at the end of winter. Decked out in black and white plumage, with a splash of red at the top of the head, the partners perform a courtship dance, beating their wings, jumping up into the air and uttering sharp cries. And when several pairs perform at the same time they look like a corps de ballet. The sight is all the more striking because this choreographic display often takes place in a winter landscape where the graphic shapes of denuded trees stand out starkly against the dazzlingly white snow. It is a sight that has inspired many a Japanese artist.

▲ FIGHTING FOR THE HAREM

In the first cool nights of autumn, the woods ring with loud plaintive cries. Stags are calling to each other. Their bellow, or troat, heralds the beginning of the courtship season. Among European red deer, it all begins with a fight. The males must form their harem, but there is strong competition to head the group of females. Fights between two males are frequent and exceptionally bloody, with antlers clashing violently. The weaker stag will yield to the winner and go off to try his luck elsewhere.

▼ THE ACCENTOR, A BELIEVER IN FREE LOVE

You would never guess that the charming little sparrow-like bird with discreet plumage that flies about in the garden in fact ignores all rules in matters of the heart. In the breeding season, if the territory is extensive and individuals are numerous, male and female accentors mate indiscriminately. The males may mate with several females and a female may be fertilised by several males. The females sit on the eggs and after two weeks pairs of birds can be seen flitting to the nests regularly to feed their young. But the offspring will never be sure of the identity of their father!

▲ IS A WRASSE MALE OR FEMALE?

Attentive and protective, the male wrasse builds nests where the female can come and lay her eggs. Some of the 500 species of wrasse that live on our planet build their nests of seaweed. Others make them from debris consisting, in the Mediterranean Sea, mainly of sea-grasses. Sometimes, just a crack in a rock will do. In addition, the nest is always well defended against any intruder that may venture near it. The wrasse really lavishes attention on his mate. This is not surprising, as he was a female before he was a male. The transformation takes place with the onset of old age and in the wrasse is always accompanied by a change of dress.

▶ WHALE GAMES

Humpback whales choose the warmer waters of the Pacific, near the shores of lower California, to meet up in winter. Their courtship is a spectacular sight: the males compete for the favour of the females amid high sprays of foam. They perform great leaps, sometimes jumping right out of the water before falling heavily back down again. And 25 tonnes of whale makes a big splash. Often, there are also violent fights to be the favoured mate. But the secret weapon of these great charmers is their song: each male has his own.

▲ SALAMANDERS SAY IT WITH DROPPINGS

The red-backed salamander of North America has very strange habits. Males attract females with their rich droppings. At the beginning of the breeding season, they remain in their holes and lay their droppings at the entrance. When a female goes past she inspects this pile to see what its owner has eaten. The remains of termites, the favourite food of the species, demonstrates the male's superiority. If he is able to find and feed on such rich food, he must be a good breeder. So he will be the chosen one. If he has eaten only ants, which are less easily digested, his chances are not as good.

▲ THE GREAT GAMES OF BIRDS OF PARADISE

Male birds of paradise are polygamous and at the beginning of the courtship season they have a lot to do. Sporting a plumage of rich colours rarely equalled in the world of birds, they all show off together in the trees. The forests of New Guinea are the scene of their spectacular demonstrations. The males call loud and clear to announce their presence and attract the females. When the females arrive, it is up to each male to display his feathers most majestically, beat his wings the loudest and perform the highest leaps and acrobatics, including hanging upside down. The female chooses her partner by crouching down in front of him.

▲ CHAMPION BOXERS

At the end of winter, hares chase each other more often, and are frequently seen racing across cultivated fields. It is the rutting season. Several males, or bucks, which often outnumber the females, will run after a female, the doe, at the same time. But this is not always enough for her to decide on the lucky male. So then the males have a boxing match. Standing on their hind legs, they box with their front paws. This is called bucking. Fights among kangaroos in Australia happen in the same way and for the same reason, to win the female.

▼ THE TERN'S ENGAGEMENT PRESENT

Assembled in groups on rocky islands or in lagoons, female terns are captivated by the demonstrations of the males. The males circle slowly, calling, then land nearby, lifting their beaks towards the sky and spreading their wings slightly. But this is not enough to convince the females of their breeding potential. So the males fly off and catch a fish, which they bring back and present to the female they desire. This little gift is often the magic formula for mating to take place. Many other birds do the same thing, like the male bee-eater, which brings a dragonfly or other insect to the bird of his dreams.

▶ **GIRAFFES FIGHT NECK TO NECK**

Firmly planted on their spread legs, two male giraffes wield blows against each other with their necks. Before courting, a male must eliminate his rivals, and the fights are often fierce. Each opponent bangs his head with great force against his rival until one of them takes flight. The winner then goes towards the female and begins a much more peaceful dance. The opening manoeuvres may last several hours. This time, the long neck becomes an instrument used to charm. The male rubs his neck against the female, giving her lingering caresses and even light blows, but affectionate ones this time.

▼ **THE BOWERBIRD'S MASTERPIECE**

To win over females, bowerbirds show an unusual degree of imagination. These near cousins of birds of paradise make good use of their artistic talents. They may lay a forest walk, a corridor of woven branches, or create complicated structures against the trunks of trees. The point is to see who shows the greatest skill. They almost all decorate their masterpieces, which are not nests, with various sorts of presents to attract their sweethearts: pebbles, shells, polished bones, crystals, shiny objects (including metal, if they find any), and even feathers stolen from their neighbours, the birds of paradise.

When one generation hunts another

In some species, to give life also means to lose it, because the whole life of each individual is devoted to the continuation of the species. Once the individual's task is completed, it dies.

▼ *These sockeye salmon, or red salmon, are on the point of spawning. The male (in the background) shows a more humped back and more severely deformed jaws than the female (in the foreground).*

It is still winter. On the Kamchatka Peninsula, in the extreme east of Siberia, there is snow for a large part of the year. Nothing seems to move in this calm, sleepy countryside, except in a small river where the water bubbles furiously as though hot springs were welling up. Abundant, exuberant, multifarious life thrives in the water, watched, from the surrounding emptiness, by large birds of prey with huge beaks. They are white-tailed eagles, and they are waiting for the salmon in the river to die.

The red salmon had come up the river a few weeks earlier on a final journey, a journey that would first give life and would end in death.

The call of the sea

Let us go back a little. A few years ago, salmon eggs were laid in this river. They hatched into young fish, which for at least a year lived in their 'maternal' environment. They had time to store away in their memory many details of the surrounding sights and sounds. Then, slowly, they began to make their way down this river, then down the larger river Zhupanova as far as the mouth, where the Pacific Ocean begins.

This journey, which took from four to six months, was perilous for the young salmon. They would have encountered predators, pollution, fishermen and dams. But having survived these obstacles, they swam out into the open sea.

The red salmon first stays in the coastal waters, then it moves out towards the east or the northeast, to the middle of the North Pacific in the direction of the icy Bering Strait. For about four years it grows and then becomes an adult salmon. In summer, it frequents the northern waters, and in winter it returns to the south in search of milder temperatures. During the whole of this period, it feeds on zooplankton and small fish.

The return to its roots

Once an adult, it is time for the salmon to return to the waters of its birth. The fish swims near the coast and, with the help of chiefly olfactory landmarks, it finds the estuary of the river it left several years ago. It is then that it begins to undergo spectacular morphological changes. Its endocrine glands

out, die and fall prey to raptors and carnivores, the patient spectators. Soon there will not be a single one left. But, safely hidden away in their holes, the eggs of the following generation wait to hatch.

▲ *At spawning time, bears come to feast on the dead salmon that drift in the river.*

The eel: a journey in the other direction

Eels have a life rather similar to that of salmon, but it follows a reverse pattern. Whereas salmon are born in fresh water, eels are born out at sea – for the European species it is the Sargasso Sea, off the Antilles. Then, after a migration of several years, they come in to shore and set off inland to conquer rivers and streams. There they spend several years, until the time when they begin the journey back to the sea as adults. By now their shape has changed so that their bodies are more elongated, their jaws are bigger, and their colour has become darker.

▲ *The eel also makes a long journey before returning to the waters of its birth to breed.*

The adult eels will return to the waters where they were born to breed and give life before they in turn die at the bottom of the sea.

▲ *The 'dance' of the salmon at the time of spawning (top). At this time the eggs are laid and fertilised (above, left). After hatching, the young (above) will grow up in their birthplace before travelling far down the river to the sea.*

cause it to take on a red tint that in the male becomes almost scarlet. In both the male and female the mouth also undergoes a curious transformation: it grows bigger, curves and gradually takes the form of an excrescence with each jaw ending in a hook, while the teeth grow longer. Similarly, the back grows more rounded and forms a hump. All this combines to give the fish quite a frightening appearance, especially the male.

Now the salmon has to swim upstream, make energy-draining attempts to leap up waterfalls and sometimes dams, and avoid the talons of a fish eagle or the claws of a bear, both very partial to salmon. By the time they arrive at the place of their birth the adult salmon are quite exhausted, especially as their deformed mouths prevent them from eating properly.

The final combat

It is the end of the journey. The females dig small hollows where they will lay their eggs. And the males begin to fight between themselves for the right to fertilise the eggs. The struggle is often fierce. But the females are not to be outdone and they vigorously defend the site they have chosen against other females in search of a good place to lay their eggs.

In between fights, the males remain motionless, reserving their vestiges of energy. Again, there is considerable risk of being hunted at this time. Before long the river returns to its normal state of calm, as the red salmon, worn

WHEN REPRODUCTION IS FATAL TO MALES

For bees, it is not good to be a male. When mating takes place, the embrace is so strong that the male often loses part of its abdomen and dies. And even if it were to survive, it would be excluded from the hive and end up dead. Being excluded from the society of his fellows is fatal. Usually, he is stabbed by the workers at the entrance to the hive and dies of the venom in their stings. This behaviour, which perhaps seems cruel, is merely one more way to ensure the survival of the species: the male has accomplished his mission and is now just another useless mouth to feed.

Pocket babies under threat

Relics of ancient times, marsupials are more defenceless than other mammals because of their manner of gestation, which exposes the foetus to environmental dangers. Although they flourished in Australia until the arrival of Europeans, many of them are threatened with extinction today.

Imagine our planet Earth, 100 million years ago. Dinosaurs rule the world. It is the end of the Cretaceous Period. Along come some curious animals, looking rather like the opossums of South America today. They are the alphadons.

These creatures were marsupial mammals, related to the modern-day kangaroos that can be seen jumping about in the Australian bush. The adventure of the marsupials was just beginning.

Suckling in a pouch

The group that forms the marsupials, the order Marsupialia, is in a sense midway in evolution between monotreme mammals (of the order Monotremata), which still lay eggs but suckle their young, and placental mammals, in which gestation is a masterpiece of adaptation and the young emerge from the mother fully developed.

Take the example of the kangaroo. About 40 days after impregnation, the female gives birth to a little worm scarcely formed, weighing one gram. The female sits down, which enables the embryo, or 'joey', to cling onto its mother's fur and climb up to her ventral pouch, called the marsupium. This climb takes a few minutes and once inside the pouch the joey immediately latches onto one of the teats there. It will finish growing in this protected environment until it is ready to poke its head out of the pouch and discover the outside world. In the red kangaroo, for instance, the joey detaches itself from its mother's teat for the first time at the age of 70 days, puts its head out of the pouch at about 150 days, and gets out for the first time from 190 days. Between 190 and 235 days, when it leaves its mother for good, this pouch remains its refuge.

Another feature of marsupials is that as soon as the new-born has reached the marsupium, females of most species mate again. A new egg is then fertilised. But the presence of the suckling young in the pouch has an inhibiting action, so that the development of the egg is delayed until

▲ *Until they are eight months old, young kangaroos go back into their mother's pouch at the slightest sign of danger.*

▲ *Like all marsupials, the female koala has a pouch in which to rear her young.*

▶ *At birth, the kangaroo 'joey' weighs one gram and is only a few millimetres long. As soon as it is born, it latches firmly onto a teat and does not let go for at least two months.*

the first baby has left the pouch. This phenomenon, called the embryonic diapause, may seem to hold things up a bit, but it has a distinct advantage. If the first baby dies, the second immediately takes its place.

▲ *The dwarf opossum-mouse is one of the smallest marsupial species in the world. It measures no more than about 10 cm in length.*

The mammal menace

Probably originating in South America, marsupials once conquered the world. But another type of mammal emerged at the same time — placental mammals. Their foetus is better protected than that of the marsupials, and they are therefore better equipped to face the dangers of life.

Faced with this competition, marsupials diminished in number. In the Tertiary Period (between 65 and 1.8 million years ago), they disappeared from many regions. It is thought that they died out in Europe about 20 million years ago. In South America and Asia, a few species survive. But it was especially in Australia and surrounding areas that they were able to flourish because this island-continent was not colonised by placental mammals for millions of years. The marsupials therefore diversified there, occupying most environments.

An uncertain future

Placental mammals arrived on the Australian continent with the Europeans in the 18th century. Until then, only the dingo, a wild dog, had existed there. The colonisers brought with them disaster, in the form of rabbits, goats, dogs, cats and rats, which rapidly became the merciless rivals of the marsupials. The new settlers, too, hunted them because some were a danger to livestock, like sheep.

Many species of marsupial in Australia died out between the 19th century and the beginning of the 20th. The most famous of these was doubtless the Tasmanian wolf, or thylacine, which was relentlessly hunted by humans until it became extinct in the 1930s.

Today, marsupials continue to be threatened, including species which used to be common, like the grey and the red kangaroos. Fortunately, there are now conservation programmes, set up to protect symbolic species like the koala and the Tasmanian devil. But in Asia, the flying phalangers, or in America, the opossums, survivors from a world in the far distant past, have a rather dark future.

▲ *These baby Tasmanian devils are a marsupial species (Sarcophilus harrisi) presently under threat.*

▲ *Humans got the better of the Tasmanian wolf. A powerful predator, it was ruthlessly hunted by European colonists in Australia until it was completely wiped out, because it sometimes attacked sheep.*

MAMMALS THAT LAY EGGS

Monotremes are primitive mammals which lay eggs, and live only in Australasia. They include the echidna, or spiny anteater (between two and five species according to scientists) and the duck-billed platypus, which has webbed feet, a duck's bill and a tail like a beaver (right). When the first platypus arrived in Europe as a stuffed specimen, scientists thought it was a practical joke. Monotreme females sit on their eggs as birds do, but suckle their young like other mammals.

A female leatherback turtle has climbed up onto the beach, dug a hole in the sand with her fins and laid her eggs. Now she is returning to the sea.

Strange parents

While some oviparous species let nature run its course, others show a wealth of resourcefulness to ensure the survival of their offspring. Many animals are prepared to go to any lengths in order to ensure the survival of their young.

On a hot summer night, the area around the marsh is abuzz with noises, calls and songs. If you were to walk along the water's edge with a torch you might come face to face with an astonished-looking little toad. Motionless and somewhat hunched up, it is carrying eggs on its back. It is the midwife toad (*Alytes obstetricans*), so-called because this species has the curious habit of carrying its eggs about until the tadpoles hatch.

What a nice mother, you may say. But you would be mistaken. This is the male's job. In fact during mating he helps the female deliver her eggs and then he fertilises them, while winding them – they form a gelatinous string – round his hind legs. He then becomes a mobile incubator, and at regular intervals even goes to the water to keep the eggs moist until they hatch.

Natural incubators

Other species do not care for their offspring as well as the midwife toad, and choose to leave things up to nature. This is the case with marine turtles like the leatherback turtle and the green turtle.

When they are ready to lay, the female turtles come ashore to bury their eggs high up on a sandy beach. After much effort spent digging the hole deep enough, they lay their eggs in it, cover them

▼ Directly after birth, young turtles undertake a dangerous journey to the sea. But even if they do not fall victim to predators on land, there are dangers awaiting them in the sea.

▲ *In the midwife toad (Alytes obstetricans), the male is responsible for incubating the eggs. He also carries them on his back for three weeks before dropping his burden in a pond to hatch.*

with sand and go back to the sea, exhausted. Warmed by the sun, this natural incubator helps the eggs to develop. And one day, the little heads of new-born turtles will emerge out of the sand and they will haul their little bodies to the sea as quickly as they can, as long as the predators lying in wait do not eat them.

Plant incubators

Birds take care of their eggs and sit on them conscientiously. However, there are some that do not obey this rule. Among them are the wild or brush turkeys that live in Australasia.

These large birds do not sit on their eggs. Instead they have an ingenious but no less demanding alternative arrangement. The male digs a large hole and fills it with leaves, twigs and other damp material. Then he covers it with a dome of sand which can be as much as a metre high. When the female arrives to lay her eggs, the male scrapes a hole in the top of the dome so that she can lay her egg where the plant matter is decomposing. And as she lays her eggs at intervals of several days, the male repeats the operation. Then the eggs are incubated in the heat given off by the decomposing matter.

When the temperature becomes too high, the male digs openings to allow the heat to escape. But he may also add leaves and branches if it is too cold. It is an operation on a grand scale. Between the preparation of the nest, its maintenance and the incubation itself, the whole process takes 11 months.

Incubating seahorses

Even in the world of fish, parents take care of their children. This is certainly the case with seahorses, for example. And here, too, the male gives generously of himself. At the time of mating, the female releases her eggs into a sort of pouch on the male, at the level of his lower abdomen. Once the eggs are in this pouch, he fertilises them and they are incubated in situ.

A radical change in the tissue of the pouch allows the offspring to receive enough oxygen. The tiny babies are born after about two months, when the male pushes them out, with some difficulty. The behaviour of trumpet fish, a species closely related to the seahorse, is similar.

In fish like the American perch or the stickleback, the male closely guards the eggs throughout the incubation period, ventilates them and chases away intruders.

In discus fish, the hatchlings take refuge in their parents' mouths if there is danger or to travel safely from one place to another. Moreover, the young feed by sucking the scales of their parents, which produce a nutritious, energy-giving substance.

▲ *It is difficult to find a better mother than the seahorse father, which incubates his offspring in his abdominal pouch.*

Mother lays and father sits

Among birds, it is generally both parents that care for the young. Often, but not always, the female sits on the eggs and the male helps to feed the nestlings. But in certain species of small waders (phalaropes and also small sandpipers), the care of the young is left entirely to the male.

▼ *This brush turkey –a mallee fowl – has constructed this natural incubator himself, from earth and plant matter.*

A PERFECT MOTHER

Who would have thought that the earwig (*Forficula*) actually possessed maternal instincts? The female lays her eggs in a very safe, sheltered place. But instead of leaving them to develop on their own, she watches over them and guards them constantly. Better still, she brushes them regularly to prevent microscopic fungal spores sticking to and growing on them. When the young are born, the female continues to watch over and protect them from danger. In some species, she even feeds the babies herself. And her care does not stop there. Studies have shown that female earwigs may take charge of the eggs of another female, even those of different species.

The dead body of a small animal...

... is dragged into a hole...

...and buried underground.

Substances secreted by the burying beetle reduce it to a compact ball,...

... to be discovered by the larvae guided by the chirping of their parents.

The spirit of sacrifice

In penguins, especially the emperor penguin, the degree of care, given first to the egg and then to the chick, is immense.

The emperor penguin breeds in the middle of the southern winter, deep in the interior of the Antarctic continent, where it is not easy to find food and where temperatures are as low as –20°C. As soon as the single egg is laid, the female begins a long journey, sometimes travelling as far as 160 kilometres, to go back to the sea and find food to restore her energy. During this time, the male, who gained weight before the egg was laid and so has large energy reserves, incubates the egg while he sits on the ice. He puts the egg on his feet and covers it almost completely with the skin of his lower abdomen. Kept in this way at the same temperature as the body temperature of the adult, the embryo can develop. Hatching takes place by the time the female has returned, full of food.

Then it is the male's turn to go back to the sea. By now he is pretty hungry as he has been fasting for more than 100 days. Both parents will repeat the journey to and from the sea in order to feed their chick until it is independent and weighs more than they do. The extra weight will allow it to fast for as long as about 30 days, until it reaches the sea to feed itself.

Male emperor penguins have another trick for surviving the difficult task of incubation. They assemble in large groups and have a rotation system whereby each of them, in turn, is in the middle of the group or on the outside. This allows them to reduce heat loss greatly.

A macabre pantry

Burying beetles are creatures with very strange habits. At breeding time, the male and female go off in search of the body of one of the small animals they usually eat, like a mouse, or shreds of meat from a larger carcass.

When they have found a little corpse they begin by excavating a hole. Then they lie on their backs, slide underneath the animal, and propel its body into the hole by 'pedalling' with their feet – a Herculean task.

Once the corpse is in the hole and buried, the beetles secrete onto it a chemical substance which hastens decomposition so that the body is soon just a compact ball.

Then they dig a gallery above this store of food, and the female lays her eggs in it. When these hatch, the adults attract the emerging larvae to the food supply by emitting a special chirping sound which guides them to it.

▼ *What are these emperor penguins looking at? Their feet, on which their offspring are quietly incubating.*

THE INFINITELY BIG, THE INFINITESIMALLY SMALL

The unknown world and the invisible

Cross-section of the smallpox virus. The red shapes are the cell nuclei containing DNA.

▲ *The central region of a spiral galaxy like our own. More than 200 billion stars extend from it like a multitude of arms, a formation typical . of spiral galaxies.*

▶ *A superb view in cross-section of the elliptical Sombrero Hat galaxy.*

THE MILKY WAY

A path among the stars

The Milky Way, a long light-coloured trail across the sky, is the visible part of a gigantic cluster of stars that is our galaxy. It is one of the building blocks that, together with hundreds of billions more galaxies, make up the Universe.

On a clear night, away from city lights, look up and you are sure to see a sort of white band across the sky. What you are looking at is in fact the trace of a grouping of billions of stars: the Milky Way.

For centuries it appealed to the imagination of poets. Then it became the object of scientific attention. The 'drop of milk from Juno's breast' and the 'path to Paradise' has turned out to be our view of a greater whole with a diameter of 100 000 light-years, made up of stars and nebulae, among other stellar objects, and of which our Solar System is just a part.

200 billion suns: our neighbours

The Milky Way, in other words, is nothing but our galaxy viewed from the inside and in cross-section. From our observation point, on Earth, there appear to be more stars in it than there are elsewhere. And, as we are not in the middle of the galaxy, on cloud-free nights we enjoy the benefit of a good view of the densest areas.

We live in this stellar conglomeration alongside more than 200 billion other suns. They are collected into a spiral galaxy that looks like a disc when it is viewed from the side. When our view is perpendicular to the plane of this disc towards 'the outside', there are not many stars in the way and we see the disc clearly as a milky band that stretches along the sky. On the other hand, if our view is into the plane itself we see a high concentration of stars which obscures the band.

The centre, 28 360 light-years away

It was no easy task to discover the shape of our galaxy and the precise location of its centre, especially because visible radiation is heavily absorbed by interstellar gas and dust. Over an area 1 000 light-years in diameter, taking the Solar System as the centre, a quantity of stars representing nearly 10 billion times the mass of our Sun has been investigated using a wide range of instruments.

Observation of the galactic halo surrounding our galaxy has revealed the presence of globular clusters. These spherical groups of several millions

▲ *Our galaxy seen from within (from the Earth) and in false colour. The band of the Milky Way is clearly visible, together with the bulge in the centre.*

▼ *Not all galaxies are the same shape. Below, from left to right are NGC 7742 (elliptical); NGC 1365 (barred spiral); and M51 (spiral), in the Great Bear.*

of stars are the relics of a period when the galaxy had not yet formed into a disc. By calculating their position scientists were able to deduce the exact position of the galactic centre. It is situated 28 360 light-years away from us in the direction of the constellation of Sagittarius.

Arms, a heart, and a bulge

As a spiral galaxy, the Milky Way has a heart from which arms extend in a swirl, much like a Catherine wheel. Imagine the heart as the most central part of what is called the bulge, a sort of swelling in the centre of an immense dinner plate. Flattened out, this bulge would have a diameter of 20 000 light-years and a thickness of about 3 200 light-years.

Although it is difficult to get to, the heart is such an enormous store of energy that astrophysicists cannot resist imagining all sorts of scenarios surrounding it. One particularly unsettling theory is that the high density of stars in this region could, by gravitational collapse, bring about supermassive black holes, which would suck up surrounding matter.

The arms of the galaxy are regions that abound in stars of all kinds and all ages: old ones, very bright young ones, and even future stars. Matter is very compressed in these regions because of the enormous force of gravity. The Solar System, of which our tiny Earth is a part, is situated in one of these arms. And just as the Earth goes round the Sun, the Solar System goes round the heart of the galaxy in the company of many other stars. This whole, highly organised system is itself only a part of another, even greater, whole.

The galactic panoply

Progress in the field of astronomical instrumentation has enabled us to study images of the sky in detail, especially images of nebulae. The nature of nebulae was previously completely unknown and they were thought to include globular clusters, dead stellar bodies, gas clouds and also galaxies. But with better telescopes, among the millions and millions of stars visible on the photographic plates, galaxies began to be seen more and more clearly. It even became possible to tell their size and distance from the Earth.

Terms designating galaxies describe their shape: spiral, barred spiral, elliptical and irregular. Each one contains between 100 and 400 billion stars. By measuring their distance from Earth it is possible to situate them in space and to know that galaxies live in clusters, themselves grouped in superclusters. Studies have also revealed two distinct types of star 'populations': the younger Population 1 stars, found in the disc, and the older Population II stars, which are found in the halo and in globular clusters.

We are now certain that the Universe at present contains many more galaxies than one galaxy contains stars. Each star is a basic building block among 100 billion others forming a galaxy, which, with hundreds of billions of others again, makes up the Universe. When looking at the Milky Way, the suburb of such a vast, complex Universe, we may well be overcome by a sensation of dizziness.

▲ *A simplified representation of our galaxy, seen from the side (top) and from above (bottom). Our Sun is two thirds of the way out from the centre. We live in the suburbs of the Milky Way!*

TYPES OF STARS IN THE GALAXY

1. The stars in the halo, or global sphere, which do not form part of the galactic disc. They are very ancient and essentially rich in hydrogen and helium (stars of population II).

2. The stars in the disc (except the bulge), which constitute population I. Stars of all ages are found here, from the youngest (a few million years old) to stars as old as those in the halo.

3. The stars in the bulge, of low mass and very old (like those of the halo), which have a yellowish colour.

An exceptional planet

Since their birth within the Solar System, the Earth and its sister planets have had very different destinies. Although made of the same matter as the others, our planet has distinguished itself. This is fortunate for us, because we would not otherwise be able to live here.

On July 4, 1997, broadcasting live from Mars, NASA's little robot, *Sojourner*, sent fantastic images of the red planet to media all over the world. The pictures made the front pages. Yet this was only the latest stage in the exploration of our Solar System, which began in the early 1960s. Since then, the impressive number of images and samples brought back by the numerous space missions has included pictures of the satellites and rings of Saturn and Uranus, the burning atmosphere of Venus and the 'pink sky' of Mars. These strange planets belong to two large families, and they are very different from Earth even though they were born at the same time.

One star, nine planets and many other objects

Somewhere in the Milky Way 4.5 billion years ago, a star was formed among hundreds of billions of others from a gas and dust cloud. It very quickly left the nebula to live its own life, drawing along with it grains of matter. These granules gradually gathered together (in the accretion phase) and began to form the future planets.

So began the history of our Sun and the system organised by its influence. Round the star revolve nine planets: Mercury, Venus, Earth, Mars, Jupiter, Saturn, Uranus, Neptune and Pluto, in order of increasing distance from the Sun. It is thought there may be a tenth planet beyond Pluto, but until more evidence is uncovered, no one is sure.

The first four planets are in the inner Solar System. Smaller than the others (Mercury has a radius of 243 kilometres and Earth 6 370 kilometres), they are solid, made up of silica rocks and iron. They also have a similar average density. If the density of water equals 1, then that of Mars is 3.9, Mercury 5.4 and Earth 5.52.

Jupiter, Saturn, Uranus and Neptune are in the outer Solar System. They are giant planets (the radius of Neptune is 24 750 kilometres, that of Jupiter 71 600 kilometres), and essentially gaseous in composition, containing mostly hydrogen and helium. Tiny Pluto, a little-known planet is the odd one out in the outer Solar System. It seems to have a rocky core because its density is twice that of water, but it is not known for sure.

Round these planets revolve numerous natural satellites: 66 are presently known, but there are probably others round the most distant planets. The Moon is the only natural satellite of the Earth. There are many other categories of objects within the Solar System: asteroids – small solid bodies of various sizes – grouped in a belt between the orbits of Mars and Jupiter, comets, and also the sets of rings round the giant planets.

▲ *A photo of the 'blue planet' taken in December 1972 by the crew of the Apollo 17 space mission on its way to the Moon. Clearly visible are Antarctica, Africa, Madagascar and the Arabian Peninsula, as well as the South Atlantic and the Indian oceans, covered with heavy cloud masses.*

the reason for the youthful appearance of its surface.

During the 700 million or so years following their formation, all the planets in the Solar System have undergone an intense bombardment by meteorites that have gouged great holes in their surfaces. These scars can still be seen on the Moon, on Mercury and on part of Mars, but have largely disappeared from the Earth's surface. Movement of the tectonic plates, volcanic activity and erosion – the effects of geological activity – have seen to it that the scars from past events have largely been erased. Volcanic activity has certainly fulfilled this function on Venus and Mars, where 75 and 50 percent of the surface respectively is covered by volcanic material. But today these planets no longer show signs of any volcanic activity at all.

▲ Mars, as recorded by the probe Viking 1 in 1976. Its southern hemisphere is pitted with craters that were made by meteorites. The northern hemisphere, covered in lava flow, appears unmarked.

◀ Saturn (in false colour) as seen by the Voyager 2 probe in 1981. The bands of bright colour below the level of the rings correspond to cloud masses in the upper atmosphere.

4.5 billion years old and still going strong

Earth is naturally the planet we know best. Compared with the other solid bodies, like the Moon, it is the most varied. It has a protective atmosphere and great quantities of water stored in a liquid state: three quarters of its surface is covered by the oceans. Nothing could be more different from the pockmarked desert of Mercury, for example. So why did the Earth, formed at the same time and from the same matter as the others, evolve in such a different way?

The planets are like geological machines with an engine that runs off the heat produced in the core. There are, in fact, two forms of heat: one generated by accretion, which is stored, and one given off by the disintegration of radioactive elements accumulated at the time the planet was formed. The heat sources are smaller or greater according to the size and mass of the planet, and therefore more or less quickly exhausted. The smallest objects in the Solar System are thus today geologically dead, while the Earth, the largest of these objects, remains active. In a sense, this constant activity keeps it healthy. It is even

▲ A photomontage of the planets Jupiter (in the background) and Neptune (in the foreground). The atmosphere of these two gaseous planets is characterised by the rapid circulation of cloud masses, which generates constant cyclones. The large red spot on Jupiter is a spectacular example of this.

Air!

The atmospheric conditions on Venus – a sky overcast with thick cloud and an air temperature of 450°C – are enough to make earthlings tremble. And yet Venus might have been the twin sister of our planet. Of similar mass and diameter, it has retained its original atmosphere.

After its birth, each planet went through a phase of differentiation, which in fact corresponded to the formation of the different layers (the crust, mantle and core) of its

ARE WE REALLY ALONE?

After several decades of space exploration, it seems that Earth is the only planet in the Solar System on which life exists. But has this always been the case? Nothing is less certain. There seems to have been an abundance of water flowing on Mars at one time. So were all the elements necessary for biological activity present on the red planet? If so, it remains to be seen what form they took. Today, conditions on this planet do not allow the existence of life, at least in relation to what we know on Earth. Perhaps there are fossilised traces of life on Mars. Exploration of this planet may hold surprises for us that could help us to understand how life appeared in the Solar System and therefore on Earth.

▼ *The Solar System viewed from the outside. In the foreground of the drawing are the orbits of Pluto, Neptune, Uranus, Saturn and Jupiter. Beyond the orbit of Jupiter, the asteroid belt constitutes the boundary between the outer and inner Solar System, where Mars, Earth, Venus and Mercury orbit.*

internal structure. This phase was accompanied by a great emission of heat and an intense evolution of gas from the crust. The discharge of gas produced the original atmosphere of the inner planets, which were either able to retain it or not, depending on their mass.

So Mercury, a small planet, lost its atmosphere which dissipated completely in space. Mars, a little larger, managed to keep a small part of it, consisting almost entirely of carbon dioxide. Earth and Venus, on the other hand, being much larger, retained much of their atmosphere. But on Venus the atmosphere is also largely made up of carbon dioxide, and was for a long time maintained by volcanic activity without undergoing renewal.

Enclosed under a blanket of permanent cloud, Mars has an atmosphere which does not allow heat to escape. This creates a greenhouse effect which excludes any possibility of life.

In the case of the Earth, because of a particular chemical process, a large part of the carbon dioxide in the atmosphere came to be fixed in the rocks. The work of 'purification' was then continued when photosynthesis was started … and the result is the atmosphere we have today which is indispensable to life.

Ninety-two simple structures

Whatever form it takes, all matter is made of the same fundamental thing in varying quantities: atoms. To date, 92 variations of atoms, called elements, have been recorded.

Just imagine. Practically everything in the Universe – the Earth itself, human beings and plants – is made of the same basic material! And this material is not recent. As far as we know, it began to exist shortly after the Big Bang, 15 billion years ago, at the time of the separation of light and matter, and at the same time as the Universe began to expand. Conditions were then right for particles, the basic elements of matter as we know it, to come into being.

The right ingredients

Before then, the Universe was just a soup, impossible to describe even with the tools of modern physics. Time and space did not have the meaning we give to them. Temperature had infinitely great values. There was no distinction between light and matter. From this original soup (also called the primordial soup), quarks, the basic constituents of matter as we understand them today, united to form neutrons (uncharged particles), electrons (negatively charged particles) and protons (positively charged particles).

In the first three minutes of the orgy of energy that was the Big Bang, the lightest atoms formed.

▼ Splitting nuclei in order to understand more about how they are made is an obsession of particle physicists.

◄ It is the stars in the middle of galaxies that enable matter to become complex and all the elements to be produced.

The simplest, hydrogen, is made of a single proton around which one electron revolves. A near cousin, deuterium, is formed from one proton and one neutron and again has one electron orbiting it. Finally, helium consists of a nucleus made of two neutrons and two protons, with two peripheral electrons. Shortly after the Big Bang, no complex structure yet existed. But the basic ingredients of the Universe had been formed: about 90 percent deuterium, 10 percent helium and a little lithium.

The alchemists' dream

Under the influence of gravitational forces, this disordered matter gradually built up to form galaxies, within which gravity again played a part to form future stars.

The heart of each star works like a nuclear reactor where the fusion of matter takes place. New elements then develop based on hydrogen. At that point the star is a vast crucible in which matter becomes more and more complex. The fusion reactions produce, in succession, carbon,

◀ A simulation of the proton, the basic constituent of all atomic nuclei. According to the present model, it is made of three quarks.

of the electron, and his classification is still rather empirical. Almost 50 years were to pass before there was a more complete classification, produced by the Danish scientist Niels Bohr, in 1922. Today, all the simple elements from hydrogen to uranium have been identified, and no others in the Universe have come to light.

But scientists press onwards. They are trying to drive matter into a corner and are exploring the inside of the atom in search of the elementary particles that make it up – quarks, electrons, gluons and bosons.

Today they have a standard model of what constitutes matter and the forces that ensure its cohesion. This model shows 17 particles, the building blocks that are necessary to make up all atoms. Actually, it shows 17 times two, because each particle has an antiparticle: the proton has its antiproton, the neutron its antineutron, the electron its positron. Photons (particles without mass, which transport light) are a separate group, because they seem to be particles and antiparticles at the same time.

silicon, iron and many other elements. At the point at which iron is created, fusion reactions are no longer possible. The creation of heavy elements (those between iron and uranium) will be able to take place only in the final explosion of the star. So it is in the course of the evolution of the stars that matter goes from the light elements to the heavier ones, from elemental simplicity to complexity.

The appearance of simplicity

Today, 92 chemical elements have been identified, studied and classified. The first real ordering of the elements was proposed by the Russian scientist Dmitri Ivanovitch Mendeleyev, in 1869. But he failed to take into account the presence

IN HONOUR OF HELIOS

By breaking down the light coming from the stars, a spectrum is obtained in which lines of different wavelengths can be seen. These lines are the signatures of the elements making up the object observed. In the 1880s, spectra taken during an eclipse of the Sun revealed the presence of a new element, until then unknown. In a reference to the light source that had revealed it (the Sun, *helios* in Greek), this element was called helium. (Right: a helium atom with its electrons.)

Take a little gas and dust...

A by-product of the Sun, the young Earth was a planet on the boil, producing steam which created an atmosphere without which life would not have been able to begin.

▲ *Like the other planets of the Solar System, the Earth was created from the aggregation of matter present in the solar nebula after the formation of the Sun. It developed from the successive accretion of planetesimals and planetoids which were in its neighbourhood, and gradually grew bigger over hundreds of millions of years.*

In the beginning there was a cloud of hydrogen and helium, and tiny little grains of matter which gradually condensed and gave birth to the Sun. The rest of the cloud (less than one percent) was drawn into a wild dance with this young star and quickly flattened into a disc, called a solar nebula. There, the nine planets formed, one of which was the Earth. That was about 4.55 billion years ago.

A turbulent birth

In the nebula, there was total upheaval. It is difficult to understand exactly how the planets we know today can have come from the simple elements it contained. Were great gaseous protoplanets formed as a sort of first attempt, possessing a solid central core? Did larger and larger particles cluster together from the dust present in the nebula? In the 1970s, the Russian astronomer Victor Safronov put forward the theory that the planets were formed in three stages.

During the initial phase, which was quite short (about 1 000 years), a multitude of solid planetoids, which means planets in the making, with a diameter of between one and five kilometres, formed from dust in the nebula. This little world must have been quite turbulent and collisions in its midst very frequent.

In an intermediate phase, owing to collisions among themselves and the phenomenon of

◀ *Meteorites are fragments of rock or metal from space, which have struck the Earth and survived entering its atmosphere.*

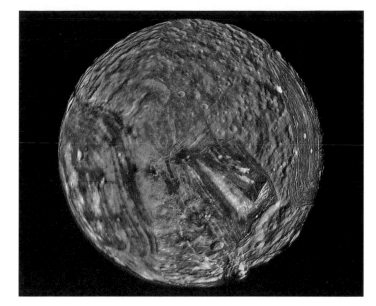

accretion – like dustballs that form when you sweep up dust and find that it sticks together – the planetoids developed to form the 'embryos' of planets.

The last phase was characterised by the growth of the embryos. They grew larger and larger (from 1 000 to 2 000 kilometres in diameter), their mass increasing and with it their force of attraction. They could then 'swallow' planetoids passing within the range of their influence, until they became planets. These last two phases must have taken about 100 million years.

A tempestuous youth

The early Earth had little in common with the blue planet that fills astronauts in space with wonder. Constantly bombarded by planetoids in violent random motion, it had an ambient temperature of 2 000° C, due partly to this bombardment, partly to its own contraction and partly to the natural radioactivity of the elements of which it was made.

In this mass of fusing magma, the exchange and transfer of matter took place and the interior of the Earth became differentiated. The heaviest of the elements, like iron, concentrated in the centre and made up its core, while the lightest elements rose to the surface to create the mantle and the crust.

The most abundant elements that make up the Earth – iron, oxygen, silicon and magnesium – were therefore distributed in a heterogeneous way. Iron, for example, represents 35 percent of Earth's mass, but the crust contains little (iron is its fourth element in order of importance) and is mostly made of silicates.

▲ *A ferrous meteorite. Iron meteorites, made of an alloy of iron and nickel, are probably pieces of planetoid cores that were shattered by collisions in the asteroid belt.*

A steam engine?

At the time of the differentiation of the Earth's elements, the very young planet was boiling away. The heat was so great that the lightest elements contained in the rocks escaped as gas. This discharge of gas created an initial atmosphere, which must have been made of carbon monoxide (CO), carbon dioxide (CO_2), molecular nitrogen (N_2) and water vapour (H_2O).

It is quite possible that some of the water had been 'brought in' by extraterrestrial elements, such as comets. These would have come from the regions farthest from the Sun. Comets are made essentially of ice, which would have melted as they hit Earth, and also contain elements like carbon and nitrogen.

In this atmosphere, carbon dioxide gas (which was also given off at the height of the phase of volcanic activity) must have been greatly preponderant. It would still be today, as it is on the planets Venus and Mars, if in the course of time it had not been dissolved and precipitated at the bottom of Earth's oceans in the form of calcium carbonate.

Doubtless, this substance later aided the beginning of life in the oceans, which in itself was to play an important part in the transformation of the atmosphere.

▲ *Miranda, a satellite of Saturn, illustrates the process of accretion, shown by its irregular surface.*

▲ *The most ancient remains of the terrestrial crust (4.2 billion years old) were found in a mineral, zircon.*

WATER, AIR .... LIFE!

Finding the beginnings of life is a great adventure which has enthralled humans for a long time. We still do not know for certain how life first began, but various hypotheses have been proposed. The first molecules no doubt came into being in the oceans. The consequent discharge of molecular oxygen into the atmosphere (from which until then it had been absent) was a decisive stage in the continuing existence of life. The dissociation of oxygen by solar radiation led to the formation of an ozone layer capable of absorbing ultraviolet radiation, radiation that would have destroyed organic molecules, in particular DNA, and would therefore have prevented the development of any life on the surface of the Earth.

Journey to the land of extremes

Galaxy 10²³

Fasten your seatbelt and hold on tight. We are going to take you into a world that will leave you bewildered and giddy, of previously unknown dimensions, where units of measurement have many rows of zeros... From a Universe containing an unknown number of galaxies, themselves formed of hundreds of billions of stars, the journey will go on to discover the most infinitesimal particle at the heart of all matter, the quark. This voyage to the limits of our knowledge will be overwhelming, since you may have some difficulty taking in these surroundings. What you will notice, however, is that from the infinitely large to the infinitesimally small, nature seems to obey simple laws.

CONSTRUCTION SETS AND STRONG GLUE

The point of departure: the Earth, a speck of dust lying on an arm of the Milky Way. With eight other planets, it revolves round the Sun, its star. This structure, the Solar System, was cast as a small troupe in the great ballet of our galaxy along with a few hundred billion other stars. The galaxies regrouped into clusters, the clusters into superclusters, the superclusters into the Universe. Each of these groupings is linked to the others by a system which is simple but as efficient as superglue: gravitational force.

And now, let us travel the other way, using supermicroscopes and even particle accelerators, to look at much smaller objects. Practically everything around us is made of molecules. A molecule itself is made of atoms which are made of nuclei around which electrons revolve. Within the nucleus are neutrons and protons. And within each neutron and proton are three quarks, the present building blocks of all matter. This small world is perfectly organised and held together by certain particles (gluons, photons and bosons), which act as 'force carriers'. Their role is to cement or stick the building blocks together to form the neutrons, protons, atoms, molecules, etc.

ZEROS AND A PEA

0.000000000000000001 m (metre). That is how the size of a quark could be written. But to avoid the tiresome task of writing an interminable line of zeros, a short cut for this notation has been invented. Rather than write 100 000, you can break down the figure into 10x10x10x10x10 and count the zeros, or alternatively write 100 000 as 10^5 (10 to the power of five). As you can see, this is an immense advantage when it comes to writing 10^{16} or 10 000 000 000 000 000, or even 10^{18} which is 0 000 000 000 000 000 001m, the size of a quark. (1mm = 0.001m or 10^{-3}m.) The exponent always shows the number of zeros, but do not forget the decimal point. All these numbers can be multiplied: for example, 4.5mm becomes 4.5×10^{-3}m, while 563 km can be written as 5.63×10^5m.

But it is still difficult to visualise dimensions such as these. To take the Solar System as an example, the Earth has a diameter of 12 600 km, the Sun a diameter of 1.4 million km. When we reduce all this, we get a pea beside a football. Lost in a space ten times as big as a football field, this little pea could equally well represent the nucleus of an atom.

Quark 10⁻¹⁵

Atomic nucleus 10⁻¹⁴

Atom 10⁻⁹

Aids virus 10

Planetary nebula 10^{15}

Sun 10^{10}

DOES IT COME TO THE SAME THING?

A sphere, some satellites and a strong force of attraction: what does this entity describe? The system of the Universe or the system of an atom? Apart from a (gigantic) difference in scale, the similarity is amazing. And the forces that structure matter suggest an organisation of the quark in the Universe as we can observe it in our own Solar System. Nature, moreover, is swarming with examples where the similarities linking the largest and the smallest are strong. Take one of the florets of a cauliflower, for example; it looks just like the whole cauliflower in miniature.

Earth 10^7

Moon 10^6

Man 1

Mite 10^{-2}

Background image:
group of galaxies 10^{26}

75

The building blocks of life

Every living organism, whether microscopic or gigantic, is made of cells. A true miniature factory, the cell takes from its environment what it needs to make its energy.

What do a yeast and a blue whale have in common? Both are made of cells, the basic units of life. The functioning of the cell is determined by a genetic programme contained in a molecule as complex as it is long – deoxyribonucleic acid, or DNA – and is directed by proteins. These proteins are controlled by and manufactured according to the same genetic programme.

In the living world, there are two types of cellular organisation: prokaryotic, in organisms where the DNA is free (for example, single-cell bacteria and blue algae), and eukaryotic, in which the DNA is enclosed in a nucleus (this applies to all other organisms, from single-cell yeasts to the giraffe).

A fairly crude organism...

The cell of a prokaryote is a relatively simple structure. It is enclosed in a membrane only a few millionths of a millimetre thick, called the plasma membrane, which protects its contents against its environment.

Made of a mixture of lipids and proteins, this 'skin' has pores and controls the exchange of molecules between the inside and outside of the cell. The inside is filled with a viscous substance called cytoplasm, which is very rich in proteins and in particles called ribosomes, which are collections of enzymes – proteins that activate biochemical reactions – and RNA, or ribonucleic acid, which is a derivative of DNA.

The function of the ribosomes is to manufacture proteins from the amino acids that are linked together in an order determined by the genes of DNA.

All the synthetic processes of the prokaryotic cell take place in its cytoplasm. It is a fairly crude machine and not very efficient.

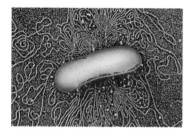

▲ *Chemical treatment has caused this bacterium to release its long DNA molecule which is usually coiled up in the middle of the cell.*

... and an elaborate model

On the other hand, eukaryotic organisms are much more complex. Their cellular functions are very compartmentalised, and are carried out within the cells by specialised organelles. Some take charge of producing energy, others see to photosynthesis, and so forth. Like all cells, eukaryotic organisms are separated from their environment by a membrane. Inside, in the cytoplasm, swim the various organelles (the largest of which is the nucleus), themselves also having membranes which partition off compartments where particular molecular reactions take place.

It is in the recesses of a specific network of membranes called the endoplasmic reticulum that all the proteins and certain molecules are made. They are then stored in cavities in the reticulum until they are locked into transitional vesicles. In these watertight parcels, they are transported to another group of organelles called the Golgi bodies, and there the 'raw' proteins are modified in order to become operational. When this is completed, the proteins leave the Golgi bodies in new packaging, called secretion vesicles. They are delivered in this form to the outside of the cell.

▲ *A eukaryotic cell (here, of beer yeast). Red indicates the nucleus; green, the mitochondria; and brown, the cell wall, making up a sophisticated factory with the machines being the organelles that manufacture proteins and energy.*

THE DIVERSITY OF CELLS

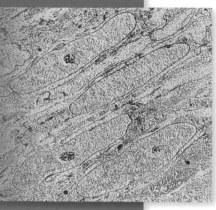

The functions of eukaryotic cells are highly compartmentalised, which allows these cells to specialise, giving preference to the work of certain organelles. For example, cells that make hormones (and therefore proteins) are characterised by a marked development in the part of the reticulum that carries ribosomes, the rough endoplasmic reticulum, and also by a considerable number of secretion vesicles. In contrast, muscle cells (right), which need energy, are extremely rich in mitochondria, the 'fuel' cells.

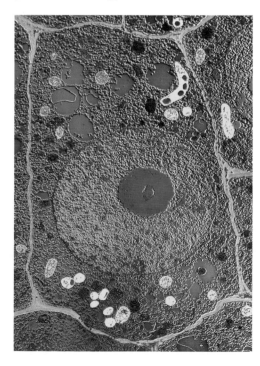

function. To renew this fuel, the mitochondria break down glucose and lipids, which they do with the aid of oxygen, and make more ATP. This process is called cellular respiration. It is a most effective mechanism. The oxidation of a single molecule of glucose is enough to form 36 molecules of ATP. In comparison, the prokaryotic bacteria that are responsible for the fermentation of alcohol manage to produce only a meagre two molecules of ATP from a single molecule of glucose.

▲ *Fungi (above, the spores of* Aspergillus*) are neither plants nor animals. They are nevertheless eukaryotes.*

A biochemical mini-factory

All this machinery requires energy, which is produced by a set of 'fuel cells', the mitochondria. These are small organelles that are virtually autonomous, containing a small amount of DNA. The mitochondria activate a molecule, ATP (adenosine triphosphate) that, as it dissociates, provides the energy essential for the cell to

▶ *Plant cells (here, maize) are surrounded by a second, almost rectangular wall, made of cellulose (in yellow).*

A separate group: plant cells

Plant cells are surrounded by a rigid wall made of cellulose, as well as their membrane. Moreover, to produce their energy, they have an extra system of organelles known as chloroplasts. These cells contain molecules of green cholorophyll and carry out the vital process of photosynthesis. In this process they transform light into the energy needed for the reaction between carbon gas and water. The sugars produced are stored in the form of starch, which allows the cells' mitochondria to manufacture ATP. It is a self-perpetuating cycle, which makes plants, unlike animals and fungi, totally autonomous. Their cells manufacture their own fuel and regenerate it.

But plant cells do not have the organelles that animal cells have, acrosomes, which allow them to move about. These microfilaments not only ensure the movement of material within the cell, but they also play an essential part in the separation of chromosomes during cell division (mitosis). When the acrosomes project beyond the cell wall in the form of a flagellum, they enable the cells to move. Without them, there would be no spermatozoa.

ANIMAL CELL

PLANT CELL

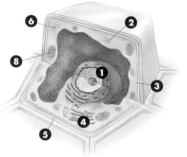

▲
1. Nucleus
2. Lysosome (digestion-excretion)
3. Cytoplasm
4. Golgi apparatus (transfer and concentration of proteins)
5. Plasma membrane
6. Mitochondrion (nutrition-respiration)
7. Centrioles
8. Chloroplast

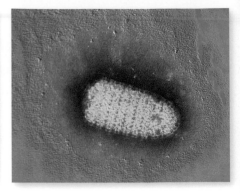

Small can be beautiful ... or deadly

Microbes are present in the whole biosphere, from the bottom of the sea to the inside of our bodies, from deep in the earth to the atmosphere. These tiny organisms are essential to life.

▲ *A living example of the Aids virus budding on a lymphocyte, a blood cell that normally ensures our immune defences. We do not die directly from Aids but from an illness that the immune system, weakened by the virus, is incapable of fighting.*

Since Pasteur, we have been quick to blame micro-organisms, or microbes, as being responsible for all our ills. Yet the majority of them are useful to us. What they all have in common is their minute size and the fact that they are made of a single cell.

Strange forms of life

Bacteria were the first form of life to appear on the Earth, 3.5 billion years ago. Their organisation is very simple. They have no nucleus but rather a nucleoid, which is a single molecule of DNA – their genetic material. They ruled our planet alone for a long time and alone they developed every process of life, from breathing oxygen to processing chlorophyll – some feed by photosynthesis – and including a form of sexual reproduction (the exchange and recombination of their genetic material). Most bacteria develop while playing a

part in the life cycle of matter. It is thanks to them that complex molecules are broken down into simple ones.

The kingdom Protista contains micro-organisms which are neither plant nor animal. These are single-celled but eukaryotic, unlike bacteria, and are thought to be the descendants of

LIVING SCULPTURES

In Hamelin Pool bay, in Australia, strange rock concretions are found, called stromatolites, which are formed by colonies of blue-green algae. These algae are photosynthetic cyanobacteria, which develop in shallow, very salty seawater and are formed from sediment, organic material and calcium carbonate crystals. Stromatolites were very common in the early ages of the world, but were known for a long time only in the form of fossils. Some go back more than 3.5 billion years, and are the first visible sign of life on Earth.

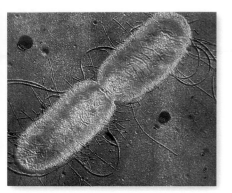

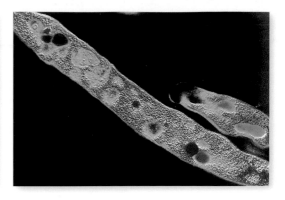

a complex cell formed by the fusion of several bacteria. Their genes are isolated in a nucleus, and organelles, the remains of formerly free bacteria, carry out specialised tasks. Some protists are plant-like, for example the *Pleurococcus*, an alga that makes the trunks of trees green. Others, such as the amoeba, are like animals in that they move and 'hunt' food.

Viruses, viroids and prions are not made up of a cell, but consist of molecules of genetic material wrapped in protein. They reproduce themselves only by entering a living cell, because they have no metabolism of their own. Living as parasites, they cause numerous diseases.

Billions of microbes per cubic metre

The number of microbes on Earth is too great to count. It is thought that there are as many in 2.5 grams of fertile soil as there are people on the Earth, that is, about six billion. The largest bacterium is about one millimetre long, and certain protists like foraminifers are more than a centimetre long. But the great mass of microbes is invisible to the naked eye. Most of them are between 0.1 and 0.001 millimetres in size. The record is held by the viruses, some of which are no bigger than 0.000006 millimetres in diameter! Their small size allows them to infiltrate everything.

Some micro-organisms, remembering a distant time when the atmosphere did not contain oxygen, live in environments where there is none. Bacteria deep in the soil, those that abound in our intestines and predigest what we eat, and yeasts responsible for the fermentation of wine or beer, are some of the anaerobic species.

Extreme records

The archeobacteria have survived since the early life of Earth where primitive conditions prevail. Some prosper in the saline crusts of salt marshes or in hot springs full of sulphur. Others proliferate in volcanic lakes of almost pure acid, or on burning hot rocks where the temperature can reach 113°C. And what about those in volcanic springs several thousand metres beneath the ocean, under enormous pressure?

Long live the microbes!

Humans depend on microbes for their food supplies. Without bacteria and yeasts, there would be no wine, beer, bread, sausages, tofu, cheese, yoghurt or sauerkraut. Without the bacteria and yeasts in the stomach of cows that help them digest grass, there would be no milk. Without the soil microflora that recycle dead organic matter so that plants can assimilate it, there would be no fruit or vegetables. And without our intestinal flora we would be unable to digest our meals.

Even the microbes that cause diseases are useful. Those like the Thüringen bacillus, which attacks crop-eating insects, are effective natural remedies. Microbes which attack us are essential, too, in small quantities. Doctors have noticed that in a world which is too sterile, our immune systems, which are made to fight off attacks from the outside, stop working properly. This is how so-called auto-immune diseases appear, like allergies and certain types of diabetes.

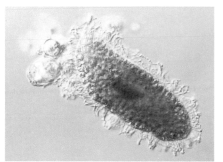

▲ The paramecium is a protozoan, a 'pre-animal' made of a single cell. Its cilia enable it to actively move in search of prey. No bigger than a drop of water, it is a fierce predator.

▼ *Below: green algae owe their colour to the presence of chlorophyll, which uses energy from the sun, water, carbon dioxide and mineral salts to synthesise sugars.*

▼ *Background photo: only micro-organisms can live in the waters of the Dead Sea, where salinity is as high as 30%.*

◀ *It is not even 1 mm long, and yet the bee mite has menaced European hives for a long time.*

Useful or harmful?

They may be little, but there are so many of them that in the end they carry a lot of weight. The part played by tiny animals in many ecosystems is of prime importance, and also in the propagation of many diseases.

▲ *The human louse is an insect that feeds on the blood of its host, causing itching.*

European beekeepers were almost destroyed. The larvae and nymphs of their bees were dying. For years their hives had been infested by a mite by the name of *Varroa jacobsoni*, which was less than a millimetre long. Eventually, this parasite withdrew. But its cousin, *Acarapis woodii*, which is even smaller, continues to give bees scabies, or mange. Mites, whether they are in hives, cheeses or mattresses, are invisible. But given their number, they carry an enormous amount of weight in Earth's biomass.

Dust-dwellers

Mites are very small indeed, but they are there just the same. In a mattress, you will find between 2 000 and 15 000 in each gram of dust. This is awkward for asthmatics, because the waste matter of dust mites, which feed on dead skin, causes allergies. Like the 7 000 other species of mite, which are rarely more than a millimetre long, the dust mite is a recycling specialist.

Near cousins of the spiders, mites too are members of the zoological group, or phylum, Arthropoda – which includes crustaceans, insects, myriapods and arachnids. Mites form an essential link in the planetary ecosystem. With more than 300 000 individuals in the soil per square metre,

▶ *A parasite of humans and animals, the tick feeds on blood. It is a carrier of various infectious diseases.*

they perform the phenomenally huge and vital task of transforming waste matter of all kinds into excrement. This feeds bacteria, which break down the organic matter further into its simplest elements.

Mites are also found in cheese – it is the cheese mite that, attracted by the fermentation, eats the rind – and as parasites.

Many people have experienced the bites of harvest mites, large red mites that cause itching on the legs. But ticks and the terrible itch mite (*Sarcoptes scabiei*) that causes the horrible irritations of scabies, are also acarians, or mites. Lastly, adolescents know them perhaps better than anyone, even if they cannot see them. Demodex is a mite that is a parasite of human hair follicles and is the frequent cause of their pimples.

Residents of hair and feathers

Fleas and lice are wingless insects. They are not very big, being between 0.8 and 6 millimetres long at most, but they are well equipped for life as a parasite. Fleas are related to flies and mosquitoes. They lay the same type of whitish larvae that are legless but highly mobile. Left to themselves, they leave the host on which they were born to feed on organic matter elsewhere. They need to settle in the hair of a mammal or the feathers of a bird in order to metamorphose. Failing that, they can wait several months in their silk cocoon before becoming nymphs, then adults.

The flea is perfectly adapted to its life as a parasite. It has lost its wings, which are useless, but its legs have become longer and more efficient: they enable a flea to jump 20 centimetres high and 40 centimetres in distance. The sides of its body are drawn in so that it can pass between the hairs or feathers of its host. It has non-functional eyes and a mouth made for sucking blood.

▲ *Harvest mite. Abundant in grasses in summer and autumn, when it comes into contact with the skin of animals, this mite sticks its rostrum into the skin and causes itching and sometimes an allergic reaction. The larvae and adults both bite, the first less often than the second, but more deeply.*

Fleas consequently cause itching and, more seriously, are carriers of diseases: the rat flea was responsible for the bubonic plague in Europe; the rabbit flea is the principal carrier of myxomatosis.

The louse, with a flat body, is more closely related to crickets and bugs but it, too, sucks the blood of its host mammal and can transmit disease. Almost all mammals can be infested. The louse clings onto the hairs with its hooked feet. For a long time, the first task of teachers in primary school was to look for louse eggs, or nits, in children's hair. Humans are mainly affected by three species: the body louse, which can carry diseases like typhus; the head louse – the schoolchildren's nightmare; and the pubic louse, which because of its inwardly curving legs is also known as crabs or the crab louse.

Other lice, those of the order Mallophaga, infest mainly birds, ridding them of particles of skin and feathers. But they can kill them by exhausting them. It is a bad end: the lice survive for only three days after the death of their host.

▲ *The pubic louse. Also called a crab, this cousin of the body and head louse infects young people particularly, who catch this unwelcome guest from sexual contact.*

A tiny giant

It is possible to be a dwarf, almost too small to be seen, and yet gigantic. This is what the ribbon worms of the phylum Nemertea are. With a diameter of a few millimetres, these largely marine invertebrates can go completely unnoticed. Hidden under a rock, they are rolled up in a harmless heap. But when they uncoil themselves, they become several metres long. The largest, the giant ribbon worm, can measure up to 60 metres!

Ribbon worms are fearsomely efficient predators. They have hundreds of eyes, some of which have a crystalline lens, and special organs which sense vibrations and smells and help them patiently to track their prey. Then they shoot out their gigantic proboscis and catch their prey in a sticky mucus, which is poisonous in some species. Other species have a spike at the end of the proboscis with which they stab their prey.

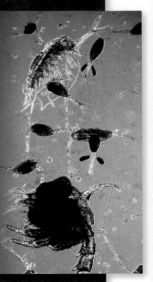

Under our beach towels

Beaches owe their stability to the invertebrates living between the grains of sand. To plough them deeply in order to clean them condemns them to be eroded. Among these teeming but invisible animals are those of the phylum Gastrotricha, which are not more than 0.3 millimetres long. Hermaphrodites, they lose their male genital organs as soon as conditions become better, and then practise parthenogenesis (the eggs develop without being fertilised by spermatozoa). Kinorhyncha are not much bigger: 0.5 millimetres long on average. Their retractable mouth catches organic waste. In particular, they are viviparous: like us, they give birth to young that are already formed. Loricifera look like pots of dead flowers, with their large mouths surrounded by tentacles.

There are many other mini worms, micro molluscs, tiny shabby crustaceans, a whole little world that lives, hunts, gets eaten and dies out of sight. The granular texture of beaches and their richness in organic matter depends on the relationships between these animals.

▼ *Some 50 000 species of acarians in the world have been counted. But the origin of these tiny animals from a branch of the arthropods remains a mystery. Some researchers say its origin may be the fluke, such as the parasitic worm Trematoda, which has methods of strong adhesion like those of the acarians.*

SETTLERS AND MIGRANTS

Conquering the world

*Pink flamingoes fly over the
national park of Banc d'Arguin
(Mauritania).*

Swimming and floating

In order to float and swim under water, animals adjust their volume and density. In addition, their streamlined shape enables them to shoot forward with amazing ease.

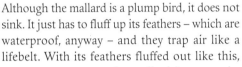

Although the mallard is a plump bird, it does not sink. It just has to fluff up its feathers – which are waterproof, anyway – and they trap air like a lifebelt. With its feathers fluffed out like this, the mallard increases its volume but decreases its density, which is then less than the density of water. And so it floats.

The proper use of a lifebelt

When the duck swims along in the water it floats with perfect equilibrium, displacing an amount of water corresponding to its weight. A duck weighing one kilogram displaces one litre of water. When it wants to dive, it gives a vigorous thrust with its back and dives under the water. The pressure of the water drives out the air trapped in its feathers. In effect, the duck has let the air out of its lifebelt. Its density has increased and approached that of the water, and it has only to make a slight effort with its webbed feet in order to remain under water. As soon as the duck comes up to the surface, it fluffs out its feathers. Once its lifebelt is full of air again, it can swim effortlessly.

Depth setting

Fish do the same. They use their swim bladder to adjust their volume and density. This organ is a sort of buoy filled with gas, which the fish can fill or empty voluntarily. At any depth, fish are thus able to regulate their volume and density, and therefore their buoyancy.

The cuttlefish has no swim bladder, but, instead, a spongy cuttlebone full of air. The sperm whale makes use of an

▲ *This female mallard floats because it is less dense than the water. Archimedes' principle prevents it from sinking. To dive, it has only to expel the air from its feathers.*

organ called the spermaceti, a large pocket full of oil situated in the whale's head. Nasal passages pass through the spermaceti. As the whale dives, its nasal passages fill with cold water, which lowers the temperature of the oil, making it denser and allowing the whale to dive to great depths. By making the oil cooler or warmer, the whale alters the volume and density of the oil and hence its ability to float.

A technique made to measure

Next it is a question of moving forward. Aquatic animals must compromise between the strength of their thrust, the resistance of the water and their energy. They glide along, as it were, in order to swim without tiring themselves. This is why fish have evolved in three different ways. There are long-distance swimmers (tuna), those with lightning acceleration (barracudas), and others expert at manoeuvring (butterfly fish). But most fish are not specialised.

A spindle and a sickle

In terms of energy efficiency, large pelagic fish (deep-sea fish) are the most remarkable, especially the tuna. Its body, flattened into the shape of a perfectly symmetrical spindle, is highly streamlined and slides through the water without resistance. Its

▼ *The cuttlefish is a real helicopter. Its cuttlebone ensures its buoyancy and its ribbon-like fin ripples and enables it to hover.*

▲ *Butterfly fish live among coral. Their oval bodies can tilt at the slightest thrust.*

tail, a very thin sickle with a wide span, is attached to its body by a relatively thin stalk and can propel it along for days at a speed of about 15 kilometres an hour without using much energy. They are poor at manoeuvring and sprinting, but are tireless when they bear down on their prey.

The mako shark, another large pelagic fish, is distinctly less efficient. It swims by undulating almost the whole of its body, whereas the tuna moves only its tail. It therefore spends more energy and moves more slowly, but is capable of much greater acceleration.

A tank and a space capsule

This is nothing, however, compared with a fish like the barracuda or the pike. These fish manoeuvre like a tank, but are capable of lightning bursts of speed because their massive tail is joined to the body on a very broad base. However, with a single flip of the tail they expend most of their energy.

The butterfly fish lives in coral, a complex environment where moving about is particularly difficult. Its egg-shaped body is very short and can easily tilt with the slightest thrust from its pectoral fins, which beat very fast, allowing it to move like a space capsule that can manoeuvre on three axes without going forward or back.

110 kilometres an hour

All sea animals, faced with the resistance of seawater, have evolved the same streamlined shape. But it is the billfishes that have adapted the best. In order to eliminate turbulence when they move, their upper jaw has lengthened into a spear. Water resistance is minimal and their swimming speed maximal: the marlin can reach a top speed of 110 kilometres an hour, and an average speed of 50.

▲ *From left to right: shark, swordfish and barracuda. Three types of hunting fish, three methods of using hydrodynamics, three ways of managing speed.*

WALKING ON WATER

Certain animals can walk on water. These include insects like the pond skater or water-strider, the water boatman and marsh treader, as well as a large South American lizard the basilisk, which has been nicknamed the Jesus lizard. Unable to alter their density, unlike fish or water birds, they manage to spread their slight weight over as large an area as possible, which decreases the pressure they exert on the surface of the water. This surface is like a skin between the water and the air. Surface tension makes it sufficiently rigid to bear a lesser weight. Even an animal with a weight like the basilisk is therefore able to walk on water, because it has feet with very long toes that spread the pressure over a large enough area.

Lords of the skies

Being able to fly is primarily a question of anatomy. And the best-adapted anatomy is clearly that of birds.

'struts' which makes them strong. As for the muscles, they have to be particularly powerful to enable the wings to beat. This is why birds possess, like humans, a sternum, a triangular bone in the breast. However, in birds it has a projecting plate, the keel, to which the pectoral muscles are attached. It is these muscles that are responsible for flight. When they contract, they pull the wings down. Then other muscles – the pectoralis minor – contract in turn and by the pulley action of tendons make the wings go up again.

Arms to fly with

The wings of birds are really forelimbs that have evolved. They are formed from the bones in the arm and forearm, whereas in other flying vertebrates (bats, for instance) they are made from the bones of the hand.

The length of the arm, and thus of the humerus, varies according to the type of flight in the species. In gliding birds (such as albatrosses and storks) this bone is particularly long. But in fast-flying birds, like swifts and hummingbirds, it is the forearm, including the ulna and the radius, that is highly developed. Finally, the 'fingers' play a very

▲ *Like all large raptors, the bald eagle soars more often than it flaps its wings, by making good use of rising thermal currents.*

High in the cold blue autumn sky a squadron of wild cranes flies overhead in faultless formation, honking. They are flying south to spend the winter. Their great wings alternate regular beats and smooth glides in a vision of grace and elegance combined with efficiency. They are already disappearing over the horizon.

Flying seems so easy for them, so different from the immense efforts of humans, who have managed to get off terra firma only with the help of expensive and sophisticated technology. Flying animals – particularly birds – are endowed by nature with cutting-edge technology.

Light bones, powerful muscles

In order to fly it is first necessary to be light. Together with muscles, the skeleton makes up a major part of body weight. But the long bones of birds have air pockets. They are hollow in the middle, which makes them light, but they are also equipped with a system of

GLIDING IS NOT TRUE FLIGHT

Certain species of frog, lizard, snake and squirrel that in general live in tropical forests are equipped with a membrane of skin between their four limbs, which allows them to move about by gliding. It is because of this form of locomotion that these animals can go quickly from tree to tree. Flying fish (right) possess large pectoral and pelvic fins that they spread like wings in order to jump out of the water and glide for short distances. This enables them to escape from predators like tuna, only to become the prey of birds.

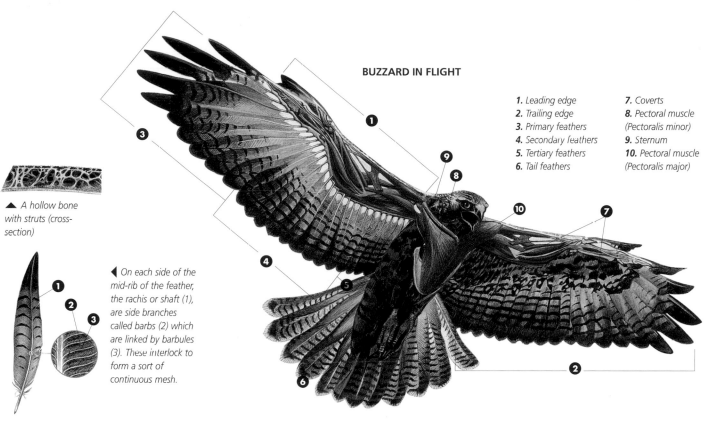

BUZZARD IN FLIGHT

1. Leading edge
2. Trailing edge
3. Primary feathers
4. Secondary feathers
5. Tertiary feathers
6. Tail feathers
7. Coverts
8. Pectoral muscle (Pectoralis minor)
9. Sternum
10. Pectoral muscle (Pectoralis major)

▲ A hollow bone with struts (cross-section)

◀ On each side of the mid-rib of the feather, the rachis or shaft (1), are side branches called barbs (2) which are linked by barbules (3). These interlock to form a sort of continuous mesh.

important role. The bones are partly fused and two of them are absent (the forefinger and middle finger). Situated on these fingers are the main flight feathers, known as the primaries.

Feathers of all sorts

In general, there are about ten (between nine and twelve) primary flight feathers, which grow at the tips of the wings. Along the forearm are the secondary flight feathers, which are slightly smaller than the primaries. They both contribute to flight, when the wings are beating and when they are not. On the front part of the wings are the coverts, small feathers which overlap like slates on a roof, for aerodynamic purposes. The tail feathers are there to act as a rudder, to lift the bird, balance it and also function as a brake when the bird lands.

The lightness of feathers also guarantees ease of flight. Thus the mid-rib or rachis that constitutes the central shaft of the feather is hollow. Filaments called barbs are attached to it along both sides, with branches called barbules, which interlock, zipping the barbs together and protecting the bird against the cold and wet.

A hundred ways to fly

Wings vary according to the species, and their size and shape depend on the bird's type of flight. The longer and finer the wing,

▲ This mouse-eared bat has, like all bats, a membrane for flight and two claws on its thumb, which enable it to hang when at rest.

the faster the bird will fly. Large, broad, rounded wings like those of the buzzards allow the bird to glide in search of prey and to migrate with ease. Ducks are equipped with short but pointed wings, so they fly rapidly by beating them.

Hummingbirds have very short wings, but possess a metabolism that allows them to beat their wings 90 times a second and also – a little like dragonflies – to hover and even fly backwards!

A flying squirrel

Birds and insects are not the only ones to fly. Bats are flying mammals. However, they do not have wings, but a skin membrane, the patagium, that links the front limbs to the ankles and often includes the tail, together with four very elongated fingers that support the patagium in flight. Flying squirrels, which actually glide, also possess a patagium, but without fingers.

▼ All it takes to gather nectar. Hummingbirds, native to the Americas, are the only birds able to fly backwards.

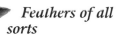

They crawl, jump and run

In order for a vertebrate to move on land, it needs legs to raise it off the ground and a spine in order not to collapse under its own weight.

▲ *Verreaux's sifaka, a lemur from Madagascar, owes its agility to its long limbs.*

▲ *Frogs have legs on springs. Longer than their body, they enable them to advance by jumping.*

▼ *The mole has the advantage of stumpy, almost rigid legs which make it a peerless digger.*

Ichthyostega was not good-looking. It had a scaly skin, skull and long tail, which gave away its fishy origins. And unable to support its weight for very long, its four squat limbs hardly raised its body off the ground. They allowed the creature only to crawl for short distances, its belly dragging on the earth. This clumsy vertebrate was one of the first to set its four feet on dry land, 340 million years ago.

Wriggling and jumping amphibians

This was the Carboniferous Period (360–286 million years ago), when amphibians reigned supreme, and even in this early time they found solutions to the question of how to move about. The salamander, for instance, which is still extant, evolved limbs that were no longer parallel to its body, but transversal. Its arms and forearms form a right angle perpendicular to the animal's body, and the femur and humerus are parallel to the ground, as are the hands and feet. Only the radius and ulna and the tibia and fibula are vertical. They lift the body and give it a jerky, sideways movement when it walks. The salamander's method of motion is efficient, in common with all amphibians, but it does not allow them to reach great speeds.

Frogs have, in addition, adapted themselves for jumping. Their very elongated hind legs form a z-shaped spring. Once in the air, the animal lands on its small front legs, which transmit the energy of the landing to its shoulders, which are fused into a solid sternum that allows them to absorb the shock. The martens (weasels and polecats), mammals which jump on their prey, have adopted a similar spring-loaded structure. Their very long, flexible skeleton rests on short front legs that are normally tucked in. With an anatomy like this, a marten can hide at ground level and then, by pushing on its long hind legs, leap forward in a fraction of a second.

Legless lizards

The skeleton of snakes is totally different. These reptiles have reverted to wriggling with their whole body like fish. Their locomotive equipment has disappeared, including the girdles (see box below), of which only vestigial traces remain in boa constrictors. Their ancestors are burrowing lizards, which adapted to very particular ecological niches where they had to slide into loose earth and small holes in order to survive. In the course of their evolution, their legs eventually disappeared.

Upright for speed

With mammals, limbs become vertically elongated, an amazing innovation. Long limbs from the waist down are freer to move and so allow faster movement over greater distances. The bones in contact with the ground (the feet, paws or hoofs) face forwards in order to increase the speed

▲ *In any soft environment, the most economical mode of locomotion in terms of energy is crawling. And in order for the crawl to be fast and efficient it requires that nothing projects from the body. The skeleton of a snake is limited to a spine and ribs, and vestigial limbs.*

THE SAME STRUCTURE FOR ALL

Amphibians, reptiles, birds and mammals have the same basic structure: a spine (1), two girdles and four limbs or vestigial limbs. The pectoral girdle and the pelvic girdle (2) are each formed of three bones. One of them is articulated with a vertebra. The other two form, with the first, the edge of a socket into which the limb fits, formed from three articulated segments. The first segment, called the humerus in the front limbs and the femur in the rear limbs (3) is in direct contact with a girdle. The second is a double structure, comprising a radius and ulna in the front, and a tibia and fibula at the back (4). The third is subdivided into carpals (wrist), metacarpals and fingers, in the hand, and tarsals (ankle), metatarsals and toes (5), in the foot.

▲ *Impalas are the prey of carnivores. Fortunately, nature gave them an average escape speed greater than that of their predators.*

▲ *The stone marten lies in wait for its prey. Like the frog, it is endowed with a skeleton that enables it to make very sudden leaps.*

potential. The bones of the pelvis become fused into a single coccyx which strengthens the propulsive mechanism. In species built for running (lion, antelope and so on) the size of their tracks also becomes smaller. The reason is to limit the surface of the paws or hoofs in contact with the ground and, again, to achieve speed. Thus, in human terms, dogs and cats walk on their fingers, whereas ungulates (boars, gazelles, stags, buffaloes, etc.) put their weight on their nails. Horses gallop on only one nail.

The cheetah owes its speed to the elasticity of its spinal column, which acts like a spring and makes it easier for the animal to run, as do its two girdles (pectoral and pelvic). The cheetah can thus reach 110 kilometres an hour for a few seconds, but over a long distance it will never equal the North American pronghorn antelope, the fastest vertebrate on earth. Much more muscled than the cheetah, it also has a heart that beats more quickly and its blood, with higher levels of haemoglobin, is supplied with oxygen from enormous lungs. The pronghorn is therefore able to gallop at 65 kilometres an hour for some 10 minutes.

An adaptable skeleton

The capacities of mammals do not end there. It is because they have shortened and thickened bones in their long fingers that moles can dig in earth so easily. Lemurs leap because of their very long tarsals, kangaroos and rabbits jump using their long metatarsals. Many primates are adapted to life in

trees. Their long arms with long fingernails and opposable thumbs allow them to get a firm grip on any branch. In fact, the skeleton of mammals allows for all possible adaptations, which largely explains the evolutionary success of these animals, like that of the dinosaurs, for they too stood firmly on their upright limbs.

▲ *In cheetahs, the shoulders and pelvis can pivot forwards and backwards on their respective vertebrae. This prevents their leg movement from being hindered and allows them to sprint more easily.*

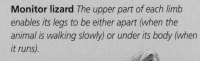

Wolf The wolf has long, not very flexible limbs, so that they can move in a plane parallel to the axis of the body, which enables it to run fast with little effort.

❶ ❷ ❸ ❹ ❺

Monitor lizard The upper part of each limb enables its legs to be either apart (when the animal is walking slowly) or under its body (when it runs).

❶ ❷ ❸ ❹ ❺

Frog The flexibility of the joints and the extra length of the tibia enable it to leap in the air when its back legs are extended.

❶ ❷ ❸ ❹ ❺

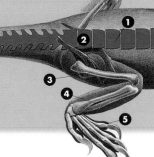

A nomadic existence

Certain species do not migrate, strictly speaking, but wander about in unpredictable ways in search of better food and living conditions.

◀ *Seasonal migration is synonymous with risk for the wildebeest of east Africa. Predators lie in wait, there are raging rivers to overcome, and there is the possibility that the new pastures will be yellowed by drought.*

▶ *The American bison used to move about according to where grass was growing. Today, they often live enclosed in reserves.*

▶ *Like wildebeest and certain gazelles, zebras have to cover a lot of ground to satisfy their food requirements.*

The Serengeti Plain, in Tanzania, is the vast setting of one of the most extraordinary sights on Earth. Even the most world-weary tourists who go there at migration time do not tire of the spectacle of tens of thousands of zebras and wildebeest galloping in continuous lines which stretch as far as the eye can see. The herds are all going north. It is May, and there is not much grass left on the plain. It is time to travel to other pastures, which generally means those of the Masai-Mara.

Roaming with the rain

Wildebeest, zebras and gazelles are herbivores dependent on the rainy season that turns the savanna of east Africa green. They live in immense herds always on the move in search of the best grazing. In the dry season they go far north into Kenya. Then, with the return of the rains to the south, they head back to Tanzania. Although these movements can be foreseen, they are never identical from one season to the next. In a wet year, the animals will cover only short distances. But if the dry season is long they will roam over increasingly greater distances.

This journey is not without its dangers. Rivers have to be crossed, and predators like lions, hyenas and African wild dogs lie in wait for any sick or young animal that may lag behind.

In spring in central Asia, the saiga antelopes migrate northwards from the region of the Caspian Sea, in order to give birth to their young. Then the animals travel down south again, as long as there is sufficient rain and enough grass. If not, the saigas have to go a lot farther to the west. Millions of animals used to undertake journeys like these, but their numbers have dropped severely because of indiscriminate hunting and human settlements on their pastures.

The American bison also used to migrate in enormous herds. Today, their numbers are not as great as they used to be and, more importantly, the wide open spaces in which they were able to move about freely have shrunk considerably, so that this species has become mostly settled.

Errant stilts

There are also birds that wander. But it is very difficult to predict, from one year to the next, where they will choose to nest. The Australian white-headed stilt nests in desert regions. If the rain comes on time and covers the desert with temporary lagoons and ponds, thousands of breeding pairs – as many as 150 000 – will nest there, side by side. But the following year, if the ground is dry, not a single stilt will be found in this area. The European white stilt habitually nests in southern Europe, but if there is a drought in Spain, for instance, it can move farther north and set up house even in Great Britain and the Low Countries.

▲ *The Bohemian waxwing of northern and eastern Europe flies long distances westwards if the berries it normally feeds on are in short supply.*

▲ *The crossbill, which feeds exclusively on pinecones, can completely invade regions where it does not normally nest if it is faced with a shortage of these fruits, which is a cyclical occurrence.*

Unpredictable invaders

Species which nest in northern Europe, North America or Siberia and depend on berries, beechnuts and pinecones for food, like the nutcracker and the jay among the Corvidae family, and some species of tit and nuthatch, can depart suddenly in autumn when it has not been a good year for fruit in their usual breeding area and food becomes scarce. They then migrate over long distances towards the south or west in search of more plentiful feeding.

The crossbills, small passerines which are indigenous to Scandinavia and Siberia, feed exclusively on pinecones. When these are not abundant they fly en masse to western Europe where food is sufficiently plentiful for them to breed in midwinter. It is thought that they then return to their home regions, but it is not known if they all do so.

Flocks of two-barred crossbills, a related species which usually nests in northern Canadian forests, have been seen arriving in the heart of the United States, and some that normally nest in Siberia have been spotted in the Baltic states and even the Low Countries.

People living in Scandinavia, the Low Countries and Germany sometimes see groups of Bohemian waxwings arrive suddenly from Scandinavia and western Siberia. If the winter in the birds' usual territory is particularly icy, they have no doubt failed to find the berries they love to eat and so have travelled westwards. These superb birds were for a long time considered an ill omen (and were given their nickname of plague-birds) heralding a severe winter with its attendant misery.

A tidal wave of locusts

The desert locust of the Sahara region and countries to the south generally leads a solitary life. At this time it is completely harmless. However, when rainfall combines with a certain degree of warmth and causes the grasslands to flourish, this insect undergoes an unusual transformation. From being a solitary insect, it gradually becomes gregarious, and its reproductive cycle speeds up. The females then lay their eggs from one to four times every six days. Each time, they lay between 35 and 70 eggs. Development of the embryos takes only 10 days, development of the larvae 25, and the adults live for 35 days, so a new generation appears every 45 to 60 days.

The swarms become bigger and bigger – they sometimes contain tens of millions of individuals, between 40 and 80 million per square kilometre. They move very fast at a great height (up to 3 000 metres in the air) and over long distances in the direction of north-west Africa or east Africa, sometimes reaching India. Some have even crossed the Atlantic and reached Brazil. They devour everything on their way, including crops. Each locust can consume its own weight in vegetation every day. Farmers often have to take radical measures, like insecticides, to overcome this plague, which unfortunately recurs frequently. Scientists now use satellites to pick up locust activity so that they can predict where and when the plagues are likely to break out.

▲ *Desert locusts have population explosions of complex origin. They move right across Africa south of the Sahara, devouring all the crops on their way, laying them waste in a few minutes.*

DEVASTATING COLUMNS

Army ants, especially those of tropical regions (*Dorylus* and *Anomma* in Africa and *Eciton* in America) are always on the move, which means that their anthills are always temporary. They adopt a formation in serried columns that can be as long as one kilometre and as wide as 20 centimetres. The workers, with their fearsome jaws, are at the head. No plant or animal can withstand them, and they leave a desert in their wake.

The odyssey of small creatures

The search for more favourable climatic conditions or for larger supplies of food leads some species of insects to travel great distances in certain seasons.

▲ *These ladybirds of the American south-west gather to spend the winter at high altitudes, where the cooler climate discourages the fungus that lives on them as a parasite.*

A few years ago, inhabitants of the Iberian Peninsula discovered they had a new guest, a large butterfly with beautiful orange wings and black markings: the monarch butterfly had crossed the Atlantic, after calling in at the Azores and the Canary Islands.

Canada to Mexico: the monarch's marathon

This fragile insect is capable of covering thousands of kilometres. It is not surprising then, with this constitution, that it crosses oceans. Undaunted by the Pacific, it has reached China and Australia.

In summer, in North America, the caterpillars of the monarch feed on asclepias, or milkweeds, and absorb their toxic sap. This makes them indigestible to vertebrate predators. The butterflies that hatch in autumn fill themselves up with nectar and store it in their abdomens, in the form of fat. When they have accumulated a sufficient supply, they gather in massive groups and fly south to Florida, California and even Mexico, some 3 000 kilometres from their point of departure.

The butterflies spend the winter perched in trees. Where milkweeds are present in certain regions, they breed. In spring, the hibernators go back up north with the newly hatched butterflies, flying along routes where milkweeds grow. This long double journey, similar to the migration of birds, is quite exceptional among insects.

▼ *Great travellers, the painted ladies are among the most widespread butterflies in the world.*

Staying cool

Other butterflies fly north when the arrival of the dry season heralds a lack of moisture. This applies especially to the painted lady butterflies that leave Africa to fly to the far north of Europe and North America, even to the frozen Arctic, where they eventually die.

Similarly, certain dragonflies leave for moister climates but do not return afterwards to their point of departure as the monarchs do.

Other insects, like the ladybirds from the Mediterranean basin or California, or the corn bug of western Asia, perform seasonal migrations in both directions, but over relatively short distances. They fly from the warm, flat regions to cooler, mountainous areas where the climate may allow them to rid themselves of any parasites they may have picked up.

▲ *Flights of monarch butterflies are spectacular and sometimes include millions of individuals that fly past in a group for hours at a time.*

Perpetual motion

With the exception of plants, living organisms are in constant movement. A whole range of motion exists between the strictly sedentary life of lichens or oak trees and the long-distance migrations of the black swift that travels a million kilometres during its lifetime. This movement includes seasonal journeys, latitudinal (from north to south or the other way round), longitudinal (from east to west or inversely) or altitudinal voyages, and nycthemeral migration, which follows the change from night to day.

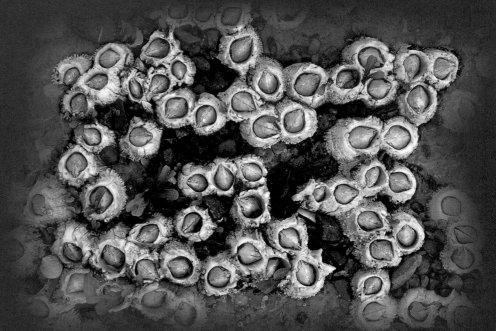

▲ INVOLUNTARY TRAVELLERS

The shellfish of the *Balanus* genus, such as the barnacle, are little crustaceans. When young they exist in the form of larvae which swim about freely. As adults they cling onto a substratum, often rock, where they stay for the rest of their lives. They can therefore be considered as strictly sessile (permanently sedentary) organisms. But some substrata do not remain in one place: pieces of wood floating in the sea, on which the barnacles often choose to live, for example, and also whales, whose backs offer a lasting refuge to colonies of these crustaceans. So these sedentary animals indeed travel – but involuntarily.

SEDENTARY FOR LIFE

It is difficult to be more sedentary than a mangrove in a swamp. In certain species the seeds produced by an adult develop a pointed shape. When the seed is ripe, it falls into the swamp and sticks in the mud at the foot of the tree it came from. It will produce a new plant on the exact spot where it fell. Of course, all plants are sedentary, in the strictest sense. But it is not always the case with their seeds, which are sometimes winged and can, depending on the wind, travel over immense distances. In this way, coconut palms, the seeds of which are often carried out to sea, have colonised most of the tropical islands on our planet.

▶ AND YET THEY MOVE!

In the world of marine invertebrates, there are species that you would think are immobile, like sea anemones (right) on a coral reef, or, in temperate regions, limpets, winkles and scallops, which seem firmly fixed to their rock base. But this is not the case. When danger threatens, these animals are capable of moving short distances, the anemone by performing a sort of St Vitus's dance, the limpets and winkles by creeping along on their single, very developed foot, and the scallops by 'snapping their beaks', that is, by opening and closing their shells.

▼ LEAPFROG MIGRATIONS

In certain species of birds, like the chiffchaff, how far they travel varies according to the competition among their fellows. If chiffchaffs nesting on the shores of the

Mediterranean find enough food there in winter and so decide to settle for the season, those in temperate Europe (the British Isles and the north of France), which migrate south in search of better conditions in winter, have to travel as far as Spain and north Africa, because the chiffchaffs of the Mediterranean shores have already occupied some of the overwintering sites. Finally, the chiffchaffs of Eurasia that nest in the forests of Scandinavia and Siberia migrate in winter over much longer distances, as far as tropical Africa or India, in order to avoid, in their turn, competition from their more southerly cousins. This is called leapfrog migration.

▼ DOWN FROM THE MOUNTAIN

Animals which live near the summits of mountains – mammals and more particularly birds – embark on seasonal altitudinal migrations. In summer they live at more than 3 000 metres above sea level. In winter, as cold weather and especially snow arrives, they move down to more temperate levels.

So the ibex (below) and chamois leave the high meadows in autumn for the upper edges of the forest, where there is still grass. At this time, too, wall creepers, tits, chaffinches and most other mountain birds move even lower, into the valleys. Some species, like the water pipit, go down so far that they are even seen by the sea.

▶ NIGHT TRAVELLERS

In the oceans, plankton embark on what are called nycthemeral vertical migrations, which means that they occur according to the alternation of night and day. In the daytime, zooplankton (like certain copepods) go down into the depths of the sea and at night come up to the surface again. It was first thought that this movement was connected to light levels, as these organisms prefer to live in a dimly lit environment. But it is really food which determines this movement. Zooplankton species rise to the surface at night because food is more abundant there and they run less risk of being eaten at that time. At dawn they descend again to the depths to avoid being eaten.

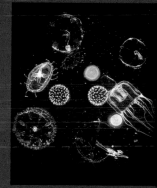

THE MIGRATION OF BIRDS

The same route every year

Some species of birds systematically undertake journeys of considerable length, twice a year, at the same time. Their search for better feeding conditions comes at a high price, in terms of the dangers they face.

▲ *In Europe, each autumn, swallows fly south towards tropical Africa.*

It is the end of summer. In a village in Germany the swallows have begun to assemble. They create an amazing spectacle, gathered in large groups on the electricity wires in the evening, apparently making long speeches. This year's young and their parents can be seen, having nested in various sites in the village during the summer. One fine morning, in a few days' time, they will have disappeared and will not be seen again until the next spring.

Winging their way over seas and deserts

The barn swallow (*Hirundo rustica*) is the most common swallow. It nests in northern Asia as well as in Europe and North America. All swallows migrate, but those that face the journey with the greatest number of obstacles are without doubt the swallows that breed in Europe.

If we were to follow the German swallows we would see that they first cross part of their home country, then France, the Pyrenees and Spain, before swooping over Gibraltar or, for the boldest, crossing the Mediterranean without going via the straits, sometimes at the cost of drowning because of

exhaustion, before reaching the Maghreb in north Africa. They must go on and cross the Sahara, where oases are few and far between and food is hard to find. The losses among the flock are considerable, especially of young birds, which are less experienced than the adults. Even if they have all accumulated large stores of fat before setting out on this perilous journey, they only have to run into a sandstorm or bad weather and they are forced to land anywhere

◀ ▲ *The arctic tern is one of the greatest travellers in the animal world. It flies from the Arctic, where it nests, to spend the northern winter in Antarctica.*

▶ *The albatross.*
These large birds
(which can exceed
1.3 m in length
with a 3 m
wingspan) are
truly the
'marathon
runners' of the
seas, able to cover
considerable
distances over
the ocean.

they can in the desert. If they cannot find enough sustenance to restore their energy, then death is sure to follow.

An uncertain return

The survivors then fly over the region south of the Sahara, which is a little less forbidding. Some swallows stop there, others continue farther south into tropical Africa and even beyond the equator. Birds born in England have been found spending the northern winter in South Africa. There they have some respite for a few months, finding warmth and insects under the African sun.

But as soon as the northern winter ends, they have to think of their return. It is the same journey, with the same dangers. From February, a few scouts appear in the south of Europe, but most of the contingent arrive at the end of March and the beginning of April. Birds that nest in the most northerly parts of the European continent are still flying over western Europe in the middle of May. This double journey is made at great cost: in the course of these annual migrations, only half the swallow population survives.

Long-distance birds

The long annual journey of the sooty shearwater is an ongoing quest for waters richer in food supplies. This seabird with long, fine wings nests in the southern hemisphere, from Cape Horn to the islands in the south Atlantic. After breeding, which takes place during the southern winter, the birds migrate north. Some fly over the Pacific, others over the Atlantic.

The 'Atlantic' birds follow the North American coastline as far as the south of Greenland, then make a loop towards the east which takes them to the coasts of Europe. Then, following these coastlines, they slowly make their way south once more, towards the nesting sites that they will reach in the months of November and December.

THE LONG ANNUAL MIGRATION OF BIRDS

Arctic tern

Lesser golden plover

Sooty shearwater

White stork

Isabelline shrike

Wandering tattler

▲ *Most European white storks migrate to north Africa at the end of summer, when their young have flown the nest.*

Perpetual flight and a world tour

Once the breeding season is over for the great albatross, the adults travel round the Antarctic for months, covering tens of thousands of kilometres and settling on the water only to feed on waste matter floating on the surface. The young birds, however, journey for a longer time, for nine or ten years until they are fully grown – this species has a life expectancy of several decades – because the adults prevent them from staying on the birth sites that they occupy in the breeding season. The exiles sometimes take advantage of this period to seek out prospective sites and in due course find a sub-Antarctic island where they, in turn, may breed once they are sexually mature.

The arctic tern, related to seagulls, completes a still more spectacular journey. This bird nests in summer in the polar regions of the northern hemisphere. In autumn, it flies southwards, via the Pacific or the Atlantic according to the population, to regain its winter quarters in sub-Antarctic waters. After having spent the winter there (for it is summer at the South Pole), the tern bravely returns to the far northern tundra to begin a new breeding season. The resulting journey is a total of up to 40 000 kilometres that it has flown over the seas, or once round the world!

Nature abhors a vacant space

But just why do these birds migrate so far at the risk of their lives? Insect-eating bird species leave temperate regions in winter for milder climates where there is always food in the air. Grass- or seedeaters move from northern regions which become covered in snow or freeze over, and spend the winter in temperate lands where food is accessible.

Species as small as warblers, flycatchers and nightingales thus complete journeys of between 3 000 and 4 000 kilometres to reach tropical lands. When storks, pelicans and certain raptors, for instance, leave their summer quarters in central Europe or western Siberia to fly south, then geese, ducks, cranes, thrushes, chaffinches and buntings take their place, coming from the north of Siberia or Scandinavia.

And what about the robin, seen in European gardens all year round. Does it migrate? Sometimes. In some areas of Europe, birds leave their nests and territories in winter and travel farther south, but then one of their more northerly relatives takes their place! So in winter, the robin whose red breast makes such a splash may be really a cousin of the summer resident, come in out of the cold.

▲ *Robins are either sedentary or migrate over quite a large distance, according to the regions where they live.*

CONTRARY MIGRATIONS

Bird migration follows unchanging patterns. But it can happen that migrating birds seem to go astray. In young birds unaccustomed to migration, a curious phenomenon has been observed. Instead of migrating in the direction defined for the whole of the species, a few individuals take the opposite direction. Thus in Siberia, where most of the migratory passerines leave in a south-easterly direction (towards tropical Asia), the young depart towards the north-west. Warblers, dwarf flycatchers (right) and buntings are some of the species subject to these 'opposite' migrations. Is it due to a mistake in the gene code, or are these birds scouts in search of future breeding areas? Their contrary migration remains a mystery.

Stowaways

The inadvertent transportation of many animals – cockroaches, shell-boring molluscs or disease-carrying mosquitoes – by humans sometimes has disastrous consequences.

▲ *Travelling equipment. The cockroach protects its eggs in a 'box' which it carries on its abdomen.*

The suburbs of Narita in Japan, Roissy in France and Queens in the United States have something very unexpected in common: they are malarial areas. Although they are not near the tropical marshes where Plasmodium, the protozoan responsible for this disease, flourishes, these places are home to the international airports of Tokyo, Paris and New York. And in their pressurised holds aeroplanes carry in comfort the anopheles mosquitoes that pass on the parasite.

Diseases travel with them

The risk of contamination in airports and the inhabited areas that surround them is slight but real nevertheless, particularly in the heat of summer. Flights coming from the tropics are more frequent during the holidays and mosquitoes are surviving on them for longer periods. In Paris, in the time spanning a quarter of a century, 28 cases of airport malaria have been identified, or more than one case a year. To limit the danger, strict measures to disinfect aircraft are applied.

In the 14th century, another involuntary traveller caused havoc throughout Europe: the black rat. It was taken on board the holds of ships in Asia and unloaded, together with its fleas, in the large ports of the West. Its fleas, which bite both rats and humans, carry the germ of bubonic plague. The Black Death killed 25 million people in western Europe, a third of the region's population.

Invaders without boundaries

Cockroaches are found in great numbers on all continents in the proximity of human beings. It is impossible to know where they originated. But it is thought that they arrived in cargo from hot countries, because in the countries of northern Europe they survive only in heated apartment blocks where the atmosphere is sufficiently humid.

Tropical bivalve molluscs that bore into submerged wood took up residence in the wooden hulls of ships from the end of the Middle Ages, at the time of voyages into warm seas. Disembarking in the Netherlands, they attacked the posts of the dykes that protected the polders. By the 18th century they threatened to destroy the defences that the Dutch people had patiently constructed against the advance of the sea.

This involuntary dispersal of living species by humans in the course of their travels has a future. According to some scientists, life now exists on the Moon, in the form of bacteria taken there by space vehicles.

▲ *Some bivalve molluscs are very fond of wood.*

▲ *The blood this female anopheles mosquito is sucking will be digested over several days. But the parasite she contains, which causes malaria, is not digested and can be transmitted any time she bites.*

▶ *It is impossible to tell the number of wrecks of ships whose hulls have been destroyed by the countless holes made by wood-eating molluscs.*

Travellers great and small

The search for food leads most land-based mammals to migrate. Every year, the caribou – or reindeer – make an immense double journey between the forests of the north and the tundra. African elephants similarly take advantage of the rainy season to look for better pastures.

▲ *Born in the Canadian far north at the beginning of the summer, this young caribou will travel south with its mother once autumn comes, to spend the winter in more forested areas.*

May in the Canadian far north. At least, according to the calendar it is May, because here in the region of the northern forests it is still winter. In the distance, on the plain, a long unbroken line moves slowly northwards: the caribou are returning.

1 000 kilometres in the snow…

The snow is still deep and food is becoming scarce, yet the animals have round bellies. This is not unusual, as these are pregnant females going up into the arctic tundra where they will give birth at the beginning of summer.

A few kilometres away, the rest of the caribou are migrating in the same direction. They will all soon reach the shores of the Beaufort Sea, somewhere between Canada and Alaska, the end of a 1 000-kilometre journey.

Travelling like this in the snow is not easy. The lead animal is regularly replaced by the one behind it. From time to time the herd stops and, scratching with their hooves, the animals uncover the meagre vegetation that has survived under the snow. The caribou walk day and night.

… for two months of plenty

By June the summer pastures are near. Already some females are stopping to give birth and resting for several days in the same place with the young. The others go on.

Summer will be short in the tundra, but the pastures will be lush, with many flowering plants. This is rich and plentiful food, much more nutritious for the females and their young than the endless lichens of the forest.

◀ *For reindeer, the rutting season begins at the end of summer. They are then in the most northern part of their migration territory.*

In July and August, when the arctic vegetation is in full bloom, the caribou continue their journey, avoiding areas where there are too many mosquitoes and flies. Loosely arranged groups form, for the males and young females have joined the mothers. As the summer wears on, it becomes an idyllic time: there are fewer mosquitoes and the caribou are therefore able to graze undisturbed, in peace at last.

A ceaseless quest

But now the herd must leave. At the end of summer, some groups have started the journey back. With the first falls of snow, the autumn migration reaches its peak. The autumn migration corridors are less clearly defined than those in the spring, for the snow is not so deep and it is still easy to reach food.

The males start to get restless: the rutting season is beginning. Each wants to assert his dominance and some harmless pushing and shoving results. It is also the time of year when the males, females and young are all together in one great group.

▼ *African elephants move to where they find grass and good feeding places.*

▲ The long migratory journey of caribou is fraught with danger: facing predators (especially wolves), crossing peat bogs and frozen rivers, they run a constant risk of losing their lives.

When winter returns, the animals will again have reached the forest areas farther south where food is more easily accessible. Each year, this migration from the northern forests to the tundra and back, added to their other movements, means that the caribou cover more than 4 000 kilometres, a record in the world of terrestrial animals.

Travellers in the rain

Unlike the caribou, African elephants are not proper migrants. They do not systematically move from one region to another but travel around within their own geographical area where their food supply takes them.

During the dry season, the herds (consisting of females and young) gather round water holes. At the time of the rainy season, they scatter, travelling in search of better pastures. The need for salt, which is extremely important for elephants and must be eaten regularly, is another reason for their constant movement. Also in the wet season, the males, who are generally solitary, approach the herds to mate. They wander from one group to another making chance encounters.

It is only 22 months later, and after a long distance has been covered, that the females will give birth to a calf. This is the longest gestation period among mammals.

Driven by hunger

Although bears are not long-distance migrants, they can undertake seasonal journeys of a considerable length. The record is held by a black bear that was tracked for some months. Urged on by hunger, it travelled more than 200 kilometres in search of food before returning to its favourite den to hibernate a few weeks later.

Journeys by bears are a function of their diet, as they are herbivorous in spring and then omnivorous in summer and autumn, eating berries and small animals. But, unlike the caribou, their feeding usually involves local wanderings, over fairly unpredictable distances. This is true mostly of males rather than females.

The polar bear migrates in search of food according to the movement of the ice. It needs stretches of open water, where it can catch seals. The polar bear moves to arctic latitudes as the ice recedes in summer and returns south to the edge of the tundra in autumn, a total migration of several hundred kilometres. In one outing to look for its favourite prey, ringed, bearded and harp seals, it may travel 20 kilometres

▼ The American black bear. This species travels only when it needs to, as food becomes harder to find.

Sea voyages

Many sea animals migrate long distances to breed, after making sure they have been well fed elsewhere. How do they find their way in the sea?

▲ *In June the leatherback turtles come to lay their eggs on the beaches of New Guinea. They bury their eggs in six holes 45 cm deep, a task which takes them about 10 days to complete. As soon as they have finished, they are off again on a long sea voyage.*

What can the green turtle, a resident of the shores of Brazil, be doing on the beaches of Ascension Island, a tiny British possession out in the middle of the Atlantic? It is laying eggs. Why is it so far from home? Because it is there that it was born, like many generations before it. When the species first appeared, millions of years ago, Ascension Island was much nearer Brazil. The island moved farther from the Brazilian coast as the African and South American plates moved apart.

So do green turtles really have such a sense of their native soil that they continue to travel 1 500 kilometres across the ocean to reach it? The answer remains a mystery.

Fat stores

The green turtle travels long distances in order to feed and reproduce. Like the whale, the plaice or the hammerhead shark, it feeds in one place in order to lay eggs in another place far away, as it has done for centuries. Since breeding makes considerable demands on energy (searching for a mate, mating, developing the eggs), the search for food is the primary motive for animals to

migrate when food becomes scarce, or to follow their prey if they too set out on a journey. The subsequent store of fat enables them to undertake longer journeys to their breeding grounds, making use as much as possible of the currents if climatic conditions are favourable.

Jellyfish hunting

The hawksbill sea turtle is one of the greatest marine migrants. Many of these turtles are born in the Pacific Ocean, south of Japan. After spending between two and six years in the sea, the young turtles cross the Pacific following their favourite prey, jellyfish, along the border between the nutrient-rich cold waters of the north and the poorer, warm waters in the southern parts of the ocean. Where these meet, the cold water, being denser, sinks below the warm water and takes with it to the bottom a large quantity of plankton as well as jellyfish. Their journey over, the turtles find themselves in lower California, on the west coast of Mexico, where some lay their eggs. The others return to Japan by the same route.

There are seven species of migrating salt-water turtles, among which are the leatherbacks. They lay their eggs in a place that thereafter they leave immediately, even though they have had a long and exhausting journey to get there. In certain

A BUILT-IN ROUTE MAP

The sea is vast and unchanging. But it is streaked with currents. By travelling along these currents, many marine species, including turtles, return year after year to their natal land. In the same way, sea trout find their way from the coast to the river where they were born. But how do they find the right place? It is a mystery. They probably possess a mental geographical diagram, the nature of which is not known, a sort of map which most migrant marine species seem to have at their disposal. Is it topographical, magnetic, olfactory or visual? We do not know. But it is certain that it gives marine migrants a true sense of direction.

▲ *In the breeding season, Caribbean lobsters sometimes travel in single file, the antennae of one touching the tail of another. Hundreds of individuals sometimes travel several kilometres in this fashion.*

▼ *Beluga whales (below) assemble in summer in an estuary (bottom, the mouth of the Cunningham River in Nunavut, Canada) in order to moult and give birth. They then travel south to feed. Mating occurs while they are at sea, during the winter.*

species, notably the hawksbill sea turtle, the young return to the place where they were born in order to lay their eggs, in turn.

The call of the river

Most sea trout are also great travellers. After hatching in a river, the young later migrate to the sea, which is much richer in food. Years later, the mature adults return to breed where they were born, then they die of exhaustion.

Young sea trout born in waterways in Great Britain swim downstream to the North Sea. From there, carried away by the currents, they approach Iceland and Greenland. They will return as adults to breed in their native river.

A song for the road

Almost all whales migrate. The movements of humpback whales are the best documented. In each hemisphere, these animals form vast groups that are different from each other. There is rarely any contact between them, and each is identifiable by its own sound language. Each pod completes long journeys, but, strangely, none crosses the equator.

Those pods of humpbacks that feed in Antarctica in summer, swim up towards Colombia, Brazil, Angola, Mozambique, north-western Australia or the south-west Pacific in winter in order to breed. The populations in the

◀ *Hammerhead sharks in the waters off Costa Rica. Perhaps they are seeking waters warmer than 22°C, which might guide them in their migration.*

◀ *Some populations of humpback whales breed off north-east and north-west Australia. By then they are nice and fat, having fed all summer in Antarctica.*

northern hemisphere feed in summer in the Bering Sea, and off the coasts of Greenland, Iceland, east of Canada or in the Norwegian Sea before migrating in winter to breed in the Antilles, off western Mexico, Morocco, the Azores or in the East China Sea.

Many other marine animals migrate, from Florida lobsters to the giant cuttlefish in the Pacific and Indian oceans, including hammerhead sharks, tuna and plaice. Although we know the reasons for their journeys, we still wonder how they find their way.

Aiming for true north

Whether it is due to a magnetic sense, innate or acquired knowledge of a route, orientation by the sun or the stars, or memories of sounds or smells, animals and particularly birds do not put a foot wrong when it comes to finding their way.

▲ *Trumpeter swans nest in the north of Canada and return every year to the same sites for the winter, following an unchanging migration route.*

▼ *Orientation experiments have been carried out on starlings and have led to a better understanding of the mechanics governing their migration.*

A fine drizzle and low cloud cover the countryside. It is a gloomy November evening. There is no sign of the moon or stars in the sky. Nothing is visible to the eye. Yet regularly, high-pitched sounds pierce the silence of the night. It is the sound of redwing thrushes migrating. These birds have been nesting in northern Europe and are on their way to the south of the continent where they will spend the winter season. How do these migrants manage to find their way when the sky is so dark and visibility so poor?

A magnetic sense

It is a favourite sport of pigeon fanciers to catch a few pigeons, take them a long way away from their lofts and see which of them arrives home first. To take the experiment further, some have put lenses of frosted glass on the pigeons' eyes so that they cannot see anything. The birds arrive back at their lofts just the same.

Pigeons have cells in their brains which contain tiny particles of magnetite. It is thought that with the help of this mineral they are attracted by the earth's magnetic field, to the north in the northern hemisphere and to the south in the southern hemisphere, which may explain how the birds find their way. But the neurophysiology of this magnetic sense is still largely unknown.

A geographic sense?

When you know the longitude and latitude of the place where you are and the place you want to get to, you can trace the route you need to take. This is called bicoordinate navigation, or true navigation. Sailors operate in this way. And it looks as though certain animals do too. In the 1960s, an experiment was carried out with starlings that were nesting in the Netherlands. They were caught and taken to Switzerland. When they were released, the young birds did not take a north-westerly direction to return to their winter quarters in the south of Britain and the west of France, but continued their way to the south-west as far as Spain and the south of France, as though they had left from the Netherlands.

On the other hand, the adults reoriented themselves spontaneously towards their winter quarters by taking a north-west direction, as if they were using a compass and a map.

In the 1980s, an American researcher ringed two golden-winged warblers (small passerines native to North America) that were spending the winter in a thicket in the middle of a forest in Venezuela. The following year, the researcher found the birds in the same bush. Similarly, swallows return each year to the same nest, in the same house in the same village.

Our present state of knowledge does not enable us to say positively that animals have the capacity for bicoordinate navigation. But at least we can imagine that they have a real sense of geography.

NAVIGATION BY ELECTRICITY

In certain rivers in the Amazon and equatorial Africa that have very muddy water, the fish have developed a strange method of finding their way (right, striped razor-fish). Their muscle tissue emits numerous low-voltage electrical discharges, creating an electric field that the fish feels with sensors under its skin. When the electrical waves encounter an obstacle, the resulting deformation of the electric field is immediately felt by the fish. It reacts at once, by fleeing, or attacking its prey.

Innate gift or acquired skill?

It is known that in certain birds like cranes and swans the first time the young migrate it is as part of a family. This is a form of apprenticeship with the parents still present to teach the migration route to their young.

Swallows often migrate in groups and one imagines that some of the young travel with the adult birds. Otherwise, how could individuals born in Europe and never having done the journey to Africa find their way and arrive at the right place? Perhaps they possess an innate knowledge of the road to be taken?

Other observations suggest the importance of the innate in migratory processes. Just before the great departure, the birds show a particular sort of agitation, called migratory agitation. Its duration is, amazingly, proportional to the distance to be travelled: if the bird is agitated for a long time, it is the prelude to a long journey.

When a nocturnal migratory bird is shut in a round cage and the bottom is covered with black ink, it leaves footprints all facing the same way: towards the south-west, for example, if it flies that way when it leaves its winter quarters.

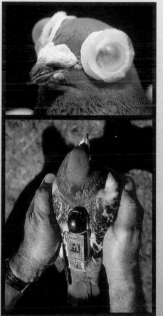

▲ *Fitting a pigeon with a recording compass or masking its eyes allows scientists to study the mysteries of animal orientation.*

The young observe the sky

During the day, migrant animals take their bearings by the sun. Their internal clocks tell them the time, and the position of the sun shows them which direction to take.

But animals also use night signs, such as the constellations, which change their position in the sky because of the rotation of the earth. People can find their bearings if they know what time it is and if they have a map of the sky. Experiments have shown that young birds raised in a planetarium that mimicked the rotation of the earth and the movement of the stars were capable of orienting themselves correctly when they were released from the planetarium, because of the fixed reference point afforded by the Pole Star, which always indicates the north. It is thought today that young birds still in the nest actually look at the stars in the night sky, as well as the position of the sun during the day, with a view to learning how to orient themselves and find their way when they embark on their first migration.

By nose or by eye: every sort of landmark

Some fish, like salmon, find their birthplace using their sense of smell (*see pp.56-57*). They memorise the smells they encounter on their journey in the same way as migrating birds do, and use them as landmarks.

The salmon are also thought to use as reference points the low-frequency sounds that wind carries over long distances. Memorising mountains, seas and topographical features of the regions through which they pass also seems to play a part in their ability to find their way back.

Insects like bees or ants make use of polarised or ultraviolet light, which is invisible to the human eye, to get their bearings. Researchers have also found that some ants seem to calculate a given distance by 'counting' the number of steps they have taken.

Experiments with insects and amphibians suggest that these animals may be able to 'see' the earth's magnetic field through photoreceptors in their eyes, which would allow them to pick up infinitesimal magnetic fluctuations.

▲ *Complex experiments are carried out with salmon to study what determines their return to the place of their birth.*

◀ *Pigeons have a well-known capacity for returning to their lofts, however far away from them they are released.*

ADAPTATION AND EVOLUTION

The price of survival

A flowering plant growing from the rocky ground of the Arizona Desert, in the United States.

Living in water

To breathe in water where oxygen levels are low, to maintain their temperature in a medium where heat is lost rapidly, and also to live there safely with lungs, aquatic animals have had to adapt their physiology.

▼ *The basking shark opens its mouth wide to breathe and feed. Its gills filter the water to take in oxygen and plankton.*

Very small animals, like some micro-organisms in plankton, simply breathe by means of the diffusion of oxygen in the water towards their cells. Those organisms that are larger than one millimetre in diameter, small species like sponges and cnidarians (corals, anemones and so on), have a system of ducts which brings water into contact with their cells where oxygen can be diffused.

A filter to breathe underwater: gills

All larger marine animals, like molluscs, worms and fish, breathe by means of a special and unique organ, the gill. It is a sort of oxygen filter with a very fine mesh and a large surface area.

The more complex the animal or the greater its need for energy, the more involuted its gills are to increase their surface for absorption. The gills have capillaries extending all over them, which can immediately distribute the oxygen-carrying blood to the cells around the body. Bivalve molluscs make water circulate over their gills by moving their cilia, or little hairs, crustaceans by waving their appendages and fish and squid simply by swimming.

▼ *The pike swallows its prey whole by suddenly opening its enormous mouth.*

A cod's life is not easy. Oxygen is much less soluble in water than in air. Its solubility in water is scarcely 0.7 percent compared with its concentration of 21 percent in air. And oxygen only diminishes further in very salty or particularly warm water. The cod, like any other fish, must therefore swallow about 10 cubic metres of liquid in order to take in the few grams of oxygen it needs each day, whereas amphibian animals like newts need to breathe in only a few litres of air.

Breathing the unbreathable

Water is vastly 'thicker' than air. Its viscosity and density are much greater, which hinders ventilation. Species that live in water spend from 10 to 20 percent more energy breathing than those that live outside water. This is a lot, especially as oxygen molecules diffuse very slowly in water.

Mineral salts, a question of equilibrium

Take the simple principle of two linked, fluid-filled containers. The concentration of substances within their liquid will always be balanced, because a substance always diffuses from an area where it is more concentrated to an area where it is less concentrated.

Now consider that water, like the body fluids of animals, is characterised by its concentration of mineral salts and ions. But aquatic animals could not live if their fluid was identical to that of their surrounding water. In fact, animals that live in fresh water need to contain more dissolved 'salt' than their habitat, whereas saltwater species can only survive with less salt than the sea.

▲▼ *The downy gills of the pincer crab (above), like those of the goldfish (below) are engorged with blood. The gill tissue is fine enough to let oxygen through by simple diffusion of water into the blood.*

Living at the ambient temperature

Water dissipates heat much more readily than air does, which is why marine animals are unable to regulate their body temperature. The cod lives with a body temperature of about 11°C imposed on it by the temperature of the water.

Only a few fast fish like tuna are capable of maintaining a temperature much higher than that of the sea, in their brains and swimming muscles. The high speeds reached by these fish indeed require an amount of energy that only a higher temperature in the motor muscles can provide. The rest of their body nevertheless remains at the surrounding temperature.

The answer to the heating problem: being big and fat

For marine mammals which are homothermal, or warm blooded – their body keeps a constant temperature – living in an environment that behaves like a heat absorber demands a great expenditure of energy. The answer is a thick layer of fat to keep important organs at the right temperature, and to have a body of imposing size. For the bigger they are, the smaller their body surface in relation to their volume, and the fewer kilojoules they lose. Per unit area of its surface, a whale loses less heat than a dolphin does.

When light is low

Only 40 percent of the light on the surface reaches one metre below in a normally clear sea. Some 40 metres deeper, this falls to 1.5 percent. Faced

◀ Sponges (left, top) and ascidians, or sea squirts (left, bottom), filter water in the same way as a basking shark does. These animals feed on bacteria, which they plaster with mucus before digesting them.

This imbalance with their surroundings means that freshwater fish experience a constant flow of water in and a loss of salts from the blood, whereas for saltwater species it is the other way round. All animals have to regulate these exchanges of salts and water in order not to die. This process is called osmoregulation. In order not to lose too much salt, the freshwater carp, for example, recovers some from its kidneys, absorbs some through its gills and takes in a lot more in its food. Its urine is, logically, very dilute. In the sea, it is the opposite. The cod excretes a small amount of urine to make up for its water loss and its gills help to eliminate the extra salts in its body.

▼ Rays are flat because they live at the bottom of the sea. Their gill slits are on the upper side and their mouth on the lower one. They find their prey largely by the magnetic field they emit.

with such a low light intensity, marine animals have not, however, become sightless. They have developed other sense organs that are capable of detecting things that human beings are often incapable of sensing or even of measuring. This evolution applies equally to fish and cephalopods (which include cuttlefish, octopus and squid), which have very efficient vision, even at great depths. They all perceive their environment through its infinitesimal changes of pressure, its vibrations, electric or magnetic fields or its smells. Sharks and dolphins probably have the widest range of sense organs.

▲ *Adaptation to a marine life does not concern only aquatic animals. Hall's giant petrel eliminates excess salt by means of a gland at the top of its beak.*

▲ *The large eyes of bigeye fish enable them to hunt at night. Their colouring, which is red in the daytime, then takes on a silvery tint.*

Withstanding pressure

Every mammal risks its life when it holds its breath to dive. This is mainly because as the animal descends, the pressure of the air stored on the surface gradually increases, until its main constituents, nitrogen and oxygen, become toxic. When the animal comes back up towards the surface, the respiratory gases dissolved in the bloodstream can form bubbles as a result of decompression (like opening a bottle of soda water) and create the risk of an embolism. In humans, this is called 'the bends'.

In order to live in the water relying on the air held in their lungs, cetaceans and seals have developed special capabilities. They have reduced their need for oxygen by making their hearts beat more slowly than those of terrestrial mammals the same size. At the same time, their blood, which is often richer

▼ *The Weddell's seal, a cold-water diver, is well equipped. Bulky and fat, it loses few kilojoules per unit area of its surface.*

SAFE IN THE WATER

In spite of all the constraints it imposes on the animals that live there, water is a welcoming environment. Life began there 3.5 billion years ago but emerged to colonise the Earth only 350 million years ago. Why was this? Because water is a cocoon; it stops destructive ultraviolet rays, its temperature and chemical composition are very stable, and the force it exerts on bodies, from the bottom to the top – Archimedes' principle – counters the force of gravity. Minerals necessary for life, like salts of nitrogen, phosphorus, sodium, chlorine and potassium, are dissolved in water and are therefore easily assimilated.

in haemoglobin, also takes in more oxygen. Certain other anatomical and physiological adaptations have allowed them to limit the forming of bubbles when they come up to the surface, to withstand levels of nitrogen that would be fatal to other mammals, or to limit these levels by reducing the volume the nitrogen occupies in their lungs. Nitrogen is then accumulated in other organs – in the bronchi, for example, in Weddell's seals – which are not part of the bloodstream. So the nitrogen has no toxic effects on the animal's tissues. Weddell's seals can therefore go down to the exceptional depth of 500 metres and remain in apnoea, in other words not taking in any oxygen, for 70 minutes.

The sperm whale descends to 2 200 metres, for a little longer than the seal. Its size is a major asset: this enables it to lose less heat and thus use less oxygen. But without Archimedes' principle to cancel out its weight, it could not afford to be so big.

Living in a cave

Cave-dwelling animals live in a poor, almost unchanging environment and tolerate difficult living conditions. There are, however, many of them and they are well adapted.

▼ *Tiny and wingless, the cave-dwelling collembola hops about on a forked appendage, the extension of its tail, which spreads out under its body.*

▼ *Adapting to the environment. The long legs of this arthropod (below) prevent it from sinking into the mud. The proteus (bottom) retains the form of a tadpole all its life.*

The axolotl, a pale rose salamander which lives in caves in Mexico, grows up and dies with the body of a tadpole. Why bother to become an adult in such a protected environment, where the jaws of predators are less to be feared than they are in the open air, and where life, in short, is so peaceful? This phenomenon, called neoteny, is the result of an adaptation to the environment of caves.

Living economically

In the confined environment of caves, the temperature is constant but the sun never enters. Plants cannot grow there to make the organic matter necessary for the cycle of life to begin. Instead, this basic material is brought into caves by surface water which percolates down from above, and is occasionally produced by certain bacteria from gases that are usually toxic to animals living in the open air. Food is therefore scarce, the number of species small and there is not much competition.

This relative security also explains the curious breeding habits of cave-dwelling animal species. The oviparous animals that live there lay fewer but larger eggs than their non-cave-dwelling cousins do. And viviparous animals produce young at longer intervals. The embryo develops much more slowly and is born later. For in such a sheltered world, the need to make sure of the future becomes less pressing.

Crowds

When it comes down to it, many animals have become accustomed to living conditions like these. Among vertebrates, bats are the most famous ambassadors for life in caves. Salanganes, the swifts whose nests are eaten by Asiatic people, are their diurnal counterpart in south-east Asia. But these two vertebrates do not live entirely in caves as the axolotl does, or its cousin the proteus, also called an olm.

There is a multitude of cave-dwelling invertebrates. Almost all groups are represented, notably the arachnids (spiders and scorpions),

crustaceans (including the famous scud, *Niphargus aquilex*, an animal so sensitive to water pollution that biologists use it to give toxicity alerts) and insects (the collembola in particular, of which there are 3 000 species). Many subsist only on the little that reaches them from the outside.

Strange cousins from the darkness

Cave-dwelling, or troglobitic, species are on the whole larger than their open-air counterparts because the lack of competition for food resources prolongs life expectancy and allows them to reach their maximum adult size.

In caves which are permanently damp, the spongy, sludgy or muddy soil, covered with excrement and corpses, obliges arthropods (crustaceans, spiders and insects) to walk on legs and claws that are much longer than normal. On the other hand, as there are many obstacles in the dark and very little to hunt, flying is not much use. So most cave-dwelling insects lose their wings.

Vertebrates and invertebrates in caves transform their food into fat more often than their relatives outside, to make up for the comparative lack of food. As there is no light, most are depigmented – some seem almost transparent – and blind. The function of the eyes is taken over by other organs (such as long antennae, more numerous sensors and so on). In addition, cave-dwelling animals are highly sensitive to vibrations, currents, slight variations in pressure and also to substances dissolved in water.

▲ *As it grows, the axolotl (top) changes its name – it is called an amblystome as soon as it reproduces – but not its appearance. The skin of bats (above) does not like ultraviolet light. So they spend their days sheltering in the dark of caves.*

Living without water

In deserts, where there is virtually no water, plants and animals have developed ingenious strategies for survival, sometimes finding water in inconceivable places.

▲ *This Californian 'beaver-tail' cactus is a succulent plant, able to retain water and thus to flower in the desert.*

In the middle of the desert in Saudi Arabia, no life is visible. It would seem that the place is completely empty, that nothing could live in such extreme conditions. But this assumption is absolutely wrong. Deserts are full of life. Of course, not at midday in full sun, but as soon as the temperature allows it, life comes out of all the holes, dens and roots and springs from under stones and rocks.

flowering plants can subsist for years in the form of seeds, and flower briefly, only when it rains for the first time. And the welwitschia of Namibia possesses very long leaves split lengthways, where dew condenses.

Other desert plants, like the euphorbias and the cactus families, survive by being succulents. Their leaves are small – some succulents have no leaves at all – and their thick skin prevents the evaporation of the water that has collected in the stems. These plants also store water in their roots or their fat, fleshy leaves.

▲ *The saguaro, a cactus in the Arizona Desert, United States, is perfectly adapted to an arid environment: it survives on very little water a year.*

Long roots and small leaves

Plants have evolved different mechanisms for surviving in the desert. Some, like watermelons, spread their roots superficially over a wide area in order to collect water over as large a surface as possible. Acacia trees have roots that go very deep down in the earth (as far as 30 metres). They thus have more chance of finding water. Certain

Protected from sunstroke

In the arid areas of Africa and the Middle East live three species of oryx and one species of addax. These animals with long fine horns are ruminants related to antelopes. They graze on succulents, roots and tubers during the coolest hours of the day. But they drink only at dawn, from leaves covered in dew, or by eating the water-filled leaves of succulents.

Their intestinal flora enable them to collect water by hydrolysis from the cellulose of plants, even very dry ones, which explains the hardness of their excrement. The addax, in particular, can be satisfied for months with the water contained only in the food it eats. Sometimes these animals dig the soil with their hooves, because they are able to detect water near the surface.

When the day warms up, these animals build up heat, understandably. But at night, the surplus kilojoules are dissipated by evaporation from their

▲ *(From left to right) The Peruvian cactus, Californian desert verbena and Namibian welwitschia all have ingenious ways of surviving in the desert.*

nasal cavities, or fossae, by the well-developed network of capillaries in their brains. This mechanism prevents them from overheating and protects them from sunstroke.

Gazelles, like the dorcas gazelle, are not as well adapted to drought as the addax or oryx, but they are capable of travelling long distances to find the grazing and moisture they need.

Everyone take shelter

The desert environment is particularly hostile to small mammals, such as gerbils, jerboas, mice, foxes, fennecs and hares. If they remained exposed to the sun reflected by the ground, the small rodents, for example, would very quickly die from dehydration. So they dig deep dens in the sand, and remain in them during the heat of the day, coming out to feed only at night. There are even some little animals which become dormant during the hottest months, in order to survive the intense heat. Their body temperature drops during this time of hibernation.

Certain species of jerboa and gerbil have large hind legs that allow them to travel along in quick

▲ *The rapid leaps of the jerboa enable it to touch the scorching ground as little as possible.*

leaps. They can thus reduce contact with the burning ground and escape from predators more easily. Their front legs are short and strong and well suited to digging in the ground or stripping the plants on which they feed.

No thanks, I don't drink!

Some of these small animals – rodents like the jirds and jerboas – are completely acclimatised to the absence of water: they never drink. Like the oryx, their bodies are capable of extracting the water contained in dry plants and this water is then saved internally, by being reabsorbed in the large intestine. Consequently, their urine is very concentrated and their droppings are very dry. And as they have no sweat glands, they do not perspire. In their holes they give off a water vapour as they breathe, which maintains a certain degree of humidity.

Desert rodents wait for the rains to arrive before they reproduce. During the short period when the desert is green, the females feed on juicy plants which compensates for the water they lose when they suckle their young.

As for the predators, such as fennecs or wild desert cats, it is the water in the bodies of their prey that supplies the liquid they need.

▼ *A fog-basking beetle (far left) and oryx (below) in the Namib Desert live in conditions of extreme dryness. The water contained in plants, or dew, is often sufficient for their daily needs.*

CHANGING THE MENU

In the desert some carnivores have become omnivorous, such as the fennec (right) which sometimes feeds on dates. The coyote of the American desert can swallow anything at all as long as it contains some water. Conversely, some herbivores have become carnivorous. A rodent from the west of North America – the grasshopper mouse – eats bugs, crickets, caterpillars, grasshoppers, scorpions and even lizards and small mammals to obtain enough water.

Just a drop is enough

Every morning a thick fog descends over the Skeleton Coast in Namibia because of the cool nights. Fine drops of water settle on the plants. This is enough for insects such as the Tenebrionidae beetles and spiders, and it also meets the daily needs of lizards and skinks.

▼ *The jird, a Mongolian rodent, takes advantage of plants to quench its thirst.*

Large beetles can also be seen moving through the sand of this desert with a dewdrop between their legs. They have created this droplet with the aid of their elytra – modified wings that form a channel on their back. At dawn the beetles position themselves with their heads dipped so that water condensing on their backs will roll down and land at their feet in the form of a drop. Other invertebrates, which soon suffer from dehydration, generally secrete a fine film of waxy matter onto the surface of their bodies, which helps to stop water evaporating. Primitive fish, like lungfish, are capable of living in a dried-up stream by burying themselves

▼ *In severe drought, dromedaries draw on the fat in their humps, which is converted into water.*

in the mud, after producing a sort of mucus that dries on their bodies and insulates them completely. They remain dormant like this, waiting for the return of the rainy season.

Fat that turns into water

Camels and dromedaries have very concentrated urine and lose only one litre of water a day in this way. Another asset is their hump or humps – the camel has two – which constitute a store of fat. In very hot weather, this fat remains intact, only the rest of the body burns energy and gives off heat. When they have no water, the fat in the hump breaks down into hydrogen, which mixes with the oxygen during respiration and produces water. These animals have only to find one waterhole and they will drink up to 120 litres in one go.

Questions about evolution

Genetic mutations normally occur over several millennia, but some can take place in less than a century or even within a few years. This phenomenon is difficult to explain.

Thirty odd years ago, blackcap birds were noticed in gardens in England in the winter. This was strange, because although this small insectivorous passerine with a melodious song nests in the British Isles and western Europe, it spends the winter around the Mediterranean. To try to understand this mystery, researchers ringed some 'English blackcaps' and tracked them – and they were found to have originated in central Europe. They had therefore modified their migration routes.

Changing course

Milder winters and a guaranteed food supply from the numerous bird-tables in English gardens are no doubt good enough reasons for the blackcaps to change their winter quarters, especially as they do not have to go very far now. And those blackcaps that overwinter closer to home are also the first to return in the spring, and have the first choice of nesting sites to breed successfully.

About 15 years after the phenomenon was observed, young blackcaps born in central Europe were put into dark cages and studied. They spontaneously turned towards the north-west – that is towards the British Isles – although they had never done the long journey. In a few generations, the migration route to the north-west had been inscribed into their genes.

From white to black

Perched on a birch trunk covered in lichen, the peppered moth, an insect which is white with black patches, becomes almost invisible to the birds that seek it out as food. But at the beginning of the Industrial Revolution in Europe, soot from factory chimneys covered entire regions and birch trees took on a dark tinge. In a period of only 30 years or so, the white form of the peppered moth – which had become so conspicuous birds could pick it out easily – had died out. Those that survived were the darker coloured ones of the same species. However, as soon as the air became less polluted and the birch trees grew whiter, the peppered moths reverted to their original coats.

Sudden changes in the environment – the disappearance of an ecosystem or rapid, lasting climatic changes – could explain an increase in the speed of evolution. Some scientists think that an anti-stress gene would in these cases switch off all the safety mechanisms that prevent mutations at DNA level.

Salmon in lakes or running water

Over 56 years, sockeye salmon introduced into Lake Washington (near Seattle in the United States) separated into two distinct species: one living near the shores of the lake, the other in the river that flows into it. At least that is what certain scientists say, invoking differences that have come about in the initial population in order to adapt to living conditions in the lake. According to others, the differences between the two species are negligible and due to crossbreeding between the sockeyes and the original inhabitants of the lake.

▲ *In a few years certain populations of blackcaps in central Europe changed their migration behaviour. Now they overwinter in north-western instead of southern Europe.*

▶ *The peppered moth, a small nocturnal European insect, adapted to pollution in England. It became darker in order to merge into birch trees blackened by the sooty air.*

Very curious relationships

The categorisation of species in the past was quite artificial because it was based simply on observable similarities. There are now better techniques for distinguishing between living organisms, notably the analysis of DNA molecules, which constantly shed new light on their connections and raising new questions. What is the link between a bird and a crocodile, a hippopotamus and a whale, or a South American rodent and an African bovine? Genetic tools reveal hitherto unsuspected affinities between species whose form and structure (morphology) give no suggestion of any possible relationship. Conversely, as a result of ecological convergence, species totally distant from each other but evolving in a neighbouring environment have strong morphological similarities. It is all a real puzzle.

◀ AN INTIMIDATING COUSIN

At first sight, rock hyraxes appear to be large rodents related to marmots or guinea pigs. Wrong! These little animals from Africa and the Middle East are near cousins of the rhinoceros and elephant. Like the elephant, they have flat black claws and continuously growing upper incisors that project forwards like small tusks. Forty million years ago there were great numbers of them in the grasslands of Africa. But they have been pushed back into rocky and wooded areas by more recent and better-adapted herbivores like bovines.

▼ A MISLEADING FAMILY RESEMBLANCE

American vultures (called New World vultures) like condors (below right) and urubus were for a long time considered to be cousins of the vultures of Africa, Asia and Europe (called Old World vultures), like the bearded vulture and the griffon vulture (below left). But advanced anatomical research has shown that they are more closely related to storks and ibises. In fact, their ancestors looked like cranes and ibises. But they behaved like scavengers, and together with the absence of true vultures this must have contributed to their morphological evolution as they adapted to this kind of diet, which meant that they came to resemble the vultures of the Old World.

THE BIRTH OF A SPECIES?

The blue tit is commonly found in European gardens. Some populations of tits live in regions where they have virtually no contact with the continental population, like the Canary Islands or the mountainous regions of North Africa. These isolated groups are already distinguished by slight variations in the colour of their plumage. Taking into account the fact that, in addition, genetic changes – occurring by chance – will inevitably take place, there will come a time when each isolated population will show enough divergence from the continental species to be considered a different species. And if ever the two populations came into contact again, they would no longer be able to interbreed.

▲ COMMON ANCESTORS

A backwater in Africa. From time to time, two ears and then two eyes emerge, and a noisy snort is heard: a hippopotamus has come to the surface to breathe. Thousands of kilometres away, in the Atlantic Ocean, a snort sends up a spray of water, then a back and a huge fin appear on the surface: a whale has come to fill its lungs with air before diving again. What could these two animals have in common? Their ancestors. Some scientists think that the hippopotamus and the whale have common ancestors, carnivorous ungulate mammals.

▶ AN ODD BIRD, THE CROCODILE

Crocodiles are considered to be the most highly evolved of the reptiles (right: an American alligator). But are they really part of this group of animals? Recent morphological studies and the discovery of fossil species seem to show that crocodiles are in fact nearer to birds (far right: an egret) than they are to other reptiles. They build nests (or rudimentary nests), lay eggs and watch over them and their young, just like their distant feathered cousins. Both crocodiles and birds – like dinosaurs – come from the great family of archosaurs that lived 225 million years ago.

Me, cold? Never!

To withstand extremely low temperatures, plants and most animals in very cold climates have reduced their size and adopted various systems of protection and insulation.

▲ *Short legs and small ears: the reduced extremities of the arctic fox help to protect it from the surrounding cold.*

In the month of June in the Canadian tundra, nature is just waking after a long winter. Great sheets of snow lie on the ground and the surrounding hills are still completely white. Here and there the tundra is lightly spotted with colour. Little flowers of all colours spring up for a short time. It looks as though there are no trees. And yet, by getting down on all fours you would be able to see dwarf willows, birches and many other species of trees on an almost microscopic scale.

A miniature garden

Harsh environmental conditions make this dwarfism necessary. The cold does not allow plants to grow normally. The wind, which is frequent and often violent at these latitudes, also prevents plants from growing tall. And the acid soil is poor, with only a thin fertile layer. Permafrost – an area of soil that is permanently frozen – is often just below the surface.

Some species of arctic buttercup and dwarf willow have to gather the sugars and nutritional elements that they produce by photosynthesis for years before they can flower. Their life cycle is therefore longer than that of their relatives in temperate latitudes.

▼ *Like many mat-forming plants used to harsh, cold conditions, the saxifrages are low-growing and have tough leaves.*

◀ *The dwarf willow, as its name suggests, is very small for a tree. It grows in latitudes where it has to endure cold, snow and wind.*

◀ *Some dwarf willows live at more southerly latitudes in temperate Europe. They are found in mountainous regions, where the climate is similar to that in the arctic regions.*

The leaves of plants battered by wind are often small in order to avoid losing too much water. And in summer in the tundra, if the weather is not hot it can be very dry. So as an extra adaptation, the stems and leaves of the plants are often hairy, which helps to store heat and reduce water loss.

In these conditions, plant growth takes place when the surface thaws, and it continues throughout the arctic summer when there is almost permanent sunlight.

At the limits of life

Still farther north, vegetation almost completely disappears. These are regions of snow and stone. However, on the rocks, mosses and lichens still grow. They are the only plants able to survive. At -10°C some species are still able to carry out photosynthesis. One lichen – *Cetraria nivalis* – has been found in Finland, which even goes on growing at temperatures as low as -20°C.

▲ *This arctic hare on the alert has its winter coat, which enables it to crouch unnoticed in the snow. During the summer it has grey-brown fur.*

The particular conditions of the arctic and antarctic regions are also found in high mountain environments in the temperate zone, where plants have adapted in a similar manner. That is why certain plant families – those of the willow, saxifrage and ranunculus, for example – are common to both environments.

Smaller for warmth

Of course, animals too have to withstand the cold. In arctic species, their morphology has similarly adapted to the environment. With a smaller body surface and shorter extremities, their surface area is reduced in relation to their volume.

The arctic hare, for example, has small ears to reduce heat loss, whereas the ears of the desert hare have a larger surface. The ears of both serve to regulate temperature, the one by reducing the area of blood exposed to the cold air, the other by increasing it, enabling its blood to be cooled.

Cuddly coat or homemade thermostat

Animals of the frozen north, such as polar bears and especially seals and walruses, often have a thick layer of fat under their skin. The fur of these mammals is also thicker and longer, which provides much better heat insulation.

The insulation of the polar bear is amazing. It has a top layer of thick white fur which separates its body from the outside air. Then below this, it has a second layer that is dark and reflects back to the body some of the energy the animal loses (in the form of infrared radiation). This double insulation means that the polar bear loses no more heat than a 200W bulb.

Other animals – notably seals – are able to regulate their blood flow according to the surrounding temperature. Their veins narrow by vasoconstriction to limit heat loss. Blood circulation to the least important organs can even be discontinued, which increases the volume of blood available to the brain and the heart.

▲ *The uncontested master of the ice, the polar bear is wonderfully suited to the cold. This heavyweight – it can weigh up to 700 kg – is an excellent swimmer, and has the gift of sharp eyesight and an acute sense of smell.*

ANTIFREEZE IN THE ENGINE

In North America, some species of reptile and amphibians of the order Batrachia possess a sort of antifreeze. This is the case with the wood frog, the painted turtle and the garter snake, for example. Sometimes in the spring there is a sudden, severe frost. The bodies of these animals may then freeze, but their vital organs are protected. In the frog, the water in its cells moves outside the cell membrane and prevents the cells from exploding as the ice expands. As soon as the frost thaws again, the animals' bodies resume their normal activities.

Soon to be extinct

Every day, several dozen animal and plant species die out on this planet, an alarming situation for which human beings are largely responsible.

▲ *In a few decades, the North American passenger pigeon, which numbered millions of individuals in the 19th century, was exterminated.*

Following a shipwreck on Bering Island in 1741, the naturalist Georg Wilhelm Steller discovered a sort of 'sea cow' related to the manatee of today. This animal, a rhytina (*Rhytina stelleris*), measured between six and nine metres long and weighed almost six tonnes. These animals were probably never very numerous, and Steller observed that they were extensively hunted. Around 1768, the species died out. The Steller's sea cow was known to science for only 27 years.

We still remember them

Around the 1830s, great numbers of passenger pigeons were still to be seen in North America. At the time of the autumn migration, tens of thousands of these birds would fly past continuously for several days. Yet in 1914, the last member of the species died in the Cincinnati zoo.

The relentless hunting of these birds, greatly encouraged by the increase in the number of firearms in the American countryside, is doubtless the main reason for this sudden disappearance. But it is also probable that this species, which nested in large colonies, perished spontaneously when its numbers began to decline. Less stimulated to breed, the birds no longer sought to build nests.

On the island of Tasmania, south of Australia, there still lived at the beginning of the 20th century a curious carnivorous marsupial, which was striped like a tiger. It was the thylacine, or Tasmanian wolf. This nocturnal predator had been hunted out of existence in Australia by the Aborigines. The arrival of colonists on Tasmania who began to raise sheep there was a windfall for it… and a catastrophe. It was a windfall because the newly extended larder seemed inexhaustible. It was a disaster because the farmers waged a bitter war against the wolf, to the extent that, when in the 1930s a decision was taken to protect it, it had died out. Some people nevertheless claim to have had glimpses of it here and there since then.

The famous dodo from Mauritius, a large bird the size of a turkey and related to the pigeon, could not fly. It therefore stood little chance of survival

▼ *Forest fires (here in the Tanjung Puting National Park in Indonesia) constitute a mortal danger to hundreds of species of animals and plants.*

A HAPPY REUNION

The Chinese crested tern, a sizeable seabird, had not been seen since 1937. In June 2000, eight adults with some young birds were sighted and photographed on the Taiwanese island of Mazu, off the Chinese coast of Fu Jian province. Measures to protect them were taken immediately because fishermen collect the eggs of seabirds nesting on the island. But they did not return the following year…

▼ *The Californian condor owes its survival to a programme of breeding in captivity.*

when colonists arrived, who hunted it mercilessly. It died out around 1750. What is less well known, on the other hand, is that at the time there existed at least two species of dodo: the one on Mauritius, and one on Rodrigues Island, to the north-east of Mauritius, which died out around 1780.

Blame man

Until a few centuries ago, the extinction of a species was an isolated event. It often involved an animal or a plant which had a small population living in a limited area, such as an island. These cases of extinction were almost all attributable to people, either directly – through hunting – or indirectly, by the introduction of domestic animals which competed with the native species and brought about its demise, or by destroying its natural habitat.

Since the beginning of the 20th century, industrial development – which followed the colonisation of tropical lands – and pollution of various sorts no longer threaten just the occasional species, but whole ecosystems. The deforestation of certain tropical regions, in the Amazon or south-east Asia, for example, affects not only the plants but also a host of animals, from large mammals to the smallest invertebrates. Now, hundreds, even thousands, of species are being threatened or wiped out. The same phenomenon occurs in

▲ *Attempts are being made to 'recreate' the quagga of South Africa using existing, related species of zebra. The last individual died in 1878.*

numerous wetlands, like deltas and large inland marshes.

The list grows longer...

Today, scientists suggest that a living species disappears from the earth every 20 minutes. The list of threatened species, at present more than 11 000, is only an indication of the danger because it is suspected that most of the invertebrates and plants that make up the larger

▲ *The dodo, a famous extinct creature. It died out soon after the colonists arrived on the islands of Mauritius and Rodrigues.*

▶ *Only a few hundred individuals of the dwarf goose remain. Victims of hunting and the disappearance of wetlands, their future is gloomy, especially in Scandinavia.*

◀ *On the island of Bali, only a few dozen Rothschild's mynahs, also known as the Bali starling, remain in the wild. The poaching and selling of these birds does not make their future appear very hopeful.*

▼ *Forest fires are an enormous danger to orang-utans. This individual is surrounded by flames in the Tanjung Puting National Park in Indonesia.*

part of the living world are still unknown to us. Out of just 9 000 bird species, it is thought that more than 1 100 will be in danger of extinction before the end of this century. And out of the 4 500 species of mammals that have been recorded, more than 1 000 are threatened.

Pesticides, ultraviolet radiation, hunting and deforestation

Amphibians (frogs, toads and newts) are also in danger. Some scientists predict that they will become totally extinct sooner or later. Pesticides have a determining influence on this development,

as some substances destroy cholinesterase, an enzyme which plays a vital role in neuromuscular function. Amphibians would be the first to be affected. Some scientists also think that the more intense penetration of ultraviolet rays, due to the decrease in the ozone layer, could cause skin problems in amphibians and bring about genetic mutations (responsible for deformities) and the death of embryos.

A greater danger, in the shorter term, threatens the orang-utan of Borneo because of runaway deforestation.

The same applies to the Californian condor, which has difficulty coping with the invasion of the countryside by electricity pylons and the use of pesticides. The last survivors were captured so that they could breed, the objective (presently being carried out) being to eventually reintroduce them into their natural habitat.

In Eurasia, the numbers of the dwarf goose, intensively hunted, are melting away like snow in the sun, while the slender-billed curlew is represented in this region by only a few individuals.

REQUIEM FOR THE DISAPPEARED

● **17th century: the moa – a sort of giant ostrich – of New Zealand, hunted by the Maoris.**
● **1799: a large antelope known as the bluebuck (*Hippotragus leucophaeus*). It was probably the first African vertebrate to die out since the beginning of the historical age. The bluebuck was hunted for its meat, used for dog food.**
● **1844: the last great penguin of the North Atlantic was killed in Iceland. The species used to breed as far south as Spain.**

● **1878: the quagga, a strange half-striped zebra, died out in South Africa.**
● **1935: the pink-headed duck, an exquisite bird which lived in India, was hunted to extinction.**
● **1954: the monk seal in the Antilles was wiped out, hunted for its fur and meat.**
● **2000: on January 6, death of the last Pyrenean ibex, amid general indifference.**
● **2001: the last Spix's macaw, living free in Brazil, has not been seen since November 2000.**

Survivors from a distant past

For reasons that we cannot explain, certain species have survived for tens of millions of years without seemingly having changed at all.

December 22, 1938. A small fishing boat anchored at the mouth of the Chalumna River in South Africa pulls in its nets. The curator of a small local museum visits the fisherman to search through his haul for interesting specimens. She spots an unusual blue-grey fin protruding from the pile of fish and takes the fish back to the museum. The massive creature is more than 1.5 metres long and weighs about 60 kilograms. And what a catch! It turns out to be a coelacanth, of the order Crossopterygii, previously known only from fossils and thought to have been extinct for some 70 million years.

A stop to evolution

The crossopterygians were very common at the end of the Palaeozoic era, about 250 million years ago. It is thought that one of the families belonging to the order, the Rhipidistia, gave birth to the first four-legged amphibians and, one thing leading to another, to mammals and humans.

So does this mean the coelacanth is our former cousin? It is not, because it is not one of the Rhipidistia but a member of the Actinistia, another family of crossopterygians which did not evolve. This primitive fish has double pelvic and pectoral fins which it uses alternately as though it were paddling. It also has a lung (only one, on the right), which it does not use but reminds us that the first bony fish had both gills and lungs.

The third eye of the last survivor

In New Zealand there is a lizard which is nearly 75 centimetres long, called a tuatara. It is the last survivor of the beaked reptiles, the Rhyncocephalia, which lived during the Mesozoic era, about 245 to 65 million years ago. The tuatara lives on islands off the east coast of North Island and the Cook Straits. Unusually, it does not have a reproductive organ. And it has a 'third eye' made by an opening in the parietal bone of its skull and covered by a transparent scale. This eye contains vestiges of a crystalline lens and a retina, but it is not known whether it can sense light. With age – this lizard, which only reaches adulthood at 20, can live for 100 years – the opening becomes covered over with a layer of opaque skin.

An ancestor many millennia old

The ginkgo biloba, or maidenhair tree, has existed since the Palaeozoic era. Although this tree was particularly abundant during the Jurassic period (which was about 208 to 145 million years ago) – eight fossil species have been found – there is only one living species left today, originating in China, which has male and female trees. Nobody knows how this survivor from ancient times was able to reach us.

Cycads, which look a little like palm trees, are also plants that came into being more than 250 million years ago. Today there are 90 species of these primitive plants.

▼ *The tuatara lizard is the only survivor of an order which is more than 200 million years old.*

▲ *This strange fish is the coelacanth, a survivor from far back in the past, discovered some 65 years ago.*

▲ *Cycads are survivors of a plant species that existed in the Palaeozoic era.*

Keeping a place in the sun

Global warming has many effects on flora and fauna. Certain species have spread, enlarging their domain, whereas others are threatened with extinction.

▲ Farther and farther north! Several factors point to the same conclusion and support the idea that the habitat of the Sardinian warbler, a species of Mediterranean origin, is moving farther north.

▼ For some 30 years, the bee-eater has not been satisfied with nesting in the south of Europe and north Africa. It is now found as far north as southern Scandinavia.

In autumn, ornithologists usually meet in the Scilly Isles, off Cornwall. These islands are a stopover for all the migrating birds, whose numbers are dwindling, whether they come from the west or from distant Siberia. For the last few years, monarchs, magnificent orange and black migrating butterflies, have sometimes joined these gatherings.

The wind from America
Each autumn, the monarchs leave North America for Mexico where they spend the winter. Some of them, however, are caught in the low-pressure areas that form off the Caribbean coast, change into cyclones near the American coast and finish their run, smaller but still active, on the shores of the Old World. Birds, especially small water birds and passerines, also get caught up in these cyclones and are carried towards Europe.

◀ The green and yellow grass snake once lived only in the Mediterranean area. It is one of a procession of species that is taking advantage of rising temperatures to move into more northerly territory.

Climatologists have, moreover, detected an increase in transatlantic low-pressure areas in autumn, a possible consequence of the changes in climate due largely to the increase in greenhouse-effect gases. In the decades to come, the residents of Europe will probably witness a greater presence of North American migrant species – some of them might even settle there.

Heading north
The same climatologists have also observed a rise in average temperatures since the beginning of the 20th century, which has had repercussions on flora and fauna around the world. In temperate zones, for example, scientists have observed that certain southern animal species are moving farther north.

In Europe, Mediterranean birds are nesting at present outside their traditional area. The European bee-eater, a brightly coloured species of tropical origin, has similarly progressively colonised the north-west of the continent. Some members of its species even nest in the south of Scandinavia.

The large green and yellow grass snake, whose territory used to be limited to the south of France, is now also found north of the Loire. The praying mantis associated with warm places is regularly seen in the north-west of Europe. And subtropical mosquitoes have now appeared in the south of the European continent.

▼ The small-flowered heart's-tongue orchid, which usually occurs in southern Europe, now survives in England as the climate warms up.

Their bite is particularly dangerous because they carry the malaria parasite.

In the plant world, too, some of the orchids – the small-flowered heart's-tongue orchid, for example – which were previously Mediterranean, are taking advantage of the more clement temperatures to extend the area where they are found, and move north.

Similar phenomena have been observed elsewhere around the globe. In North America, scientists think that the increase in the number of subtropical mammals like the brown-nosed coati and the collared peccary, which is spreading from Mexico into the south-west of the United States, is largely due to milder winters favouring the survival of these animals.

Endangered species?

On the other hand, it is probable today that the milder climate together with the resulting increase in precipitation will also have a negative effect on certain species.

Species which live in boreo-alpine regions are particularly affected. They are often vestiges of glaciation, that is, they are remnants of species that lived in western Europe at the time when these regions were covered with ice. Having taken refuge in the mountains and adapted to colder temperatures, they are now retreating as the temperatures increase. This is what has happened to the Apollo,

▲ The Apollo butterfly used to be widespread on Europe's mountains, where there was once a more marked arctic-alpine climate. Today, you have to climb to very high altitudes to see it.

a superb butterfly found in mountainous regions in Europe, which is now disappearing from the lower mountain sites.

The increasingly damp climate also makes other species draw back to drier areas. This is true of the lesser grey shrike, previously widespread in western Europe, whose habitat has now moved to the east of the continent.

Global warming also risks making the tundra recede and the coniferous forests of the taiga take its place, which would cause many ecological changes. Some forest species are likely to profit by it, while the far north species might suffer. Confronted by changes in their environment, will these species adapt?

▲ The brown-nosed coati, a small omnivorous animal, is one of the American species taking advantage of climatic warming to move farther and farther north.

AMPHIBIANS UNDER THREAT

Scientists are unanimous: amphibians (frogs, toads and newts) all over the planet are failing to flourish. One of the causes of this phenomenon is probably global warming. The ponds where they breed dry up more quickly, the water is not as deep and ultraviolet rays penetrate it more easily. More exposed to radiation, the eggs become more vulnerable to microscopic pathogenic fungi that kill them in their thousands. Numerous species are thus dying out.

Wildlife in the city

In cities, plants and animals have found board and lodging, together with a certain amount of protection. They have adapted to metropolitan life and have even become invasive in some places. But their new environment is far from natural to them.

▲ *Common in the countryside, the red admiral butterfly is also found in towns. It is not unusual to find it seeking nectar from the flowers on a city balcony.*

People who live in cities often escape to the countryside at weekends, imagining that they are getting back to nature. However, one day spent walking around a city in the company of experienced naturalists would be enough to discover that it contained animals and plants that you would never have imagined possible in a large capital.

Like all cities in industrialised countries, for more than a century London has been the home of plants and animals that have adapted to city life. Originally from the countryside, woods, forests, meadows, and even from lakes or the sea, these species have exploited a new ecological niche where they have sometimes flourished. Today, rats are not the only wildlife living alongside humans in cities.

The asphalt herbarium

If you took a walk on the streets of London and looked down at your feet, here and there you would see a green shoot daring to emerge from the pavement. It is a plantain, a humble plant that has nevertheless found the strength to grow in the gap between two paving stones. Farther on, you would no doubt see a dandelion shoot that had managed to push through the asphalt. On a wall a pretty pink flower might catch your eye, looking like a little snapdragon. It is the ivy-leaved toadflax, which needs only a few grains of soil to establish itself in a crevice. On closer inspection, a botanist would soon see dozens more different plants.

As soon as a vacant lot appears in the middle of the city where there was once an old building, it is immediately colonised by plants that grow from seeds carried there by the wind. Large yellow-flowered evening primroses, fleabane, which looks

▲ *The plantain needs only a little soil in order to survive. What it finds in the cracks between paving stones will suffice.*

◀ *Overhanging the river from the stone wall that is its foothold, this buddleja survives come what may – a good example of adaptation.*

▶ *The Virginia creeper (right) and wisteria (far right) have found these city buildings an excellent support on which to grow. Their hardiness and their relatively high resistance to pollution have done the rest.*

like daisies, common groundsel and fireweed quickly take over the ground even though it is poor in nutrients. These are what are known as pioneer plants, and they do not require more than the minimum to thrive.

Immigrant trees

Trees and shrubs also come to town. Some grow there naturally, from wind-borne seeds. This is the case with the buddleja, or butterfly bush, which looks misleadingly like lilac. Like many other plant species that have rooted in European cities, it is not indigenous but comes from China. Many trees from far-off lands have been planted in towns and cities because of their elegant or majestic shapes. But it is their hardiness or their resistance that has enabled them to adapt.

The tree of heaven (*Ailanthus altissima*) is a robust, quick-growing tree, also originally from China. It grows in any cranny in a wall or any pile of rubble in large towns in Europe and America, as long as its roots can get a firm hold. By means of its fruits, which have a small membranous wing that easily catches the wind, it quickly colonises any large built-up area. Other trees, like the European sycamore, the false acacia and the catalpa from North America, and the paulownia from Asia, have become perfectly accustomed to city life far from their native lands. Resistance to bad weather and especially to atmospheric pollution, rapid growth and an efficient method of reproduction (notably by wind-borne seeds) are major advantages for successful immigration.

▲ *Ivy-leaved toadflax. A crack in a wall and a little soil are all it needs to grow.*

The urban microcosm

In the shadow of the plants and trees microfauna have developed that are extremely well adapted to life in cities. Waste ground where flowering plants and buddlejas grow is a godsend for many kinds of butterfly. In Europe, butterflies such as the red admiral, peacock, small tortoiseshell and cabbage white, constantly frequent this type of environment and some species even lay their eggs in the middle of the city.

We are at the Place de la Concorde, Paris, at the side of the Tuileries, in the middle of an indescribable hubbub. In the cranny of a sunny wall a little spider waits. It is a zebra spider, or jumping spider, a species that jumps on its prey (midges and other small insects) as soon as it comes within reach. The spider knows nothing of the noise or the crowds and quietly, out of sight, lives its life as a predator. In a nearby city park, another spider may also be found, the garden orb-web spider, which has spun its web in a clump of rose bushes.

And then there are the creatures that live with us in our homes – the cockroaches, fleas, flies and

▼ *In winter, black-headed gulls are found in large cities. The presence of rivers encourages them to venture into an urban environment in search of food.*

▲ *After nightfall, foxes venture right into towns, where refuse offers them an inexhaustible supply of food.*

moths – and let us not forget ladybirds, which like to settle on the plants in balcony flowerpots and window boxes. Even in the tunnels of underground railways the familiar refrain of the house cricket can sometimes be heard, attracted by the underground warmth.

Spending winter in town

Birds have also found refuge in towns and cities. The average temperature there is higher than in the countryside, so that, particularly in winter, urban areas become a haven of warmth. In addition, it is easier to stay alive there than in the wild because there are fewer predators.

Common European starlings gather, sometimes in tens of thousands, in evening dormitories in the large trees of city parks. Tits, chaffinches and blackbirds occur in great numbers in European cities, while various species of swallows and swifts have invaded all large built-up areas.

They have come because food is fairly plentiful, thanks to the scraps thrown out by humans. House sparrows and pigeons know this very well, and so does the red fox, which is becoming increasingly bolder about coming into towns and cities during the night to visit the refuse bins.

The arrival of the predators

A new type of ecosystem has developed in the urban environment. Plants, which formed the first link in the chain, have been followed by insects and then by herbivores. And now the predators have also begun to roam. The stone marten, for example, frequently takes up residence in the eaves of houses. Raptors, too, such as the common kestrel and the peregrine falcon, nest on tall buildings and hunt mice, sparrows and pigeons.

The call of the water

If there is a river or a stretch of water in the town, then animals will show up, and this does not apply only to species of fish.

Birds also move upstream, sometimes coming from the sea, like cormorants and many varieties of seagulls. The black-headed gull has adapted to urban life since the beginning of the 20th century, and in winter it leaves coastal marshes to come and spend the winter in the middle of towns, feeding on rubbish tips or just living on handouts from passersby.

In certain towns around the world, canals and stretches of inland water afford board and lodging to a large number of water birds. Herons, ducks, coots, moorhens and Canadian and Egyptian geese arrive to spend the winter in the middle of large urban areas where there is a better chance of finding food and where the temperatures are warmer.

▲ *City sights: a house sparrow feeds its youngster on the windscreen of a car (top), and a bird considers its reflection in a car's wing mirror (above).*

BEARS IN TOWN

Every autumn, polar bears go shopping in Churchill, a town in the north of Canada on the edge of Hudson Bay. Before they hibernate, they need to accumulate a good store of fat. Open rubbish tips offer them a cheap means of sustenance. Several dozen animals prowl around – and sometimes through – the town looking for such food. So it is always preferable to be armed when you are going out to empty your dustbin in Churchill.

SHAPES AND COLOURS

Nature's theatre

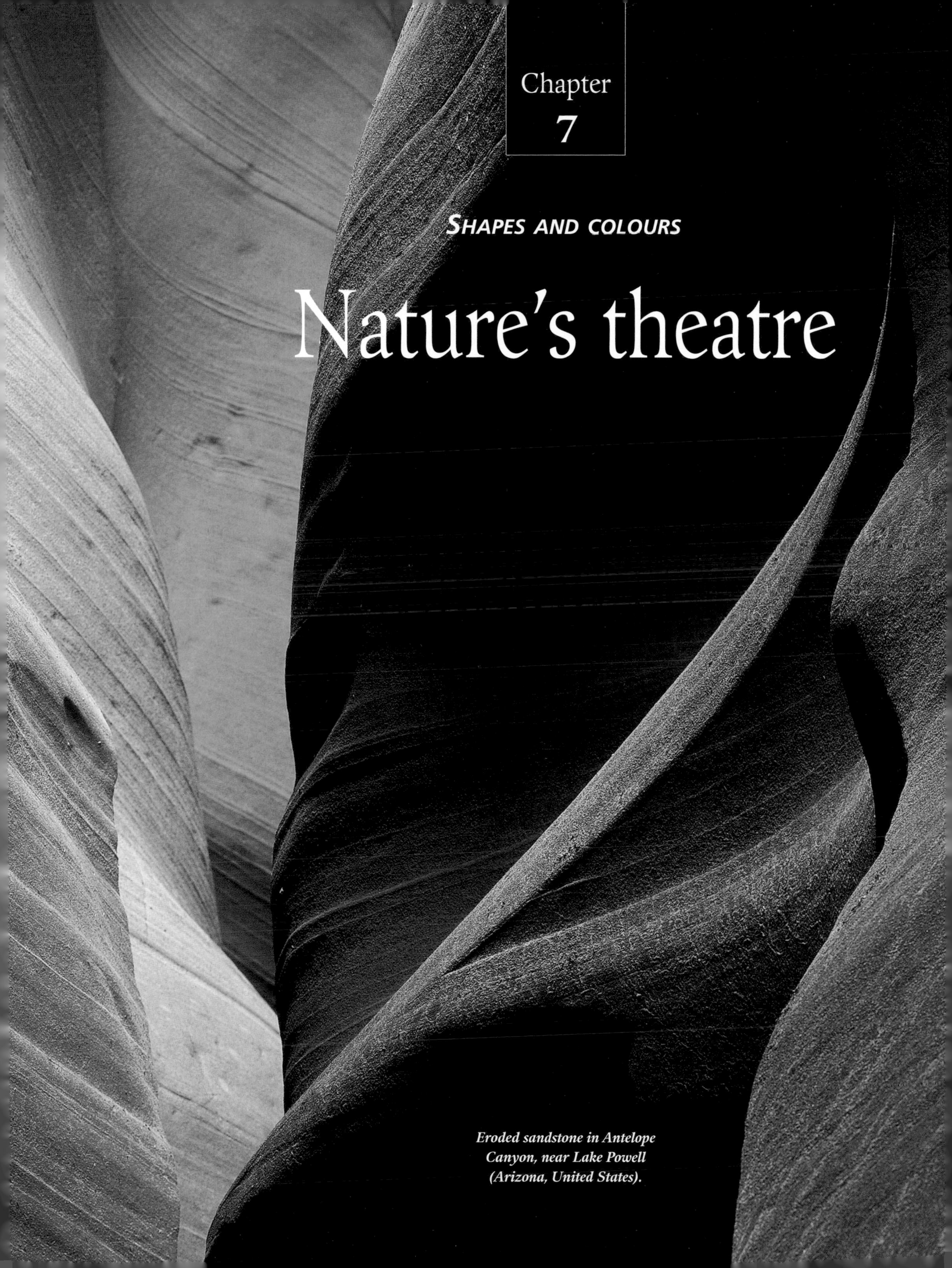

*Eroded sandstone in Antelope
Canyon, near Lake Powell
(Arizona, United States).*

The mineral palette

When rocks are not covered by vegetation, they adorn the landscape with a multitude of colours. This variety might be due to their composition, but it may also arise from their history.

Red, pink, white, grey, yellow, green and brown – the colours of the Painted Desert in Arizona are a veritable artist's palette extending over kilometres and revealing a range of tints that grow darker or brighter with each change in the light. But a simple glance at this multicoloured landscape is not enough to determine the nature of the rocks that make it up. For colour alone, as spectacular as it may be, is only one indication among others.

Black and grey tones are characteristic of basaltic rocks and carbon rocks as well as of certain schists – although they are all very different sorts of rock. The Painted Desert, however, owes its black-and-white striped slopes to an assemblage of clays and marls which contain varying amounts of organic matter, and in some places to levels of volcanic ash from ancient eruptions.

The colour of the interior

The colour of a rock has multiple origins. For instance, fragments of animal shells made of whitish calcite or elements of seaweed give many sorts of limestone and chalk their shiny white colour. The cliffs bordering the English Channel provide an excellent illustration. Here the rock owes its colour to its main constituents.

But in nature, there are often mixtures. A small amount of clay is enough to give limestone a different colour, and it might then take on any colour from white to ochre.

Any trace of metal salts, however small, also changes the colouring. Red rocks contain very small amounts of iron oxides. Thus, types of sandstone, clay and bauxite, although of very different mineralogical composition, may nonetheless show the same colours. White bauxite, an aluminium ore, is highly sought after when it contains only aluminium oxides and hydroxides.

Weathered rocks, multicoloured rocks

From the time when they were laid down, tens and even thousands of millions of years ago, sedimentary rocks may have undergone modifications to their original colour.

This phenomenon can be seen in the sandstone in the area of Lodève in the south of the Massif Central, in France. On the same rocky outcrop, the colour ranges from red to the purest green. The iron oxides (red) were reduced to sulphates (green) when these old alluvial sands were soaked

▼ *Pink, white, ochre or grey, the clays and sandstones of the Painted Desert in Arizona owe the subtle shades of their colouring to the iron oxides and organic matter that they contain.*

THE MANY COLOURS OF PETRA

The funerary monuments of Petra in Jordan were carved out of coarse sandstone 500 million years old, a beautiful rock containing a range of colours from yellow to mauve and orange, red, chestnut, grey, blue and even black. Each cliff, each façade, even the smallest excavation, is a curious mixture of colours and shapes, bands sometimes parallel to the bed of the rock, and wavy or fan-shaped curves overlapping or superimposed. These multicoloured structures are called the rings, or bands, of Liesegang. They are caused by repeated precipitation of iron and manganese oxides and hydroxides in porous rock like sandstone.

▲ *The rhyolites of Landmannalaugar, in Iceland, which stand out among the dark basalt rocks, owe their pink and ochre colours to chemical weathering by volcanic gases.*

with water for a sufficiently long period immediately after they were laid down.

Viewed from nearby, many of the clays show a mixture of ochre and pink patches, a subtle colouring that pedologists (scientists who study the formation and characteristics of soil) call marbling. These clays come from the slowly decanted silt from the water of a flooded plain. Exposed to the open air, they formed a soil covered with vegetation. Under the ground, the distribution of metal salts was altered by the flow of water.

In volcanic rocks, it is usually volcanic gases, which are chemically highly active, that are responsible for weathering and changes in colour, like the shift from white to yellow or pink that is visible in the rocks of Iceland.

Surface sheen

Sometimes buried under hundreds of metres of other sediment, rocks return to the open air after many ups and downs. There they are affected by atmospheric agents which further confuse the issue as far as colour is concerned.

More discreet than physical weathering, which sculpts their shape, chemical weathering gives rocks a patina of a different colour. This is seen, for example, when a hammer strikes a yellowish limestone layer may break off a grey chip. This colour change results when iron salts are oxidised at the surface. Very often the colour of the rock becomes duller. Basalt, very dark when broken, is

◀ *Deeply eroded, the ochre rocks in Roussillon, south-east France, are made of sand coloured by a small amount of iron oxide.*

◀ *Salar de Uyuni, Bolivia. The white colour of the rocks on these flats is the colour of their only mineral component, salt.*

greyish when it is on the surface. Certain minerals in rocks lose their bright colour just from contact with water. In towns, the effects are even more obvious. Many rocks also rapidly lose their shine with the effects of pollution.

A journey to the heart of minerals

Precious stones are minerals characterised by the strict organisation of their molecules. Their rarity is due to certain pressure and temperature conditions, which are not often found in nature.

▼ *This rock crystal, made of giant quartz crystals (silicon dioxide crystallising in a hexagonal form) comes from Brazil.*

It is difficult to imagine a crown without jewels. The symbols of wealth and power, precious stones are fascinating. But are they precious in themselves apart from the way we look at them? After all, sapphires and rubies are only aluminium oxides with a chemical formula that is identical to that of the whitish aluminium at the bottom of oxidised saucepans.

Diamond and coal, close relatives

Like all crystals that collectors vie to possess, precious stones are not rocks, which are in fact aggregates of minerals. They are actually pure minerals.

Each mineral corresponds to a natural chemical type of precise and fixed composition. For instance, the formula of the diamond, which is as simple as they come, is the same as that of pure carbon, which means that it is chemically identical to graphite and anthracite. A girl's best friend could easily be a lump of coal on her finger. Of the 3 000 minerals that exist, not all are precious. Some are even so ordinary that our roads are made of them. Granite is made of three minerals – quartz, feldspar and black mica. Limestone is largely composed of one mineral, calcite, which is one of the crystallised forms of calcium carbonate.

But to learn the secrets of minerals, it is first necessary to understand their structure on an atomic scale and know the physical conditions that have enabled them to form.

Well-organised molecules

All matter is found in one of three states: gaseous, liquid or crystal. It is the last that requires the least energy, and all mineral matter tends to this state.

In gases and liquids, the distribution of molecules is random and changes at each moment. In crystals, however, the molecules are permanently

▲ *A perfect cube of pink fluorspar crystal (calcium fluoride).*

Cubic
Garnet

Hexagonal
Quartz

Trigonal
Tourmaline

Tetragonal
Rutile

Orthorhombic
Olivine

Monoclinic
Hornblende

Triclinic
Feldspar

▲ The variety of colours and shapes of minerals is astonishing.
1. *Vanadinite*
2. *Blue tourmaline*
3. *Heliodore*
4. *Corundum*
5. *Pyromophite*

linked together in a perfectly regular and fixed three-dimensional lattice, the geometry of which depends on the nature of the molecules.

If you were to look within a crystal from one direction, you would see that the distribution of its molecules is homogeneous, but each direction has a different distribution. The properties of a crystal – such as hardness, brittleness and its refractive index – thus vary with direction.

This atomic arrangement results in regular solids that form crystals with either a simple form, like the perfect cube of a fluorspar crystal, or a complex one, like the faceted crystals. There are in fact seven crystal systems which are categorised according to their type of symmetry.

Giant crystals

Some minerals reach an exceptional size, but they are very rare because they are formed in a very particular way, requiring the presence in the rocks of open fractures or spaces linked to each other by fissures that allow fluids to circulate.

In certain conditions, for instance in proximity to a mass of cooling magma, these fluids may reach a temperature of several hundred degrees and contain dissolved mineral matter. Fractures and spaces will then be lined with crystals that will grow without constraint. This is how geodes are made, for example.

Transparency and colour: a matter of impurities

The transparency of a crystal is due to the absence of impurities. So it must contain no atoms other than those necessary for the construction of its crystal lattice, and no liquid or gaseous inclusion. On the other hand, the beautiful colours of certain minerals are linked to and depend on the presence of foreign atoms.

Take the most common crystal, quartz (silicon dioxide). It is colourless and transparent when it is pure (as rock crystal), but if, during the course of its growth, its lattice acquires a few iron atoms, it will become purple (when it forms amethyst). If quartz gains a few atoms of titanium and manganese, it will take on the fine pink colour of rose quartz.

In the same way, the red of the ruby is linked to the presence of atoms of chromium, and the blue of the sapphire to those of iron and titanium in addition to aluminium.

THE MAKING OF A DIAMOND: AN EXCEPTIONAL EVENT

The crystallisation of diamonds requires pressures and temperatures that exist naturally only at depths of more than 100 kilometres, in the middle of the upper mantle of the earth. Carbon is stored in the surface layers of the planet. Some people think that surface carbon could, through the great movements that affect the terrestrial core, be drawn down to these depths (as far as the subduction zones, for instance) and later crystallise in the form of diamonds. But they have to come up to the surface again, which happens only when particular volcanic eruptions occur in the middle of very old and thick parts of continental crust (cratons) like those in South Africa and Siberia. The magma involved in these eruptions is called kimberlite. Its very special chemical signature makes it the deepest magma on the planet (at least 200 kilometres down). The rarity of diamonds is therefore understandable.

An arsenal of seductive power

Flowers and fruit have startling colours so that they may attract animals in order to reproduce.

▲ Look at me! It is not by chance that the colour of ripe fruit like these oranges stands out against the foliage. They must be visible to the animals likely to feed on them, which will then spread their seeds.

Melons have sections so that they can be eaten by a family, and oranges have a bright colour so that they can be easily found among the leaves and picked. These very naive explanations were given by Bernardin de Saint-Pierre, author of the 18th-century novel *Paul et Virginie,* who saw in the perfection of Nature, made for the benefit of Man, as he thought, a proof of the existence of God. He was often mocked for this simplistic philosophy, but he was not wrong about the oranges.

Plants live on water, carbon dioxide and mineral elements drawn from their environment. They manufacture the complex substances they require by means of solar energy. This is harnessed by green chlorophyll, which explains why green is the main colour of most environments on Earth. There are many shades of green, but all the higher forms of plant life, or almost all, are predominantly green. The only parts of them that are not are the flowers and fruit, which often have vivid colours. Is this by chance or by necessity?

Colour messages

The sex life of plants is very different from that of animals. If there are indeed two sexes, they are often on the same plant and the same flower. The first stage of

◀ While seeking out the nectar of a sunflower, this bee unwittingly carries the fertilising pollen that will enable the plant to produce many seeds.

reproduction, the union of the male pollen and the female ovule, is totally passive, with none of the refinements of animal mating behaviour. Therefore, helpers are needed to bring this union about, and also to disperse the seed, the hope of future generations.

The bright colours are messages addressed to animals, finery designed to attract a go-between. For without an insect, a bird or a bat which will fly from one flower to another transferring pollen, fertilisation cannot take place. In this way flowers could be seen as posters advertising that there is plenty of food on the table.

Once attracted, the animals will only be interested in the sweet nectar or the protein-rich pollen. But their arrival guarantees good pollinisation, and so better reproduction. This is why natural selection has favoured these flamboyant beauties of the wild.

◀ The strelitzia owes its common name as the bird-of-paradise flower to its fan of flowers in blue and yellow-orange.

▲ This large hibiscus flower is a funnel that guides insects towards the end of the corolla where nectar is found. On the way the insect will become covered in pollen from the protruding stamens.

An offering of fruit

Plants have developed fruits to feed animals, which in return spread their seeds about. In this way plants are able to conquer new territories. In autumn, birds are attracted to the red or orange fruits of hawthorns, dog roses and pyracanthas. They will sow the seeds contained in their droppings over a radius of several kilometres, and much farther if they migrate.

If the fruit is available to an animal, the seed must not be part of its nutrition. During digestion, the casing that protects the seed withstands the digestive juices, but it nevertheless softens so that the seed can germinate. In many plants, a seed which has not been digested cannot germinate because the outer case is still too hard.

Proof from a bean

There is today no longer any doubt that the colour of flowers and fruits is designed to attract animals. Even the green bean, highly appreciated for its colour by gourmets but not by bean-pickers, since it is so difficult to distinguish from the green leaves

▲ Hibiscus flowers do not attract only insects. Their large brightly coloured flowers are traditionally used as an ornament and a ritual offering in the islands in the Pacific and in tropical Asia.

of the plant, is actually a fruit that is not yet ripe. If beans were left to ripen, they would become yellow. To assist the gardener in a hurry who wishes to harvest them, modern breeders have followed the thinking of Bernardin de Saint-Pierre and have created varieties with coloured pods – like the Purple Teepee, which has a purple skin that turns green again when it has been cooked.

If you needed any further persuasion of the role of colour in the plant kingdom, you would have to do no more than observe those plants, such as grasses, that are pollinated by the wind and not by insects. They have minute greenish flowers which are inconspicuous.

Flowers that open only at night, like wild clematis, are generally discreet and not spectacular. But, instead, they are sweetly perfumed so that nocturnal animals may find them, particularly moths which are equipped to be receptive to smells.

THE TAMBALACOQUE AND THE DODO

By 1970, the tambalacoque had become very rare on the island of Mauritius. The youngest trees dated from the period of the dodos, which became extinct in the 17th century, but which used to feed on the fruit. The seed of the tambalacoque, with a very thick skin, could germinate only after passing through the stomach of a dodo. The bird's very efficient crop wore away the seed casing and its digestive juices softened it. Without the dodos, the seed remained too hard. But with the introduction of turkeys, which have a sufficiently strong digestive system to excrete seeds that could germinate, the tambalacoque was saved.

▲ Green beans are difficult to find and pick. To make the task easier, growers have bred a variety of purple bean that becomes green when cooked.

135

Miracles of balance

Fairy castles, columns, turrets, pillars or needles –
all these amazing, gravity-defying shapes owe their
existence to the process of erosion.

The sight of weirdly contorted and seemingly fragile rock shapes with a large block of rock perched on the top never ceases to astonish. Protected by a cap of more resistant rock, the 'fairy castles' of Euseigne in the Valais, Switzerland, are well named, because such magical creations might as well be the work of supernatural beings.

When the fairies watch

Rock sculptures are often carved out from glacial moraines, because these glacial deposits contain blocks of very large sizes as well as tiny pieces of rock. Unlike running water, which sorts out rock fragments and carries them along, glaciers merely transport them to the surface or into the ice mass and then leave them there in a jumble. Afterwards, streams of water remove the smaller elements that previously escaped erosion in the shelter of the larger blocks, and leave behind fantastic shapes that stir the imagination.

◀ *On the edge of the Grand Canyon, at Dead Horse Point, Utah, the balancing acts of these rock turrets are only a passing stage in the long work of erosion.*

SPECTACULAR EROSION: THE NEEDLES OF MONUMENT VALLEY

The needles of Monument Valley, in the United States (right), the setting of many westerns, are outliers (1), the sole survivors of strata that exist elsewhere in vast plateaus (2). They are made of sandstone layers of widely varying thicknesses that erode in different ways Solid sandstone (3) rests on much narrower layers of clay and sandstone, (4) run through by many fractures, whereas the high walls overlooking them have almost no fractures. These outliers gradually recede, sometimes because of large slabs falling away, unbalanced by their base being undermined, or even by earthquakes.

In Cappadocia (in central Turkey), the gullying action of water inside volcanic tuffs (rocks formed from fused rock fragments ejected from a volcano) several hundred metres thick has created an infinity of needle shapes, sometimes topped with a block of darker lava, sometimes with a cap of welded and more resistant tuff. Volcanoes provide very heterogeneous material which is easily eroded: large blocks may be thrown up at the same time as ash, which will later be borne away by flowing water.

In the middle of this tormented landscape, a troglodyte dwelling reveals how friable these volcanic deposits are. People have dug into them to make countless houses and about 400 churches and chapels – proper underground towns.

Columns that probe the sky

Turrets and columns are also formed in more ordinary rocks, and attract attention because they frequently look like the ruins of a building. An example is the spectacular rocky needles of the American West. Among the countless shapes carved by erosion, we seem drawn to those that are remarkably like our own constructions.

Gigantic pillars

The Greek province of Thessalia is known for its extraordinary Meteora site (meteora means 'hanging in the air' in Greek). Some 60 monoliths between 200 and 250 metres high stand like a stone forest. Through the geological ages, the hard rock was attacked by running water and whole slabs of it fell down along great fissures, which explains the sheer walls and vertiginous precipices. As far back as the 11th century, monks retreated to the tops of the pillars to be closer to God and Byzantine monasteries were built there.

▼ *Fairy castles formed from volcanic ash (below, far left, near Crater Lake, Oregon, United States), from glacial moraine (southern Italian Tyrol, below centre) and in Cappadocia, Turkey (below, far right).*

The magic of colour

Light contains a mixture of an infinity of coloured rays. Matter and the atoms that make up our environment 'steal' part of the light emitted by the Sun, a lamp or any other light source before reflecting some of these rays as a colour. Therefore, everything around us has a colour. When the atoms of a banana skin absorb part of the light that shines on it, the reflected fraction is yellow in colour. It is our brain that tells us so, once our eyes have identified the colour. But what if we were to borrow the eyes of our neighbour? We would certainly not see exactly the same shades, because each human being is unique and has a different perception of colour.

▶ **THE COLOURS OF THE PRISM**
There is a shower of rain, and suddenly a rainbow spans the horizon with the arch of its palette: red, orange, yellow, green, blue, indigo, violet, with an infinity of shades between each stripe, immutably ordered. It is a purely optical phenomenon that can be reproduced in a laboratory. When a ray of light is shone onto the face of a glass prism to emerge through the other side, it is broken down into a beam of rays of coloured light. If you were to place a sheet of paper in the path of this beam, you would see red, orange, yellow, green, blue, indigo and violet bands: the colours of the rainbow. In the sky, the moisture in the air, in the form of raindrops, plays the part of the prism and breaks down the light of the sun to reveal its colours.

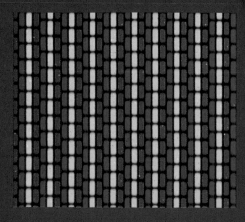

◀ **BROADCASTING LIGHT AND COLOUR**
If you went up to a television screen showing a field of snow, you would see that it is made up of a multitude of small bundles composed of three points, each point emitting its own ray of light – red, blue or green. When the reds, greens and blues are at the same intensity, our eyes see white. If the blue goes off, we see a mixture of the red and green rays, which gives a yellow light. If the red goes off, we see cyan (a mixture of blue and green light rays). The colours of light are said to be additive – by adjusting the proportion of the red, blue and green rays any other colour can be made. That is why, when we are talking of light and not of paint colours, it is these three colours that are called primary colours.

▼ THE REFLECTION OF LIGHT AND COLOUR

A banana does not emit light. If it appears yellow to us, it is because it reflects light. Let us suppose, for the sake of simplicity, that light is made up of only red, blue and green, since these three colours enable us to make all of the others. If we see the banana as yellow, it is because the blue rays have been absorbed by the matter making up the banana skin, and only the red and green rays are being reflected. In the same way, if the water of a lagoon looks blue, or cyan, it is because it is absorbing the red component of daylight and reflecting only the blue and green rays. Yellow paint (which subtracts blue light) mixed with cyan paint (which subtracts red light) will therefore reflect only green light. In paints, cyan and yellow indeed make green. And when all the components of light are absorbed, the result is black.

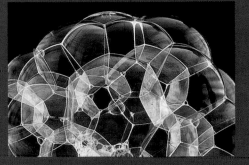

▲ OH WHAT A BEAUTIFUL BUBBLE!

The surface of a soap bubble shows wonderful, rainbow reflections. Where do they come from? When the bubble is in the light, its internal and external surfaces reflect light rays which, made of colours of different frequencies, combine. In some cases, the light vibrations cancel each other out, and only black is visible. In other cases, they strengthen each other, and colours appear. This is called interference. It can be observed on the surface of a CD-ROM, in the mother-of-pearl inside a shell, or on a puddle of oil.

▶ HOW THE EYE SEES COLOURS

When our eyes are presented with an image – a bird, for instance – the light the bird reflects goes through the lens (**1**) of the eye, which functions like the lens of a camera, and strikes the retina (**2**). The retina is lined with myriad receptors, cones which are sensitive to one of the primary colours. When the blue receptors do not perceive their colour, they are not stimulated, but the green and red receptors are, and they send a signal to the optic nerve (**3**), which leads directly to the brain. The brain translates the perceived colour as yellow. But then what happens in the eyes of the colour-blind? Because of a genetic malfunction, the red and green receptors are sensitive to the same things. A green apple and a red apple will therefore stimulate both of them, and the brain will be unable to tell the difference between the two.

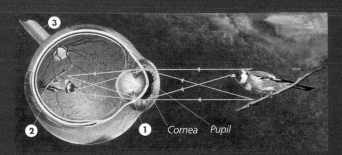

Cornea *Pupil*

EVERYTHING DEPENDS ON THE FREQUENCY

Physicists have found that all light is an electromagnetic wave that vibrates according to a particular frequency. In the range of the rainbow, the frequency increases along the colour spectrum from red (low) to violet (high). What about beyond these extremes? Light still exists, but our eyes cannot see it because our receptors are not sensitive to it. This is the case with infrared rays, for example, which have a lower frequency than red, and ultraviolet, which has a higher frequency than violet. A radiotelescope, right, tracks all light frequencies.

Nature and the art of sculpting spaces

Arches and other natural bridges occur frequently in sedimentary rocks. Tunnels and potholes occur in limestone rocks. Nature owes this amazing art of sculpted spaces to erosion.

▲ *Window Arch (Utah, United States). One of the masterpieces of erosion, this is made out of solid sandstone.*

In the west of the United States, Arches National Park contains dozens of delicately vaulted arches. What helped nature create these shapes into perfect works of art? Erosion – a process of excavating, raising and carrying away.

Arches like sculptures

Sedimentary rocks are among the most easily eroded, but they react differently to the processes of erosion according to their nature, their grain, their degree of cementing and their homogeneity. A thick block of sandstone which is relatively unfractured is the ideal material for erosion. It is nevertheless difficult to know precisely how an arch, for example, is formed from it. Since the erosion of rock is a very slow process, it is estimated that it must have taken several thousands of years, and sometimes even longer, for the rock to be carved out. During this long time there will have been changes in climate. A damp, cold climate makes it likely that the rock will split, as water in its pores and microfissures

freezes. When it is drier, sheets or flakes of rock may fall off (called desquamation), because of a series of sharp changes in temperature or the crystallisation of salts in the uneven parts of the rock. On a thin sandstone wall, a cavity might be the first stage in the hollowing-out of an arch.

From a tunnel to a pothole

In Mammoth Cave, Kentucky, in the United States, there is an extensive network of subterranean passages – more than 500 kilometres of them have been explored. They are the result of water filtering through fissures in the limestone, which widened as

the carbonates in them were dissolved. Some of these tunnels have underground rivers running through them. Some of them come to the surface as potholes and sinkholes. As the limestone dissolves, the rock opens in all directions: the passages follow the incline of the bedding layers dating from the end of the Palaeozoic era, and vertical wells form from fractures in the rock.

Some limestone arches may have been segments of underground passages before being worn down to mere bridges after a very long period of erosion. But the natural bridge called Pont d'Arc, in France, is actually a result of the Ardèche River hollowing out a thin limestone wall on either side. This was the narrow base of a meander, long since abandoned as the river has found a shorter route.

▲ *One of the great underground passages in Mammoth Cave (Kentucky, United States), hollowed out of the limestone rocks by water.*

▶ *The Pont d'Arc. An arch spans the Ardèche, in France, which has 'short-circuited' one of the river's meanders.*

THE PARADOX OF GRANITE

Balanced boulders

Granite looks as though it could last for ever. Yet in its natural state it undergoes many transformations. It is slowly worn away and turns into a loose sand called arenite, containing blocks and boulders of eroded rock that sometimes end up precariously balanced.

The great granite peaks of Europe's Mont-Blanc Massif seem to challenge time. Can water do any harm to this rock that is apparently so hard and solid? In fact, you do not need to look very far to discover that appearances can be deceptive. On a geological time-scale, this fine stone, which is used in the marble industry, is particularly vulnerable.

▼ *A jumble of granite boulders line the edge of a stream in the Tarn, France.*

Balancing in the sand

On the edge of the Massif Central in France, among granite rocks almost 300 million years old, a road cutting reveals this rock reduced to a state where it crumbles at a touch. Above blocks that are still whole, it is almost unrecognisable, its crystals are separate, dull and mixed with clay. This loose 'rock' is called granitic sand, or arenite. After the loss of trees has caused soil erosion, ravines quickly form in it and granite boulders become visible.

Because of water filtering into the fractures in the granite, certain minerals lose some of their chemical elements and become fragile and then change into clay minerals. This very slow degradation occurs more easily when there is abundant water and a warm climate.

If erosion then attacks this mantle of arenite, it lays bare first the granite boulders that are being weathered and then, underneath them, granite that has fractures running through it but is still whole. In the resulting heaps of granite rocks that are uncovered, some blocks may be precariously balanced.

▼ *In Australia's Northern Territory, erosion has uncovered blocks of granite, known as Devil's Marble. Even though they weigh several tonnes, the rocks balance finely, one on top of the other.*

The long journey of granite

Like lava, granite was first a magma. This is produced by the fusion of rocks in the Earth's upper mantle. Viscous in nature, it has difficulty rising up through the solid rocks of the continental crust and stops before it has reached the level of the surface.

Its slow cooling period, which lasts over several hundred thousands of years, enables crystals to be formed – white or pink feldspar, quartz like grains of salt, white or black mica (muscovite or biotite) that splinters into fine layers, and so on. Then the granite solidifies, and erosion begins its work and gradually makes the kilometres of rocks covering it disappear. This is why today we can walk over a granite massif.

The fractures that run through the granite and cause it to degrade are of various origins. They may be recession fissures created at the moment of cooling, or breaks due to the tensions it is subjected to in the Earth's crust.

▲ *The weathering of granite (1) can produce several metres of arenite (2). Lower down, some granite boulders have not yet been transformed (3). Water carries out its work, filtering into the fractures in the rock. If the arenite is blown away, the boulders are freed from the sand and are left in a chaotic jumble (4). Then, tors (5), walls of intact granite, appear in their turn.*

Water: artist and builder

Water dissolves mineral salts, carries them away and then deposits them. In limestone regions, this process leaves caves decorated with delicate works in aragonite and calcite. In the open air it creates banks of travertine.

▲ *The Aven Armand, in the Lozère, France, contains a profusion of formations created drop by drop.*

As their paintings and drawings show, prehistoric people went deep into caves and grottoes. Were they aware of the amazing formations sometimes found inside subterranean caverns? In modern times, it was not until the end of the 19th century that bold adventurers made cave exploration popular. Nowadays, forests of formations are visited by thousands and never fail to impress visitors with their variety, each cave having its own character. Stalactites and stalagmites, columns, rippling curtains, bouquets of crystals, cave beads, rimstone pools, lakes and underground rivers compete from chamber to chamber. Emerging from the shadows and enhanced by lighting effects, some of these strange forms have been given names according to human fancy and imagination. Time and water are the artists here.

Carbon dioxide is dissolved in water and when it flows through calcium carbonate, the chief ingredient of limestone rocks, it changes the carbonate into soluble bicarbonate. When the water arrives in a cavern having passed through these rocks from the outside, it loses part of its carbon dioxide and the bicarbonate it is carrying reverts to carbonate. The water drips slowly, leaving the carbonate behind, which builds up dripstone formations in aragonite or calcite crystals, drop by drop.

A perfect dam wall

Among the wonders of Afghanistan, the lakes of Band-e Amir offer tourists the beautiful view of

CONCRETE REVELATIONS

Underground caves, of which the formations are in a sense the archives, are of interest to many other fields besides hydrology. The growth of stalactites and stalagmites gives information about the amount of water that has entered the caves. The halt of their growth thus corresponds to very dry periods, notably during the cold periods of the Quaternary period (1.8 million years ago to today), when the ice on the surface meant that water no longer penetrated below ground. Another indication of climate is the mineralogical nature of calcium carbonate, calcite or aragonite, which tend to form a deposit when water levels are low. Isotopic dating, which complements carbon-14 methods, thus enables scientists to follow the climatic changes of the Quaternary period.

▲ *Some of the formations that decorate caves in France: (from left to right) stalactites, stalagmites and pillars (in the caves of Lacave, above left); a bouquet of crystals (in Limousis cave, above centre); and folds of drapery (in the Malaval caves, above right).*

stretches of turquoise green water lying at intervals along a valley at an altitude of 3 000 metres. Round about are mountains denuded of vegetation, where water is very scarce. The water is being held by walls of travertine, which act like a dam. These constructions, which are more than 10 metres high, withstand the pressure of the water even if the rock appears very fragile. The travertine is made of limestone, which crumbles somewhat to the touch and shows traces of vegetable matter. It has been created by the cold water from melting snow dissolving the limestone of the surrounding rocks, followed by the mixing action of the water, its rise in temperature and the vegetation on the valley floor being added to the sediment.

Building materials

In other limestone regions of the world, underground water feeds into springs that build up deposits of travertine at the place they emerge from the ground. As the outside temperature is usually higher, they have in fact left behind the limestone remaining from their journey through the caves. The rivers to which they give rise are sometimes also dammed with small dykes of travertine. When the springs dry up, the travertine is often used for building. It is light, because of the spaces in it left by the plant matter around which the carbonates precipitated, it is easy to cut and has good insulating properties.

Hot solvent

Carbonates are easily dissolved by cold water, which is richer in carbon dioxide. But deep, warm water may also contain enough carbon dioxide to attack limestone and dissolve carbonates. It may then leave them on the surface, where there is not so much carbon dioxide in the atmosphere. At Pamukkale, in Turkey, water with a temperature of 40°C and very rich in carbonates and carbon dioxide has built a succession of spectacular basins through which it continues to gush. In Yellowstone Park (Wyoming, United States), the warm waters of Mammoth Hot Springs are responsible for the creation of rounded shapes that look like the backs of mammoths.

▼ *Leaving behind its carbonates, water has built barrages and basins at Yellowstone in the United States (below left), Pamukkale in Turkey (below right) and Band-e Amir in Afghanistan (bottom right).*

Columns and prisms

When it cools down, lava contracts and remains fixed in a characteristic structure called organ-pipes. Erosion will complete these shapes and create amazing sculptures that seem to rise straight up from the Earth's surface.

▼ *Devil's Tower (Wyoming), a phonolite peak 183 m high. Its origin is controversial: is it the heart of an old lava lake, the centre of a fallen dome, or a lava intrusion in the shape of a tower?*

Spielberg fans will remember the astonishing landscape that was the setting for his film *Close Encounters of the Third Kind*: a volcanic cylinder, completely surrounded by a sheer cliff cut in a set of prisms which widened at the base. But it was not just a fantasy recreated on a film set. It was in fact a real landscape feature – the Devil's Tower – of which the inhabitants of Wyoming, in the United States, are so proud. But this amazing collection of rocks shaped like organ pipes is actually nothing exceptional. It is just the result of prismatic fracturing – a process that shapes lava as it is cooling down.

Contractions and tensions

Prismatic fracturing is characteristic of magma rocks that have cooled, whether on the surface – as is the case, for instance, of a lava flow or an old lava lake like the Devil's Tower – or just below the surface, as is the case of a volcanic chimney and its branches.

The phenomenon of prismatic fracturing is connected to the fact that lava, whatever its composition, occupies a greater volume when it is hot than when it has cooled down. It has thus been estimated that a flow of basalt lava 100 metres long, having an initial temperature of 1 200° C, will shrink by about 25 centimetres during cooling. This thermal contraction gives rise to very significant tensions in the centre of the flow.

Cooling processes

A lava flow cools at the same time on the surface, by contact with the air, and from the base, by contact with the ground over which it flows. Cooling by air is by far the more efficient of the two. Thus the top of a basalt flow several metres thick cools rapidly, as it flows along, to temperatures in the hundreds, while temperatures at its base approach 600°C and at its heart still reach more than 1 000°C.

In these conditions, the top and bottom will have already solidified – basaltic lava solidifies at 900°C – whereas the centre of the flow will still be liquid. The first tensions in the lava therefore arise at the levels of the top and the base.

▲ *Top, a close-up of basalt organ-pipes in Iceland, with remarkably regular prisms. Above, a vein of basalt intersecting the flows of the Piton des Neiges (on the Indian Ocean island of Réunion). The prism formation is almost horizontal.*

▲ The colonnade
of Svartifoss, in
Iceland. The
prisms, which here
are vertical, belong
to a basalt flow.

These tensions will be dissolved when perpendicular cracks appear in the cooling surfaces, propagating downwards from the top of the flow and upwards from the base at the same time, as the cooling process nears the centre. And they create the prisms that form the organ-pipe structure.

A hexagonal lattice

Organ-pipe prisms are geometrically very regular, which is connected not just to the vertical propagation of fractures, but also to their horizontal propagation. Once begun, a fracture propagates at an increasing speed within the cooling lava. When its speed reaches about two kilometres a second, the tension at the head of the fracture becomes such that a single fracture is not enough to release it. Two new fractures forming at an angle of about 120 degrees with the first will then appear spontaneously. These will propagate in their turn in the same way, so that they will also end up by duplicating, and so it goes on.

At the end of three successive doublings, the fissures thus formed join up, creating the outline of a hexagonal network that spreads over the whole flow in a fraction of a second.

The work of erosion

Erosion and weathering will of course make good use of this network of prismatic fractures. And any cliff made out of these rocks will naturally break along the lines of these prisms.

In Northern Ireland, in county Antrim, the Giant's Causeway is a promontory of some 40 000 columns which head out into the sea. About half of them are hexagonal, and the rest vary but include octagonal columns. They look like stepping stones heading off in the direction of Scotland.

In France, the Massif Central – which has been the site of great volcanic activity over the last 10 million years – displays many examples of organ-pipes. Among the most striking are those at Espaly, near Puy-en-Velay, and the amazing lava flow at Jaujac, which spread through the valley of the Lignon, a tributary of the Ardèche, less than 30 000 years ago.

▲ On this cliff on
the island of Staffa,
off the coast of
Scotland, the
colonnade and the
table formation can
clearly be seen.

COLONNADES AND ENTABLATURE

When the lava flow is very thick (between a dozen and several dozen metres), you can see that the systems of regular vertical prisms – one beginning at the base that geologists call the true colonnade (1) and one beginning on the surface, the false colonnade (3) – are separated by an area called the entablature (2). In this area, the prism formation is very disturbed. The prisms grow smaller and become irregular, sometimes curving and often forming superimposed fan shapes. The flow visible on Staffa island, in Scotland, is a good example of this particular configuration. The phenomenon probably has a dual origin. On one hand, it is due to the interaction and competition between two systems of fractures. On the other, as the inside remained hot, and therefore liquid, much longer than the rest of the flow, at this level tunnels of lava formed that fed the head of the flow during the whole eruption. The presence of these tunnels created thermal disturbances that completely changed the geometry of the cooling.

▲ *Three examples of tree frogs. South American forest dwellers, they are covered with glands that secrete a poison.*

Discouraging the enemy

The brilliant colours displayed by many animals are really intended to send an alarm signal to predators. An arsenal of sickening smells, together with terrifying shapes, also frightens them off.

A group of tourists is travelling through the Amazon rain forest. A young woman leans down over a magnificent tropical flower to pick up a little multicoloured frog perched on a large leaf. The guide gives a shout and rushes to stop her, because the skin of this little tree frog secretes a particularly toxic substance that in humans burns the skin. By eating it, a young predator could die. The Amerindians use this frog's venom on the tips of their arrows when hunting small animals.

Fine colours, foul smells

Frogs of the family Dendrobatidae all display shimmering colours, like many tropical species, and are as dangerous to touch as they are to eat. Some species are bright red all over, others yellow and black, yet others sea green and bright orange, in short, a rainbow of colours that stand out against the monotonous green of the forest.

Most brightly coloured species of invertebrates, together with a few equally gaudy vertebrates, send out alarm signals in this way. Their flamboyance means they are poisonous. Eating them might cost a greedy predator its life.

The black, yellow and red bands of the coral snake say a similar thing. However, the common kingsnake of the genus *Lampropeltis,* found in Colorado, in the United States, which has borrowed the bright colours of the coral snake, is completely harmless, although potential predators do not know this. This phenomenon of imitation is known as mimesis.

◄ *The fearsome coral snake. This venomous species inhabits the New World. It is found particularly in Texas, in the United States.*

The European fire salamander has parotid glands that secrete a substance which, if not deadly, leaves such a bad taste in the mouth of the predator that it will not be forgotten in a hurry. And the black-and-white skunks of North America are notorious for spraying, from their genital region, a liquid with a smell that is a horrible mixture of sulphur, a cesspool, garlic and scorched hair.

Scary masks

The back end of the caterpillars of some butterflies and moths is bigger and thicker than the front end, and bears the likeness of a mask with enormous eyes, a huge mouth or even – like the caterpillar of the Costa Rican pepper moth – a snake's head. These patterns are obviously a survival mechanism designed to deter predators.

When they are at rest, some moths are grey. But when they open their wings, for instance when a hungry bird approaches, great big round shapes, like the huge eyes of an owl, stop the potential predator in its tracks.

◄ *Which is the front and which is the back of this monstrous caterpillar of the puss moth? It is a perfect example of the art of intimidation.*

WATER IN ALL ITS STATES

Water, water everywhere...

A frozen mountain stream in the
Bruche Valley (Alsace, France).

Come rain, come hail...

Clouds store gigantic quantities of water in the atmosphere, only to release them one day in the form of precipitation that is sometimes liquid and sometimes solid.

The Gauls in the *Asterix* cartoon series were right to worry that the sky might fall on their heads. In the form of clouds or damp air, the lower layers of the atmosphere – the troposphere – hold a quantity of water large enough to cover the surface of the Earth to a depth of one metre. That is, if it reached the ground all at the same time, which of course never happens. Only some clouds, like those of the stratus and cumulus families (*see pp. 176-177*) are capable of letting rain, drizzle, snow, hail, sleet, frost and mist fall to the ground.

From microdrops to rain drops

The moisture contained in clouds takes the form of small ice crystals or droplets of water. As long as they are light enough, they remain in suspension in the air currents that pass through the clouds.

If these particles of water become very numerous, they end up clustering together to form larger ones. It takes about a million microdrops to make a drop of water sufficiently heavy to be drawn downwards. This is how rain is formed. And this phenomenon occurs in warm regions.

In temperate regions, precipitation begins with a sort of snow. The top of the clouds is almost always at an altitude where, because the temperatures are below 1°C, water is present in

◀ In summer in the tropics (here in Malaysia), anticyclones give way to depressions due to continental warming. It is the rainy season. At nightfall, there are sudden violent downpours, when large drops fall in quick succession.

◀ When it is very cold at high altitudes and the atmosphere is turbulent, raindrops form little balls of ice, the hailstones so feared by farmers.

the form of tiny ice particles. When these become numerous, they clump together and form snow crystals. As their weight draws them downwards, these crystals encounter layers of air that become less and less cold. If the temperature reaches 0°C before they reach the ground, the crystals melt, changing into drops of water, and they fall as rain.

CLOUDS AT GROUND LEVEL

Mist is just a cloud that has formed at ground level. This phenomenon occurs especially after sunset, when the Earth cools down more rapidly than the surrounding air. Wet air condenses at ground level without even having to rise any higher. Fog is a dense mist in which visibility is less than one kilometre. Water vapour may also condense in small drops on the ground and on plants, forming dew. And when the temperature falls below 0°C, dew becomes ice, and is called hoar-frost or rime.

26 000 litres of water per square metre

If the weather is calm and a precipitation cloud has no strong winds blowing through it, the droplets of water do not become much bigger. Fine, close rain, drizzle or Scotch mist, made up of droplets less than 0.5 millimetres in diameter, may then fall for long periods. This is usually caused by the presence of fairly light stratus clouds spread over a large area.

During a period of meteorological instability with strong winds, the clouds are often thick, of a cumulus type, and are buffeted by strong internal movements which favour the formation of large drops of water. The resulting rain is often short-lived, but it falls in dense squalls over relatively small areas. Friction of the air not only limits the speed of the fall of the drops – no more than 30 kilometres an hour – but also their size: above six millimetres, they burst.

The most enormous cumulonimbus clouds are capable of spilling 4 000 tonnes of water a second over dozens of square kilometres. In 1952, as much as 1 872 litres of water a square metre fell in a single day at a point on the Indian Ocean island of Réunion – almost twice as much as the city of Paris receives in a whole year.

But even this astonishing amount is not a record. Between 1860 and 1861, rain gauges – containers which measure the amount of precipitation – recorded the 12 wettest months ever in a region of India : an appalling deluge of more than 26 000 litres a square metre.

▲*Frost (top) forms when damp air comes into contact with cold ground; snow (above) forms when the ice crystals of a cloud cannot melt while they are falling.*

Hailstones weighing one kilogram

Not all the ice crystals in a precipitation cloud fall down onto the Earth in the form of raindrops. If these drops pass through a layer of very cold air on their way down, their outer surface freezes. They then change into little translucent balls less than five millimetres in diameter. This is the phenomenon known as sleet.

But if very strong winds blow through the clouds before falling, the crystals will be drawn up again to the top of the cloud several times by rising air currents. Each time, more ice particles will accumulate onto the crystals. Layer by layer, hailstones will be formed. When they weigh enough to withstand the rising currents, they will finally fall to the ground. A hailstone generally measures between 0.5 and 5 centimetres in diameter. But people have been known to pick up incredible 'cannon balls' almost 20 centimetres across, with a weight of about one kilogram, which had fallen at a speed of something like 160 kilometres an hour. These stones cause extensive damage to crops, buildings and vehicles.

If it is really very cold in the cloud and the temperatures are below freezing all the way down to the ground, the ice crystals will cluster together to form snowflakes. Seen under a microscope, they have a countless variety of shapes, depending chiefly on the temperature of the air they have passed through during their fall. They can be classified into almost 100 separate groups. But every single flake, even if it reaches more than two centimetres across, has a common characteristic: they form superb, symmetrical geometric shapes, always with six sides.

149

Rivers in the sea

Because of marine currents, the nutrients and heat in water are distributed through the oceans on the planet. The influence of currents on climate is considerable.

▲ Three very fast sea currents shown on a satellite photo: at the top, the Gulf Stream; below it on the left, the Brazil Current; and below it on the right, the Agulhas Current.

▼ This tropical garden in Cornwall, England, owes its existence to the Gulf Stream, a warm current flowing from the Caribbean.

At the beginning of the 19th century, Australian ferns were planted in the south of Cornwall. This region, situated in the south-west of Great Britain, is at the same latitude as the icy locations of Calgary (in Canada) and Irkutsk (in Siberia). The ferns, from a warm wet climate, should have died of cold, but instead they flourished. And today the Cornwall garden of Heligan, where the ferns grow among subtropical plants imported like themselves, is luxuriant.

In winter in Calgary everyone is freezing cold, but in the south-west of England people hardly shiver. Between the two places there is not only an ocean, but also, and especially, a warm sea current, the Gulf Stream, which makes the climate on the coasts of western Europe milder than that of inland Canada.

Mixed waters

If a substance is more concentrated in one place than in another, the flow of this substance will always go from the richer concentration to the poorer in order to re-establish a relative equilibrium. This is the law of diffusion demonstrated by fluid-filled containers which are linked by a tube. In the sea, this flow is called a current.

On our planet, the major currents are formed because there are differences in temperature and salinity between the masses of water.

Because of these currents, heat from the tropics can reach high latitudes and cold from the polar regions can refresh the equatorial regions. Without these currents, too, the nutrients in the silt at the bottom of the sea would never come to the surface and the supply of oxygen at great depths would be poor.

Currents allow the water masses that make up the world's oceans to mingle. They act like a thermostat which frees and retains the heat energy of the sea, and so influence the temperature of air masses and the world's climate.

A conveyor belt in the sea

Thermohaline circulation is generated by differences in temperature and salt concentrations between two masses of water, and its function is of prime importance to our planet. Acting as a gigantic conveyor belt over the whole of the surface of the Earth's oceans, it

Indian Counter-Current
North-Equatorial Current
South-Equatorial Current
Agulhas Current
5
West-Australian Current
Kuros...
E a s...

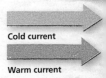

Cold current

Warm current

▲ The great sea currents grow warmer or colder according to where they flow – and change their names.
1. The Gulf Stream, which turns into the Canary Current and the North-Equatorial Current.
2. The Benguela Current, later the South-Equatorial Current and the Brazilian Current.

North Atlantic Drift

Labrador Current

Oyashio

Canary Current

Alaska Current

North Pacific Drift

Gulf Stream

1

North Equatorial Current

California Current

North-Equatorial Current

3

South Equatorial Current

Benguela Current

Equatorial Counter-current

2

Brazilian Current

South-Equatorial Current

4

Humboldt Current

...ian Current

...c Circumpolar Current

▲In the east of Canada, there is often a wave of extremely cold weather in winter. Without the warmth of the Gulf Stream, the north-west Atlantic makes the atmosphere cold and dry along the North American coast, which is chilled still further by cold winds from the Arctic, which are unhindered by any anticyclone.

◄ On the coast of Namibia, air chilled by the Benguela Current meets hot air from the Namib Desert. Before dawn, the humidity condenses into a fog which disappears as the sun rises and the air warms up.

3. The North-Equatorial Current, which will become the Kuroshio, the North Pacific Drift, then the California Current.
4. The South-Equatorial Current, later the East-Australian Current then the Humboldt Current.
5. The West-Australian Current, which will become the South-Equatorial Current and then the Agulhas Current.

circulates the water from the North Atlantic to the North Pacific oceans, and back again, about every 800 years.

Take a starting point, say the Gulf Stream, a current in the middle of the Atlantic, at the level of the equator. The surface layer of water, heated by the Sun, moves north along the east coast of North America towards Greenland. During this long journey, it slowly loses its heat through evaporation, which consequently means that its salt content increases. At the Grand Banks, it branches, one current moving eastwards as the North Atlantic Drift, the other turning south-east back towards the mid-Atlantic.

On reaching the Labrador Sea, however, it descends under the ice, because of its higher density. So this mid-Atlantic current, now cold,

travels south following the bottom of the Atlantic Ocean until it reaches Antarctica, where it grows even colder. On the way, it is enriched by water flowing out of the Mediterranean. This water, denser because it is richer in salt, passes under the Atlantic surface waters, giving rise to another current at medium depth.

151

▲ *This band of whirling white water is actually the Gulf Stream seen from a satellite. It is identifiable by the light that is refracted by the water in a particular manner, affected by the speed of the current. The Gulf Stream is, in effect, the most powerful river in the world.*

This Antarctic current moves east to the edge of the Pacific, where part of it branches north to the Indian Ocean, and the rest continues to the Pacific where it circles anti-clockwise to return to the Antarctic. In the Indian and Pacific oceans the Antarctic waters mingle with the warm tropical waters. Becoming less and less dense, they rise to the surface and, after joining up in the south of the Indian Ocean, they travel back to the Atlantic in a large-scale east-to-west movement.

The wind intervenes

The other type of circulation, which is wind-driven, is effected on the surface of the sea. The winds, coming from the land, send water out to sea. A slope is then created on the surface of the water that is immediately levelled out by a current of cold water from deeper down.

Because of the rotation of the Earth, currents moved by wind are deflected, by what is called the Coriolis force, to the right of their trajectory in the northern hemisphere, and to the left in the southern hemisphere. The deviation is about 25° near the coast and 45° out at sea.

For each current there is a counter-current of the opposite temperature which, far from the coast, replaces the water that is pushed off to the right or left by the Coriolis force. In the Atlantic, for example, the warm current of the Gulf Stream is compensated in the sea off Canada by the cold Labrador Current. In the Pacific, the warm Kuroshio (flowing from the Philippines to the North Pacific) is joined by the cold Oyashio in the Bering Sea. In this way, six large circuits of currents are formed.

The journey of surface waters

The surface currents originate with the trade winds, which are winds that blow from the north-east (in the northern hemisphere) or the south-east (in the southern hemisphere). Between the tropics, these trade winds blow water towards the west. As it moves, the water warms up. When it reaches the western shores of each of the oceans, it is obliged to turn and go along the coasts, to the left or the right depending on the hemisphere. In the north, it goes clockwise (which means towards the north), and in the south, it travels anti-clockwise (southwards).

When it reaches high latitudes, the water is then blown out to sea by the westerly winds. When they encounter the eastern shores of their respective oceans the currents go up (in the southern hemisphere) or down (in the northern hemisphere) the coasts. They thus complete their cycles.

Friction and turbulence

Deep currents come up against the mountainous relief of the ocean floor, to the point where they actually flow in places into gigantic undersea waterfalls, whirlpools that help to mix together masses of water of different temperatures and salinity.

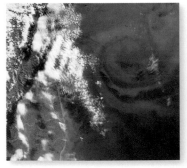

At the surface, the movement of currents generates friction where they come into contact with the masses of water over which they move. Among them are what are known as Rossby waves. As they move slowly from one edge of an ocean to the other, the waves behave like actual currents. They have a determining influence on the circulation of water on a world scale.

▲ *In New Zealand, the upwelling of cold water from the depths favours the proliferation of plant plankton (shown above in pale blue).*

Undulations of the sea

Deep undulations occur on the surface of the seas. These swells, caused by sea winds, break in rollers on the shore.

▲ *The rollers that break on the shores of Hawaii, in the United States, do not depend on the wind to exist. They are the result of the swell that agitates the open sea and picks up speed towards the coast, where it loses its shape.*

It is an extraordinary sight. Plunging down the slopes of waves as steep as valley walls, a large steel freighter caught in a storm in the open sea looks no bigger than a nutshell. Disaster films and adventure stories derive great benefit from the sea swell, with its crests '10 storeys high'.

But a swell should not be confused with a wave. It is always formed out at sea and is the consequence of an atmospheric depression, when the drop in pressure, with its rising and falling air movements and its horizontal winds, creates a considerable and chaotic agitation on the surface of the sea.

Rippling swells and boiling foam

As you move away from this area, you see surface movements begin to form wave trains that flow in only one direction and according to a regular sequence. The height of each wave – the difference between the crest and the trough, its wavelength – the distance between two crests, as well as its travelling speed, are all relatively constant. Two crests are rarely more than 300 metres apart. As far as height is concerned, a wave can reach 25 metres, for example when two swell trains meet and reinforce each other. This oscillation stirs up the water up to 100 metres below the surface.

Once launched, a swell continues, even when there is no wind. The strong depressions that rise off Newfoundland, for instance, engender

▶ *The height of a swell can reach 25 m or more. A record height of 33.6 m is said to have been observed in the North Pacific in 1933.*

long-distance swells that travel for three days as far as the Bay of Biscay, on the west coast of France, almost 3 000 kilometres away.

All this changes near the shore, when the water becomes shallower. The crests of the waves become higher and the seabed gets in the way of the propagation of the undulation, which slows down at its base while the crest continues at the same speed. The result of this imbalance is that the crest topples over into the trough in front of it. This is the roller that all surfers wait for. Such rollers are all the more spectacular when the seabed rises steeply near the coast, as it does in the Gulf of Guinea, in west Africa, where the surf is awe-inspiring.

Tides: gigantic oscillations

Tides, a phenomenon separate from the swell, are periodic movements of the waters of the sea, which rise and fall along the shores about once every 12 hours. This gigantic oscillation is due to the mass of the Moon, which attracts the waters of the Earth as it passes. The Moon's period of rotation round the Earth determines the tides.

The Sun, a lot farther away than the Moon but considerably larger, also influences this movement. The greatest movement of the tide was measured in Fundy Bay, in Canada, with a 19.6-metre difference between the high- and low-water levels.

Reservoirs of water

Holding fresh or salt water, lakes are a reflection of the great natural forces that affect our planet. Some were formed after volcanic eruptions, others arose from the work of glaciers, and still others show evidence of the deformation of the Earth's crust.

▲ *The crater of the volcano Pinatubo, in the Philippines, which erupted dramatically in 1991, is occupied by a lake.*

Make a hole in some soil and fill it with water as many times as it takes for the water to stay there. You have just made the basic replica of a lake: a hollow full of water. Thus, for a lake to survive there must be water to keep it full all year round. Mountain streams, rivers, former seas or underground water also make use of the slightest depression to create these mirrors that reflect the sky, whether they are small pools of water set among the mountains, or inland seas.

At the heart of the volcano

When a lake is called Crater Lake, you can be sure that it is of volcanic origin. With a diameter of about 10 kilometres and a depth of 58 metres, this lake in the south of Oregon, in the United States, occupies a caldera, which is a circular depression caused by the collapse of the summit of a volcano after an eruption.

Similarly, the Taal, near Pinatubo on the island of Luçon in the Philippines, possesses a vast expanse of water (267 square kilometres) also in a caldera, with a cone in the centre, on top of which is a crater occupied by a small lake. But the craters of volcanoes do not always make good containers for water. They are often built from ash or lapilli (small pieces of stone thrown from a volcano), which are too permeable.

The meeting of magma and water

In the north and south of the volcanic Chaine des Puys, in France, the craters of the Gour de Tazenat and Lake Pavin, which store surface water of an intense blue, are steep pits formed by very violent explosions. What caused these disasters? Underground eruptions that cut deep gashes in the substratum. They come about when underground water comes into contact with a column of magma (with a heat of about 1 100°C), gets overheated, turns into steam and seeks an outlet under great pressure.

But the action of volcanoes can also be indirect. By the time very liquid basalt flows have covered tens of kilometres, they may line the bottom of

▼ *Lake Superior, which looks just like an inland sea, is of glacial origin, like the other Great Lakes that lie between Canada and the United States.*

▲ *Whatever their latitude, lakes constitute large reserves of fresh water that often give rise to urban areas. From left to right: an island in Lake Victoria, in east Africa, where a village has built up; Lake Michigan, with the metropolis of Chicago on its shores; the wild shores of Lake Baikal, in Russia.*

valleys and upset the hydrographic network. They may then dam the course of rivers and create lakes like Lake Aydat, in the Puys mountains of France.

In glacial basins

With an area of almost 83 000 square kilometres, Lake Superior is the largest lake on Earth. Positioned between Canada and the United States, the Great Lakes, all of which are exceptionally large, formed behind natural barrages. At the end of the last ice age of the Quaternary period (about one million years ago), the retreating ice sheets that had covered most of the North American continent left moraines, behind which these immense stretches of water became established. The large lakes in the north of Italy have a similar origin.

The glaciers of the last ice age, too, have left numerous traces behind them: Canada and Scandinavia are full of lakes occupying basins that were hollowed out by rivers of ice.

Trenches full of water

In east Africa, Lake Victoria can rival the Great Lakes of North America, having an area of 69 500 square kilometres. A witness to the ceaseless movement of the earth's substratum, it is situated in a region known for its volcanoes, such as Kilimanjaro and Nyiragongo, and for its series of large lakes, from Lake Albert, in the north, to Lake Malawi, in the south, and including the deepest, Lake Tanganyika (with a depth of 1 480 metres). This string of lakes stretches along the floor of the western Rift Valley. The movements and disruptions of continents not only cause the folding and building up of mountain chains, but also create faults which cause the ground to collapse in the form of channels. These trenches are called rifts.

In the Middle East, the Dead Sea, which is 400 metres above sea level, also occupies a tectonic trench, along the Levantine fault. In this very arid region, intense evaporation has made this lake the saltiest on Earth.

On the other hand, Lake Baikal, in Russia, is also in a trench which resulted from a landslide. Its area (31 500 square kilometres) and its depth (1 620 metres) make it one of the largest freshwater reservoirs in the world.

THE ARCHIVES OF THE EARTH

Like the ice on icecaps, the deposits accumulated at the bottom of certain lakes, notably those on the periphery of mountainous regions that receive water from rivers and mountain streams, contain precious information about the evolution of the climate during the Quaternary period. Alluvium is deposited in the lake according to the rhythm of the seasons – there is more in spring and summer, when the snow melts, than in winter. Climatic changes in the Quaternary period, with colder periods than we have now, are reflected in the way the sediments are formed. Samples taken – core samples – are therefore minutely analysed in laboratories. Because of the pollen they have retained, they also yield information about the succession of plants that adapted to these ancient climatic changes.

The journey of a raindrop

The volume of water on the Earth is estimated to be 1 386 million cubic kilometres. This is a huge amount, but at the same time, very little. There is more water on our planet than anywhere else in the Universe – as far we can tell – but it represents only about 0.025 percent of the Earth's mass. Fresh water accounts for only 2.7 percent of the total volume of this precious liquid. It not only serves to quench the thirst of living beings, but also, and more importantly, to transmit the heat of the Sun across the globe. In the same way that a wheel transmits a force, the water cycle transports solar energy from the tropical regions to the poles, from the sea to the land, and back again.

2. CONDENSATION

As it rises, the pressure of the hot air loaded with water vapour decreases. This is accompanied by cooling, which may bring about

condensation. The water vapour then reappears in the form of spherical droplets of between 0.008 and 0.8 millimetres in diameter. As they gather together, they form clouds (left, a stormy sky in the Rift Valley, Kenya), which do not contain a high proportion of condensed water – never more than 10 grams per cubic metre of air. Condensation frees all the energy from the Sun accumulated by the water vapour when it was formed. The slight warming of the local atmosphere that results causes very strong upward currents to form. As the atmosphere is also colder where the water evaporates, convection currents appear between the evaporation zones (subtropical and polar regions) and condensation zones (the equator and medium latitudes).

1. EVAPORATION

In the heat of the Sun, the surface of the oceans evaporates. Every day, 430 000 cubic kilometres of water thus pass from a liquid to a vaporised state and join the lower layers of the atmosphere. In the course of this transformation, a considerable part of the Sun's energy, stored in the water molecules, is used up, which causes a slight local cooling of the atmosphere. The atmosphere in fact draws almost 50 percent of the energy that animates it (the winds) from the cycle of evaporation and condensation. Evaporation is greatest in the tropical regions (above, treetops in the mist in Guyana), where the Sun is strongest, on the continents, which are always hotter than the seas, and at the poles, which are very dry. Added to this is evaporation from lakes and rivers, and the transpiration and respiration of living organisms, making in all about 70 000 cubic kilometres of water. Once evaporated, water is less dense than the surrounding air and is carried upwards by the winds.

3. PRECIPITATION

The droplets in the clouds grow larger, join together and sometimes reach a size of 0.1 millimetre. This gives them a falling speed of about one metre a second. The rising currents then cannot support them any longer. Depending on the air temperature, they fall as rain, hail or snow. In a year, 110 000 cubic kilometres of water precipitates on continents, essentially on the regions round the equator or at medium latitudes (between approximately 35° and 55°). Thus, while China receives some seven percent of the total annual precipitation that waters the planet, the Amazon gets twice as much. If the air temperature is too high just above the land, the precipitated water will not reach the ground.

4. INFILTRATION

Water precipitated above land first falls on vegetation – from which it might evaporate again. The rest pours onto the ground. Part of it filters down to the hydrographic network underground – rivers, phreatic water and springs (below, underground water in the cave of Poço Encantado, Brazil). Some gathers in depressions and forms lakes and ponds. Yet more wets the soil, and the remainder returns to the sea by swelling water courses. It may also accumulate in the form of ice (icebergs and permanent glaciers), which forms a volume of 29 000 000 cubic kilometres, making the largest freshwater reserve in the world, far more than underground water (8 200 000 cubic kilometres), stagnant water (205 000 cubic kilometres), soil humidity (70 000 cubic kilometres) and water courses (1 700 cubic kilometres).

5. CONSUMPTION

The last and smallest reserve of water on Earth is found inside living things: 1 100 cubic kilometres, almost as much as the total volume of water courses. Water accounts for more than 95 percent of the total mass of a jellyfish and for almost 70 percent of that of a human being. Cells store 70 percent of this water, the rest is spread between the cells and in the various body fluids. A person must absorb at least 2.5 litres of liquid to compensate for the 2.5 litres he or she loses each day. It is little compared with cotton,

which has to drink 10 000 litres of water to grow one kilogram, and with our daily consumption (in industries, agriculture – left, irrigation of potatoes in France – services and cleansing), which is not far off three cubic metres for each person.

6. POLLUTION

Flowing over agricultural land, water collects fertilisers (based on nitrates and phosphates) and also very toxic pesticides, herbicides and fungicides. Watercourses are contaminated by waste from industries and towns, which includes heavy metals, solvents and various organic substances like dioxin. All these molecules eventually affect people after accumulating for decades in the food chain. More seriously still, they are concentrated in the phreatic water from which most of our drinking water comes. Underground water is only renewed very slowly, every 1 400 years on average, whereas it takes only a few weeks for the volume of water in a river to be replaced. The water ends up in the sea, where all sorts of pollution is concentrated (left, an algal bloom off the French coast). In spite of everything, evaporation plays the part of a gigantic purifying station, by ridding water of almost all its dirt. It is only above towns that water vapour becomes polluted once more.

157

Between hot and cold

*Whether it is hot or cold, the water that wells up from springs is the same.
It is surface water that has filtered down to the substratum and is returning to
the open air after a journey of varying length and complexity.*

Set in the middle of dark, jagged volcanic rocks, the Blue Lagoon, in Iceland, offers an extraordinary bathing experience. The water is an unforgettable spectacle of the purest blue. And it is sufficiently hot to supply a geothermic power station.

In this country, where there are many volcanoes that erupt from time to time, the rise of the magma causes the temperature in the substratum to increase. The water that filters down thus heats up near the surface and easily rises again, sometimes spurting out in the form of geysers.

◀ The Vaucluse spring, in south-east France, at low water. In limestone areas, springs dry up when the level of the underground water drops to the point where it cannot feed into them.

Gushes of hot water…

At Yellowstone National Park in the United States, there are more active geysers in one area than anywhere else on Earth. The geysers spout at odd times, or, in contrast, as regularly as a metronome. These phenomena occur when hot and cool sources of water meet and the water boils, or when gases force the water to rise (not necessarily at boiling point). Their eruption pattern depends on the type of reservoir they have. The water that spouts from geysers is rich in dissolved mineral salts. These form deposits where the water emerges, in the form of cones or rims.

The phenomenon of hot springs is not only found in volcanic regions. Water can impregnate the rocky substratum to a depth of several thousand metres, where the extremely high temperature causes it to evaporate. But before that it gets hotter as it sinks, on average by 3°C every 100 metres, except in volcanic regions where the geothermic index degree is much higher.

◀ A hot water basin at Beppu (Kyushu, Japan).

SPRINGS AND THEIR DIFFERENT USES

1. A well drawing on phreatic water
2. Granite
3. Alluvium
4. A well drawing alluvial water
5. Spring at the base of the limestone
6. Limestone plateau
7. Clays
8. Spring at the point of contact with clays via a fault
9. Fault
10. Borehole to reach deep underground water

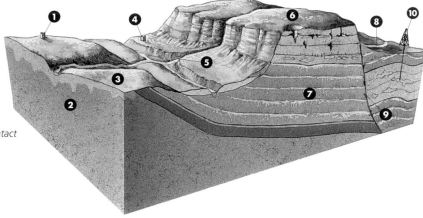

... or layers of cold water

Depending on the rocks, and of course according to the climate, there is a greater or smaller infiltration of rainwater into the ground.

In temperate regions, water filters through cracks in soils with little clay or sand, alluvial terraces along water courses, and other surface formations caused by the erosion or weathering of rocks, to form phreatic water. This underground source is easily accessible by means of wells a few metres deep. The surplus overflows in small slow-flowing springs.

On the other hand, when the underlying soil layer is impermeable, for instance because it has a high clay content, after heavy rains the waters stagnate on the surface, draining and evaporating very slowly. They also lie on the surface, in hollows, in the form of lakes, ponds and marshes.

In drier climates, surface water is not so common. The riverbeds are only full after hard, abundant rainfall. All the same, water will still be found there. The alluvium still contains enough for wells or underground channels to be dug. And there may still exist aquifers, or water-containing rocks, remaining from wetter periods of the Quaternary period, about one million years ago. These last types of water source are particularly vulnerable to overexploitation, as there is no certainty that they will be renewed.

Rivers and springs in limestone

The situation is very different in limestone regions. There, rainwater disappears as soon as it reaches the ground. Rivers are therefore rare, because they have difficulty surviving and may run completely dry, as their water descends to join water courses underground.

The fissures in limestone rocks, enlarged by the dissolving of the calcium carbonate, transform this type of rock into a giant sponge. The water that flows down through it will return to the open air in a few large springs that lie along the base of the limestone, when this rock rests on a substratum of impermeable rocks like clay.

Sometimes, faults or breaks in the Earth's crust bring aquiferous rocks into contact with other types of rock which then create an impermeable barrier. After heavy rain the resulting spring can reach an output of several cubic metres of water a second. This type of spring dries up in periods of drought or, as in the case of the Vaucluse spring in Provence, France, because an underground passage functions like a siphon, diverting the water from the spring, which becomes evident when the level of the underground water drops.

▲ *Two ways of using underground water: washing clothes in water obtained by drilling, in Syria (top), and relaxing in the geothermic warmth of the Blue Lagoon, in Iceland (above).*

▲ *Castle Geyser, in Yellowstone National Park, United States.*

HOT SPRINGS AND THE ROLE OF FAULTS

Particular geological conditions are necessary for hot water to flow deep underground. Very often, hot springs occur at – or near - the site of large faults that act as an underground drain: as rocks have more fractures at these places, the rise of hot water, less dense, is made easier. But not all faults play this role. Dissolved mineral salts settle and block the fissures. It is thus necessary for the faults to be disturbed from time to time even with very slight earthquakes, in order for the fissures in the rocks to be reopened and for the hot springs to flow once more.

The long still rivers

*The low temperatures of high latitudes and mountains transform snow into ice.
This slowly moves as a result of its weight and plays a part in modelling
the surfaces of the Earth by erosion.*

Anyone flying over the glaciers of Alaska might think they were passing over a Californian motorway without the cars. Or above an immense river with tributaries, their courses meeting and then parting without mixing, like moving walkways. The ice of glaciers is indeed far from immobile. It is formed when snow piles up, settles, and under its own weight changes into ice that moves slowly downhill, usually at the speed of a few tens of metres a year.

Ice at work: plains, cirques and moraines

A glacier carries along with it pebbles and lumps of rock of all sizes. Water that freezes in the cracks of rock walls and peaks causes pieces of rock to break off. These fragments then fall to the edge of the glacier to form moraines along the sides of the ribbons of ice. At the confluence of two valleys, moraines are found in the middle of the ice.

Altitude and latitude regulate the temperatures that determine the balance between the formation of ice and its thawing. At the head of the glacier, the waters from melting ice run off and form water courses, taking with them the finest elements of the glacial 'flour' that give it a white colour. The sand and pebbles are spread out by this water and form glacial plains, called *sandür* in Iceland.

The mark of glacial erosion is clear in many regions. The departure of the ice has left glacial cirques, or amphitheatres, rounded rocks, U-shaped valleys and moraines. The last are furrowed by flowing water that leaves in some places fairy-castle shapes that erosion has not yet worn away.

The poles, ice mountains

But valley glaciers are nothing in comparison with the icecaps near the poles. There the ice is very thick (up to 4 000 metres) and covers considerable areas. These continental glaciers in Greenland (where they cover 2.7 million cubic kilometres) and Antarctica (extending over 28 million cubic kilometres), almost completely cover the rocks. Nevertheless, the rocks rise in sharp peaks, like Mount Vinson, not far from the South Pole, which reaches a height of 5 140 metres.

▲ *Glacier Bay
National Park, in
Alaska. The
irregular flow of
the ice is revealed
by the dark,
winding threads
of moraine.*

▼ *Vatnajökull,
one of the largest
glaciers in Iceland,
covers active
volcanoes.*

GLACIAL EROSION

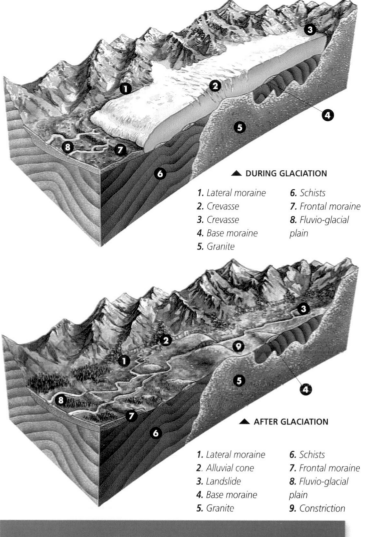

Because of its weight, ice flows towards the sea in immense glaciers. From their heads gigantic pieces break off as icebergs, tumbling into the sea when the sea is not itself turned into an ice field. These mountains of ice, of which only a small part is visible above the sea's surface, then drift with the currents and the swell towards more temperate waters where they melt and disappear. Attempts have been made to tow the largest icebergs to regions with few sources of fresh water (such as the Arabian Peninsula), but the plan has not proved feasible and has had to be abandoned.

▲ *In King Charles Bay (Spitsbergen Island, Norway), huge icebergs break off the head of the Kongsbreen glacier.*

▲ **DURING GLACIATION**

1. Lateral moraine	*6. Schists*
2. Crevasse	*7. Frontal moraine*
3. Crevasse	*8. Fluvio-glacial*
4. Base moraine	*plain*
5. Granite	

▲ **AFTER GLACIATION**

1. Lateral moraine	*6. Schists*
2. Alluvial cone	*7. Frontal moraine*
3. Landslide	*8. Fluvio-glacial*
4. Base moraine	*plain*
5. Granite	*9. Constriction*

Fire and ice

In Antarctica, in the middle of icy wastes swept by violent winds, a volcano with a permanent lava lake, Mount Erebus, is a reminder that the deep dynamics of the Earth are also present at these latitudes. The marriage of fire and ice is of no consequence here. In Iceland, on the other hand, it can be disastrous. The large icecap glaciers of Iceland cover land that is totally volcanic, torn by clefts, or rifts of the mid-Atlantic ridge. Volcanoes are scattered over it, some of them hidden under the ice. When they become active, they cause the ice to melt. Immense pockets of water are suddenly released and sweep everything away with them.

A few small mountain glaciers also flow from the top of volcanoes, like the Nevado del Ruiz, in Colombia, which sadly destroyed the village of Armero in 1985. Sweeping downhill, the water from melted snow and ice gathered in a valley, eroding away the sides. A stream of mud then burst out over the houses and covered them completely.

ICE TELLS ITS STORY

If the receding of the glaciers directly reflects the warming observed during the last decade, their history is still more enlightening. The study of ice core samples drilled out of the icecaps, called inlandsis, has revealed the climatic fluctuations of the last thousands of years. We now know that, since the end of the Tertiary period, about 2.5 million years ago, cold periods alternated with warmer periods. The landscape also bears signs of the heavy ice ages of the Quaternary period, about one million years ago. For example, erratic blocks, like the one in Central Park in New York, which was transported by the Canadian ice inlandsis, were left behind by glaciers. Erosion could not wear down the largest blocks of the moraines.

The winding streams of life

Rivers and their tributaries are the arteries of the planet. They drain the 'surplus' water away to the sea and take an active part in the vast work of reshaping the Earth's surface.

▲ *In the Amazon, the boundaries between river and land are blurred where the plentiful water of the River Amazon flows very slowly across the alluvial plain.*

At the heart of the Amazon estuary in Brazil, Marajó Island is surrounded by the waters of a river that flows into the Atlantic at the rate of 200 000 cubic metres a second. Though the Nile is longer, the Amazon is larger because of the volume of water it carries. Seventeen tributaries supply it with water collected from the whole of the northern half of the South American continent (an area of about 7 000 000 square kilometres).

Almost 7 000 kilometres long, often more than 10 kilometres wide, several dozen metres deep and navigable from the sea to 1 600 kilometres from its mouth, the Amazon is indeed the most powerful river on Earth. And for centuries, it has been wholly responsible for shaping the lifestyle and activities of the people who live on its banks.

River of excess

The Amazon is a spinal column, from which branch off tributaries that are themselves really large rivers. They mostly flow down from the Andes (like the Negro, Purus and Madeira). Others come from the central Brazilian plateau (Tapajós and Xingú), in the south, and to a lesser extent from Venezuela and the Guiana Highlands in the north. The junctions of the river and one or other of its tributaries, like the confluence with the Río Negro, carry such a lot of water that they form a sort of inland sea.

▶ *The Yangtze Jiang is one of the largest rivers in the world. It flows from Tibet to the China Sea and is navigable in central China.*

The Amazon is situated on both sides of the equator in a region with a hot wet climate, where the precipitation is between 1 500 and 3 500 millimetres a year. Water from the streams in the Andes, which are supplied with water from melted snow, is supplemented with abundant run-off, because the soil in the Amazon rain forest does not absorb all the rainfall. Thus water converges from streams to small rivers, then from these to the tributaries of the Amazon, the main artery for drainage. The Amazon flows into the ocean through an estuary full of islands, more than 60 kilometres wide at its mouth.

Running water

Flowing water does not only empty surplus water from the land into the sea. In the course of its journey, it also works at shaping the Earth's

surface. This sculpting process may be either discreet or flagrant, continuous or intermittent, and deals with huge volumes, counted each year in tens or hundreds of millions of tonnes of soil. Even the most calm-looking water is active: it transports elements in solution, like calcium bicarbonate from crumbling limestone rocks.

Water carries anything that is loose or loosens easily – sand, clays and surface soils – so that the rivers often assume a particular colour. Of the tributaries of the Amazon, the *ríos negros* (black rivers) contain iron and organic matter whereas the *ríos blancos* (white rivers) carry silt, which makes them cloudy (the Madeira and the Amazon, for example). Where the Río Negro meets the Amazon, the ribbon of the river remains bi-coloured for many kilometres before the two rivers mingle completely.

Rough passage

Rivers of equatorial regions with a low relief, which carry only fine particles in suspension, are not capable of effective erosion when the bedrock is

◀ The Colorado River, seen here in Marble Canyon, United States, dug away for millions of years to create its jagged path through the middle of these ancient rocks that date from the Palaeozoic.

◀ On the Zambezi, seen from the Zambian side, the Victoria Falls show how incapable some rivers are of digging a sloping riverbed, because they have no large-scale material to transport.

◀ The waters of the Río Negro, coloured by fine elements in suspension, have difficulty mixing with the waters of the Amazon, which are clearer.

▲ *Carrying pebbles along with them, rivers (here the Semine, in the Ain, France) carve out 'giant's saucepans' from their rocky bed.*

hard. This explains why, in Africa for instance, their course is often interrupted by waterfalls or rapids, corresponding to areas where rocks best withstand weathering.

The work of erosion is much more clearly seen in mountainous regions, where the slopes are steeper. The beds of mountain streams are often strewn with large pieces of rock that move, slide and topple over at periods when there is a lot of water and are worn away by repeated friction against the pebbles. When the streams flow out into the bottom of a valley, all this coarse material is left behind and forms huge fan-shaped heaps called alluvial cones. When they enter a lake, the same thing happens: a small delta will be formed, the first stage in the filling-in of the basin.

Large-scale work

Deep gashes, gorges and canyons – the work accomplished by certain watercourses – can be spectacular, even if it takes thousands of years to carry out, unnoticed on a human scale. Steep valleys are generally carved out of solid rock. Running water can attack them only with the help of rock fragments. Sand, gravel or pebbles then undermine them further by friction. Sufficiently strong whirlpools carry the debris away and together they create potholes, or what are sometimes called 'giant's saucepans'.

By the same process, watercourses also wear away the banks that overhang them and broaden the river bed at the sides, thus making beautiful large curves. However, it takes the intervention of other sorts of

erosion phenomena for the valley to broaden out more than this. Landslides, weathering, crumbling and breaking of the rocks will gradually widen the carved edges originally made by flowing water.

Free or captive

The great rivers of the vast alluvial plains are not fixed in place as are those that run along the bottom of gorges. They may change their course, because their meanders are free to go where they like within their main bed, as does the Okavango River in Botswana.

The variation in course of rivers is sometimes even greater. Changes of place or even of partner may occur, as movements of the soil or rises in the water level drive rivers to capture their neighbours. The river then may deviate and incorporate some of the water from another water course. This is how the Moselle, in France, joined the Meuse before becoming a tributary of the Meurthe.

▲ *The inland delta of the Okavango, in Botswana. In the alluvial plain, the river twists and turns in changing patterns as it is free to meander.*

RIVERS DRY AND RIVERS IN SPATE

The Amazon and Congo rivers have no problem with their water supply. The runoff from considerable surface areas, or pouring basins, and the abundant rain of the equatorial regions ensure a good rate of flow. In temperate regions, watercourses are sufficiently fed by precipitation, even if it is not as heavy. In mountainous regions, the melting of snow and ice makes up for the relative dryness in summer. But many regions suffer from a more arid climate. In these cases, the general rule about water flowing as far as the sea no longer applies. The beds of the wadi in the Middle East (right, in Dubai), and the fiumare in the south of Italy see abundant water flowing only a few times a year. Most of the time completely dry, they serve as communication routes. Very occasionally rain will fill a channel or a hollow for a short time before it drains away and evaporates.

CELESTIAL AND ATMOSPHERIC PHENOMENA

The magical world of the sky

*Sunset over the Timor
Sea, off Darwin,
Australia.*

A gigantic nuclear reactor

The Sun is the star of our Solar System. It emits enormous amounts of energy of such intensity that we feel the effects of its heat and light even from 150 million kilometres away.

With a diameter of 1.4 million kilometres, the size of the Sun is such that the planet Earth could fit into it one million times. Its gravitational force attracts – and keeps in its orbit – objects that are more than six thousand million kilometres away. Like all stars, the Sun is an enormous thermonuclear boiler that is continually burning large quantities of hydrogen, the most basic element in the Universe. The fusion reaction between hydrogen and helium is its primary source of energy.

A living star

The Sun does not escape the general rule that governs the life of all stars. This rule states that there must be equilibrium between two major forces: gravitational force (the star's tendency to contract towards the centre of its

◀ *This enormous prominence bears witness to the Sun's activity. Its height above the surface of the Sun is greater than the distance between the Earth and the Moon.*

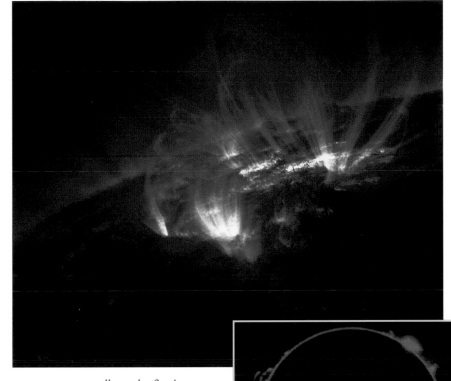

sphere) and thermonuclear energy, which is released by the fusion of hydrogen atoms (and governs a star's tendency to expand). The second force is just a consequence of the first, which in turn is counterbalanced by the energy of the second. When this equilibrium is established, a star can exist (*see pp. 10 -11*). Because of this, the Sun has been able to shine for 4.6 billion years in a relatively stable manner. Its reserves of hydrogen ought to allow it to continue to generate energy for at least a similar number of years to come.

A burning core

The temperature and pressure conditions at the core of the Sun are impressive: 15 million degrees Celsius and a pressure equivalent to 160 times what we experience on Earth.

It is in the heart of this furnace that the thermonuclear reactions take place. Every second, 500 million tonnes of hydrogen are converted to helium in this way, and four million tonnes of matter are transformed into energy. This liberated energy races towards the surface of the Sun and first diffuses into the radiative zone surrounding the core. It is then transmitted out to the periphery of the star in vast swirling movements that characterise the convective layer.

A powerful magnetic field

The solar photosphere, beyond the convective layer, is a zone of approximately 200 kilometres in thickness, where matter is quite transparent, which allows the Sun's rays to escape freely. Its average temperature is about 6 400 °C. Studies of the photosphere reveal that its surface is irregular. In fact it consists of a mosaic of bright areas known as granulations, which have a breadth of about 1 000 to 2 000 kilometres. These are the last manifestations of the convective layer.

Sometimes, darker patches – called sunspots – appear in the photosphere, corresponding to regions that are between 2 000 and 3 000°C cooler. Here, there are magnetic fields with an intensity 25 000 times greater than the Earth's magnetic field. These sunspots usually last a few weeks. The biggest ones can last for several months. They are often surrounded by faculae, clouds of bright, burning gas. Observation of sunspots has led to a better understanding of the Sun's cycles of activity. By studying their position and movement, it has also been possible to estimate that the Sun's period of rotation about the axis between its poles is approximately 30 days.

The solar corona

When a total solar eclipse occurs, it is possible to see a ring like a pink fringe several thousand kilometres broad, around the solar disc. This is the chromosphere. On occasion, magnificent prominences shoot out from this layer. They are jets

▲ *Solar prominences. They are seen with the help of NASA's TRACE detector (top); and with a coronograph (above), an instrument for studying the solar corona.*

▶ *Sunspots, corresponding to cooler areas, are sometimes of considerable size: the Earth would fit into the point marked (1).*

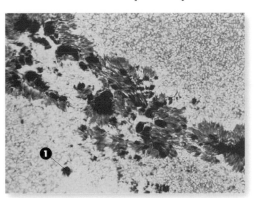

❶

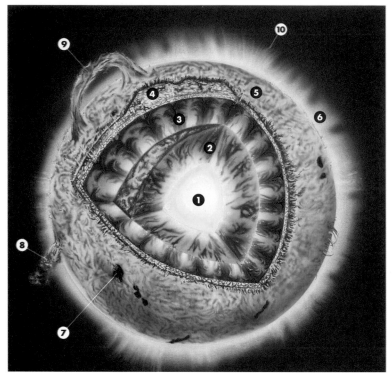

▲ *In the core (1) hydrogen is changed into helium. The radiative layer (2) facilitates the transmission of this energy towards the convective layer (3). The photosphere (4) and chromosphere (5) convey it to the corona (6).*
7. *Sunspot*
8. *Solar flare*
9. *Prominence*
10. *Solar plumes*

at different times. This is called solar activity. It is seen in various phenomena (spots, faculae, prominences and flares) and varies according to a cycle of about 11 years.

During a period of solar activity the number of sunspots increases, from almost 0 to 100 (the solar maximum). As the cycle ends, the number of sunspots then decreases, from 100 to almost 0, and then the cycle begins again. At its maximum, the auroras are very bright and the solar perturbations are sometimes perceptible even as far away as the Earth's surface.

The nature and cause of the 11-year cycle is one of the great mysteries of solar astronomy. While we have much detailed information about the solar cycle and the process that creates it, it is still not possible to provide a model that would enable us to predict the formation of sunspots based on physical principles.

This problem is comparable to our inability to predict earthquakes. It is unfortunate, because in the long term the irregularities in solar activity may have repercussions on Earth, even bringing about considerable variations in climate. More than ever, our little planet depends on the moods of the Sun.

▲ *Travelling between Earth and Venus, the Soho probe observes the Sun's outbursts of fury. Above, a chromospheric flare seen from the probe. A screen shields the solar disc and makes the corona visible.*

of gas that sometimes look like curls or bushes and are associated with sunspots. Their shape is sculpted by the Sun's magnetic field. When they erupt, they burst from the surface at a speed of nearly 1 000 kilometres a second, and are sometimes more than 500 000 kilometres high.

Beyond the chromosphere, the corona extends over more than 10 solar radii. It has a very high temperature and has long jets of hot gas running through it. Over a greater distance, the corona sends through the whole of interplanetary space a constant flow of energy called the solar wind. Its average speed near the Earth is 400 kilometres a second. Its interaction with the Earth's atmosphere creates superb effects which look like folds of drapery, and are further enhanced at the time of solar flares. These are the polar auroral displays, called the aurora borealis in the northern hemisphere and the aurora australis in the southern hemisphere.

Changing moods

The Sun has a magnetic field that permanently influences the photosphere, the chromosphere and the corona, but in different ways

CHRONICLE OF A PREDICTED DEATH

A humble star, the Sun has an estimated lifespan of ten thousand million years. Presently in the middle of its life, it should die in about five thousand million years' time. First there will be more irregular thermonuclear reactions, producing fluctuations in the light from our star. To stabilise itself, the Sun will begin to grow very much larger to the point where it will swallow the Earth! After a short burst while it changes fuel from hydrogen to helium, it will finally die a rapid and violent death: its peripheral layers will be volatilised and will fly off into space at great speed. An enormous bubble of matter will dilate in space and in the central part there will remain only a little very bright residue called, for this reason, a white dwarf. The Earth, the other planets, satellites and asteroids will all have been completely destroyed or sterilised by this orgy of energy. It will also be the end of the Solar System. Right, the planetary nebula Dumbell is on its way to becoming a white dwarf.

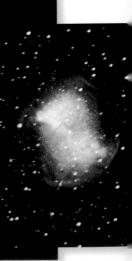

When light deceives us

Mirages are optical illusions due to the refraction of light on layers of air, one of which is much hotter than the other. It is a frequent phenomenon, particularly in deserts and the polar regions.

It is a well-known image: lost in the desert, overcome by heat, two men finally see a sheet of water, hasten up to it and dive in – only to find themselves at full length in the hot sand. (Maybe you have recognised Thomson and Thompson in an episode from Tintin in the *Land of Black Gold*.) And yet they had really seen that water. Was it a hallucination? No, it was just a deformation of reality known as a mirage.

Pools of water on a hot road

To understand the nature of a mirage, you first need to understand the trajectory of light. In interplanetary space, light travels in a straight line. But when it encounters a different medium, like air or water, which it does within our atmosphere, it changes direction slightly. You can do a quick experiment with a spoon in a glass of water to show this: at the line between the water and the air, the spoon appears to be bent.

This sort of situation can come about in air, when the temperature changes suddenly. So in summer, on roads heated by the sun, the layer of air in contact with the ground can reach 10°C higher than the layer a few centimetres above it. This variation in temperature changes the refractive index of the air and acts on light. The light rays coming from an object beyond the horizon will curve and deviate upwards. It is then possible to see in the distance a shiny sheet in which some parts of the scenery seem to be reflected. The silhouette of a tree, for instance, appears to float upside down in a pool of water. This is an inferior, or downward, mirage. The water is in fact only an image of the sky and the inverted tree, the image of a real tree a little further on.

▼ Boats on a lake? No, they are just vehicles on a hot road.

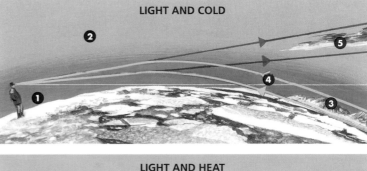

LIGHT AND COLD

LIGHT AND HEAT

Corsica in the sky

A steamer sails on the high seas. And yet, from the shore, some people can see it already, even though it is still beyond the horizon. Again, this is not a collective hallucination but simply a mirage called a superior, or upward, mirage.

In fact, in the polar regions or out at sea, the lower layers of the atmosphere are cooler than those lying above them. The deviation of the light rays then occurs downwards. The image appears high up and it is possible to see objects normally out of sight beyond the horizon.

The inhabitants of Nice are accustomed to this phenomenon. From their beaches they can sometimes see the island of Corsica in the sky, at least in a mirage.

▲ **Light and cold**
1. Cold air
2. Warmer air
3. Object out of sight
4. Light rays deflected towards the ground
5. What is seen: a reflection the right way up

Light and heat
1. Warm air
2. Cooler air
3. Object out of sight
4. Light rays deflected upwards
5. What is seen: a reflection upside down

THE·MOON

A satellite with influence

At first sight, they could not be more different. Yet the Earth and the Moon form an inseparable couple. The history of our only satellite is in fact closely linked to that of our planet.

▲ *The Moon was born when the debris from a collision between the Earth and a very large planetoid recombined and went into orbit round the Earth.*

A dead star, with no atmosphere, where the temperature from light to dark varies by 100°C; a star whose surface is pitted with craters, mountain ranges and wide expanses of cold lava; a star that looks desolate and greyish and yet is the object of all eyes, indeed an inspiration for romance.

Observed for thousands of years, studied and coveted, our satellite – the Moon – is the only celestial body on which a human being has ever set foot. Between July 1969 and December 1972, 12 men had the privilege of living there, 384 000 kilometres from their home planet.

An explosive birth

The Solar System was very young and the Earth was being formed. Like the other planets, it suffered intense bombardment by meteorites. And then, on one occasion, a celestial body the size of the present planet Mars crashed violently into it. Debris from the two bodies was mixed together and went into orbit around the Earth. Little by

little, the pieces clustered together and formed the present Moon, a relatively large satellite since today it represents about a quarter of the diameter of the Earth.

This catastrophic scenario has been confirmed by the analysis of moon rock brought back by the various missions (human and remote) to our satellite. The Moon is indeed made of light

◀ *The footprint of Neil Armstrong on the Moon. On the night of July 20 to 21, 1969, this astronaut became the first human being to set foot on the Earth's satellite.*

▲ *The phenomenon of earthshine occurs shortly before or after the new Moon. Beyond the crescent, which is the part of the Moon directly illuminated by the Sun, the rest of the Moon can be made out. This is light from the Sun reflected by the Earth bringing this area out of the shadow.*

elements that come from the outer crusts of the the two spheres that collided. It is therefore slightly less dense than the Earth.

Bombarded by meteorites

Some 382 kilograms of moon samples have been brought back to Earth. The information from this valuable evidence, together with investigations carried out on the spot (sometimes using a robot), have given us a better understanding of the Moon's history.

We know today that it must have had quite a short active period. In the beginning, the heat was so great that it melted the rocks. The Moon was then covered with a magma that gradually cooled and crystallised. Bombarded with a rain of meteorites, the crust fractured in places allowing magma to flow out. These large expanses correspond to what have been called the seas, forming the dark areas on the Moon that are visible with the naked eye from Earth. The moonscape was in fact entirely shaped by fallen meteorites, forming craters, folds and massifs. But whereas the Earth's landscape is continuously changing, the Moon has hardly changed for nearly four thousand million years.

A billiard game of lights

Although the Moon does not produce light, it is clearly visible from the Earth. In fact it reflects the light of the Sun. And if its form seems to change, it is because, in its course round the Earth, it is lit up in different ways. When the Sun, the Moon and the Earth are aligned, we have a new moon. Seven days later, the first quarter can be seen, and on the fourteenth day it is full moon, which

◄ *A few days after the new moon, a thin crescent is visible after sunset.*

◄ *On the seventh day, the Moon, seen from the Earth, is half lit up: this is the first quarter.*

◄ *On the day of the full moon, the Moon rises when the Sun sets; it then remains visible all night.*

◄ *This thin, luminous crescent is visible only a few days before the new moon, when the Sun has not yet risen.*

A MUTUAL ATTRACTION

The Earth and the Moon are linked by the force of gravity. And, just as the Earth acts on the Moon, the Moon exerts a strong influence on the Earth, in particular on the oceans. Water masses are pulled out of shape by the phenomenon of tidal movement, which can be exceedingly spectacular. The record for a high tide is in Fundy Bay, Canada, with a height difference of more than 19 metres. The action of the Moon also affects the upper layers of the Earth's crust. Each time the Moon passes, it is possible to measure tide-like terrestrial movements of nearly 30 centimetres. The Earth is more malleable than we think.

corresponds to the alignment, in order, of the Sun, Earth and Moon. On the following days, the Moon appears later and later, until the third quarter. And the cycle of 29 days, or the lunar month, comes full circle at the following new moon. Three or four days before or after the new moon, the Moon looks like a thin crescent, but the rest of the sphere is faintly visible. For while the Moon reflects the light of the Sun towards the Earth, the Earth does the same for the Moon. The reflected light illuminates the remainder of the satellite and enables us to see this diaphanous aspect, called earthshine.

In the pressure-cooker of the atmosphere

Because of the Earth's curvature and the angle of its axis, the Sun's heat reaches unevenly over the Earth's surface: the equator is hotter than the poles. But temperature seeks to become uniform within an environment, so the imbalance sets in motion a gigantic process for the redistribution of energy, chiefly by mobilising the Earth's atmosphere, using vertical air flow, evaporation, clouds, horizontal winds, anticyclones and depressions. All these phenomena, which condition climate, develop in the lowest and least stable layer of the atmosphere, the troposphere, which is about 15 kilometres thick. It is this general atmospheric circulation that, like a colossal heat pump, is responsible for up to 80 percent of the transfer of the energy that circulates from the equator to the poles.

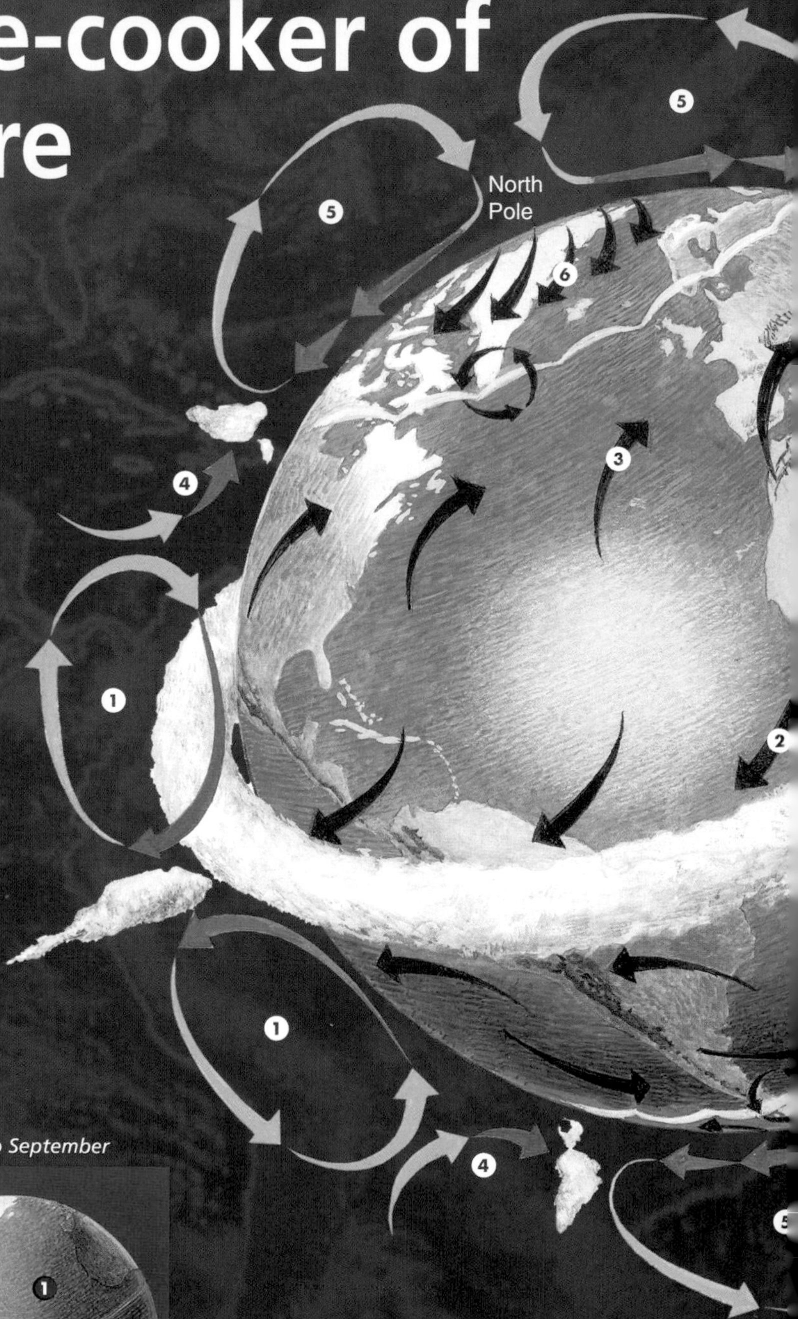

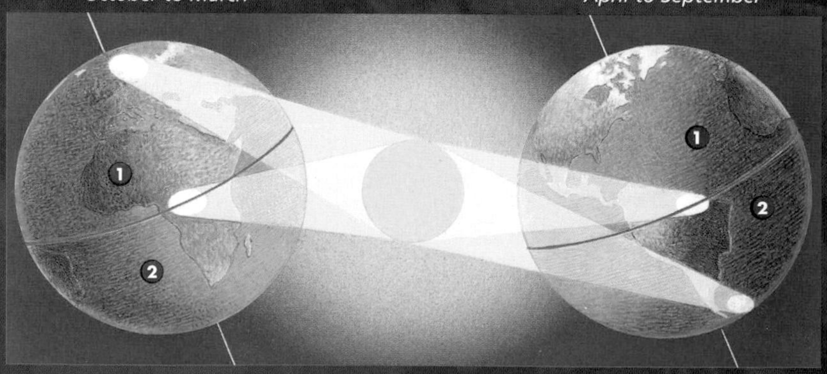

October to March

April to September

▲ **1- Unequal under the sun**
The Earth revolves round the Sun, but also rotates on its own axis, turning all points on its surface to face the Sun. However, at the equator, the Sun is always vertical at midday. The equator therefore receives more heat than the poles do, where the Sun strikes the ground at a very low angle. Moreover, as the Earth's rotational axis is inclined in relation to the plane in which it revolves round the Sun, the northern hemisphere (1) receives more sunlight than the southern hemisphere (2), from April to September, and the southern hemisphere receives more sunlight from October to March.

▲ **2- Uplift and subsidence**
The surface of the Earth – land and sea – receives and reflects the Sun's rays and so heats the surrounding air. This air expands and becomes less dense – lighter – than the neighbouring layers. So it rises, laden with moisture. This is uplift. But as it gains height, it cools at a rate of about 1°C every 100 to 200 metres. It then releases its water, which condenses in the form of clouds before spreading out horizontally. At a height of about 10 kilometres, air movements give rise to strong regular currents which sometimes reach speeds of up to 400 kilometres an hour. These are jet streams.

◀ 3- THREE CELLS AND WINDS

The largest of the convection cells is the tropical Hadley Cell (**1**). It begins in the region of the equator, where the heat supplies strong rising currents, and is an area permanently full of cumulonimbus clouds, which regularly release torrents of rain. The cool dry air of the cell sinks down again in the region of the 30th parallels (south and north). This explains why most of the world's great deserts are at these latitudes.

When the Hadley Cell arrives in the lower layers of the atmosphere, the flow of air from the cell divides. Part of it goes towards the equator and forms fairly stable winds, the trade winds (**2**). The rest goes in the direction of the poles, and gives rise to the system of westerlies (**3**), predominant in the temperate latitudes in the northern and southern hemispheres. This is the temperate Ferrel Cell. Its humid air masses, which in the vicinity of the polar circles (**4**) pass above the cold dry air coming from the Polar Cell (**5**), complete their circuit in the upper layers of the troposphere. In the two polar regions, the predominant winds blow from the east (**6**).

▼ 4- DEPRESSIONS AND ANTICYCLONES

As it rises, hot air causes a drop in pressure at the level of the Earth's surface and forms a depression. The atmospheric pressure here is lower than the average (1 015 kPa). 'Pa' is the abbreviation for 'pascals', the unit of measurement of pressure exerted on a surface. Atmospheric pressure decreases as one nears the centre of the depression, where it is lowest (points of equal pressure are linked by imaginary lines called isobars). On the other hand, cold air descending causes an increase in pressure and creates anticyclones (a pressure

greater than 1 015 kPa). At the equator, therefore, there is a string of depressions; near the 30th parallel a string of anticyclones; and at the polar circles, depressions. This system is, however, disturbed by the continents. The permafrost in Siberia creates a local anticyclone, and the warm ground in deserts creates tropical depressions. These pressure systems are stable, even if they vary in place and intensity according to the amount of sun, that is, seasonally. The climate of Europe, for example, is influenced by the Iceland depression and anticyclones from the Azores and Siberia.

5- THE CORIOLIS FORCE

Between the pressure systems, in-draughts are systematically created. The winds travel from high to low pressure areas, which means from the centre of anticyclones to the centre of depressions. This atmospheric circulation is also influenced by the Coriolis force. Created by the Earth's rotation about its own axis, it deflects all movements to the left in the southern hemisphere and to the right in the north. Its intensity, zero at the equator, increases as it moves away from it. Under the influence of the Coriolis force, a wind that would normally blow between the centres of high and low pressure areas is moved off course and blows round the pressure systems, clockwise round a depression and anti-clockwise round an anticyclone in the southern hemisphere, and conversely in the northern hemisphere. This principle is also seen at work in the large flux of the convection cells: the trade winds, for instance, blow to the south-east (and not to the south) in the southern hemisphere, and north-east (not to the north) in the northern hemisphere.

South Pole

The air that has risen becomes drier, colder and denser (shown in blue, above and right), and so finally sinks down again to the Earth's surface. This is subsidence. Here it heats up again (shown in orange) and again becomes humid, rises again, and so on. These enormous loops of air, which affect the whole troposphere, are convection cells and determine the circulation of the main winds over the planet. They are formed according to latitude, symmetrically in relation to a line near the equator – the intertropical convergence zone – which oscillates between 10° of latitude south or north according to the season, where there is almost no wind.

Depression

Anticyclone

Isobars

Highly coloured arches

When the Sun lights up a curtain of rain, a stunning phenomenon of light occurs: a superb many-coloured arch stretching across the sky.

Of all nature's wide range of special effects, the rainbow is probably the most spectacular. Although it is very common, familiar and the inspiration of many descriptions, a rainbow still makes people marvel. But the creation of this phenomenon and the materials it requires – a little rainwater and sunlight – are very simple.

The alliance of water and light
To be sure of seeing a rainbow, you should turn your back on the Sun and look at a sheet of rain. This downpour is made of millions of droplets of water, the size of which depends on local circumstances (the temperature and pressure in the cloud, and the height and size of the cloud).

The light emitted by the sun goes through the drops, which you could think of as many little spheres. As it travels from the air through the water drop and from the water to the air again, the sunlight – white light – is refracted. This causes it to split into a multitude of colours, from red to dark blue. After being reflected within the transparent drops of water, this light comes out again at an angle of 42° in relation to the angle at which it went in and returns to the observer in the form of a circle which we always see straight on.

But not all the colours are deflected in the same way. Red, for example, is deflected less than blue is. The arch formed is therefore blue towards its inner edge, and red on the outside. Spectacular and bright, this is the primary rainbow.

The play of light and water becomes more complicated when a second reflection occurs in the raindrop. The angle of exit is then 52° and produces a second rainbow which surrounds the first. But the extra angle of reflection causes an inversion of the colours in the second arch. Now the red is inside the arch and the blue outside. Less intense in brightness, this is called the secondary rainbow. The region between the two rainbows, where not much light is scattered, is known as 'Alexander's dark band'.

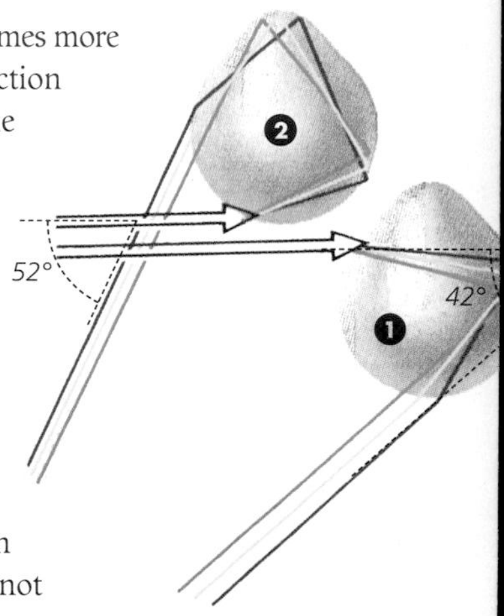

In good light
As in the theatre, a show's success depends largely on the participants. In order to produce beautiful rainbows, the sun must not be too high above the horizon. The lower it is, the higher the rainbow will be, and the better it will stand out against the sky. And the larger the raindrops, the larger the spread of colours. Drops up to one millimetre across give very fine, bright optical phenomena, whereas the microscopic droplets of fog form arches that are hardly visible. These very faint arches may also be caused by another light source, like the Moon which, not being as strong as the Sun, can produce only pale effects.

▲ When light (white arrows) goes through the middle of a raindrop, a single reflection (1) produces a primary rainbow. A second reflection (2) produces a secondary rainbow.

▼ Great, wet, empty spaces (such as in Finland, seen below) produce spectacular rainbows.

THUNDER AND LIGHTNING

The angry sky

In storm clouds, strong tensions between the positive and negative poles cause lightning, and the expanding air makes the sound of thunder.

▲ *Storm clouds act as fearsome accumulators of static electricity. If the accumulation becomes excessive, the electricity is discharged and the sky is split with a blinding flash of lightning.*

The storms of hot summer days can be felt before they arrive. The weather is sultry and windless. Then the sky darkens as large dark clouds gather. And soon the rain comes streaming down, accompanied by thunder and lightning.

Hot, humid air rises in large quantities. At about 15 kilometres above the Earth's surface, it cools suddenly. The water vapour in the air partly condenses into ice particles, forming storm clouds. During their ascent, these particles, which rise at a speed of almost 100 kilometres an hour, collide within the growing cloud. The resulting friction causes the transfer of electrical charges. Falling particles, which are the heaviest, gain a negative charge, which accumulates at the base of the cloud. Rising particles, the lightest, gain a positive charge, which is carried to the top of the cloud.

A 100 000 amp discharge

At the edges of the cloud, the negative and positive poles attract each other, like the poles of a magnet. The tension becomes so great that these particles finally produce an electric arc. This violent discharge takes the form of a streak of lightning, a short electric current under very high tension. It can occur as sheet lightning, within the cloud, or forked lightning, between the cloud and the ground. The first lightning bolt comes from the middle of the storm cloud. Then small discharges appear, leaping in zigzags down towards the ground. This 'stepped leader' prepares the ground for true lightning, by making the air an electrical conductor as the discharges travel through it. When the discharges are about 10 metres above the ground, they make way for the bolt of lightning, a real one, massive and blinding. In fact, it is a return stroke, and leaves the ground in order to flow upwards to the cloud. The electric current that circulates for a fraction of a second is of considerable intensity: 100 000 amps, or 200 000 times greater than a filament light bulb and a million times more luminous. The local temperature rises as high as 30 000°C. The air near the lightning expands so rapidly and violently that it causes an explosive sound, which is thunder.

▲ *Thunderbolts are most likely to fall on tall, pointed objects, so a tree is often a target. That is why you should never shelter under a tree during a storm.*

MR FRANKLIN'S LIGHTNING CONDUCTOR

I t was in the 18th century that the American scientist, Benjamin Franklin, set up the first lightning conductor on the roof of the house of Benjamin West (right). Bolts of lightning are attracted to points, so a metal rod was installed on the roof of the building with a copper wire leading from it into the ground. This rod duly received the surrounding lightning and the electric current that was generated was absorbed by the huge mass of the Earth.

CLOUDS

A ballet at high altitude

The formation of clouds, which contributes to the water cycle on Earth and to the redistribution of energy, is a phenomenon essential to the balance of climates.

▲ *Cumulus clouds, the most familiar of all clouds, seem to go on for ever in the sky. There are dozens of different sorts of cumulus.*

If you have ever looked out of an aeroplane as it flies through thick masses of cloud, it might have seemed to you that it was drowning in a sea of cotton wool. But this apparent solidity is deceptive. The clouds are only vast areas of fog at high altitude. They may nevertheless contain enormous quantities of water. A cumulonimbus 25 kilometres wide contains nearly one million tonnes of water in suspension.

A story of water vapour

Contemplative souls may fantasise over the shapes produced by the ever-changing dance of cloud forms. But scientists have officially distinguished 10 major types: altocumulus, altostratus, cirrus, cirrocumulus, cirrostratus, cumulus, cumulonimbus, nimbostratus, stratocumulus and stratus. This strange litany of names is based on Latin, the language also used in the classification of animals and plants.

Streaks, sheets, veils or fleece – clouds have very distinct shapes and characteristics, the formation and evolution of which take place at different altitudes in the atmosphere according to the local terrain and meteorological conditions. However, a high-flying cloud and a puff floating above the daisies begin in the same way, in the form of air loaded with water vapour that is heated and rises.

The amazing metamorphosis of water drops

All surfaces of open water – ponds, rivers, lakes, seas, etc. – evaporate naturally and release water molecules into the atmosphere. This process represents an enormous amount: about 500 million tonnes of water a year.

This wet air may rise as a result of the heating of the ground. But it may also rise because a mass of cooler (and therefore denser) air slides underneath it, pushing it upwards. Or again, the

AEROPLANES, CREATORS OF CIRRUS CLOUDS

Jet aircraft, which fly at a height of about 30 000 metres, leave in their wake long white streaks, which are in fact high clouds. The gases leaving the exhaust ducts contain a lot of water vapour which condenses instantly in the sub-zero temperatures at this height. They therefore form a thin cloud of ice particles, like those of the cirrus family, which appears at the outlet of each engine. In the end these trails of condensation grow thinner and disintegrate.

▲ From left to right: sheets of stratus cloud constitute fog which may come down as far as the ground, or they may form a low grey cloud layer; altocumulus clouds form regular lines that look like cotton wool; cirrus clouds float at high altitude.

air may move upwards because while in motion it encounters a mountain that obliges it to rise.

As altitude increases, atmospheric pressure and temperature decrease. These factors cause the water vapour to condense into tiny droplets of water, or ice crystals if the temperature is low enough, and a cloud forms. This condensation comes about only if there are particles in suspension in the air, such as dust or spray, etc. If the air were absolutely pure, clouds would not be able to form.

The droplets cluster together in the middle of the cloud as it takes shape. Not until about a million of them have gathered together will they form a single drop of rainwater or ice.

Reading the weather: what the clouds say

Elegant cirrus clouds, long filaments of ice particles, always float in the cold layers of the stratosphere, more than six kilometres high. They do not greatly affect the levels of sunlight, but their appearance is generally a sign of rain and the end of a period of fine weather.

Chubby cumulus clouds gather like fleece in countless evocative shapes at medium altitude, about five kilometres above ground level. They can be seen forming on hot days. When the sun disappears, they fade away, deprived of their supply of warm moist air.

Giant cumulonimbus clouds rise in pillars with an anvil-shaped top, which can reach 15 kilometres in height and result in storms.

The lowest of all clouds, the stratus, form sheets at about 500 metres above the ground. They and the nimbostratus give us those solid lead skies that are grey as far as the eye can see, and sometimes bring snow. These low-lying clouds sometimes come right down to ground level in the

form of mist or fog, which occurs chiefly on cold days. In these cases, the water vapour in the air condenses directly as it comes into contact with the Earth's surface.

Formidable heating machines

Clouds contribute to the Earth's water cycle by spreading water in the atmosphere which they thus humidify, and over the ground, by precipitation.

That is not all. They are also giant heaters. Hot air, carried upwards, releases part of the energy it has acquired in the lower layers of the atmosphere during the course of the condensation of its water vapour. The formation of just one cumulonimbus mobilises an amount of energy equivalent to that required to meet the energy needs of every household in the world. By thus spreading energy among various points in the atmosphere, clouds play a primary role in the climatic system of the Earth. They also contribute to the reflection back into the atmosphere of some solar radiation, which helps to limit the greenhouse effect over the globe.

▼ Storm clouds, cumulonimbus store up considerable masses of water, which will fall as heavy rain.

Very special effects

Day and night, the dome of the sky is a stage on which superb scenes are played, with the Sun in the leading role. But often the show is merely one of visual effects, the subtle play of perturbed sunlight in the atmosphere.

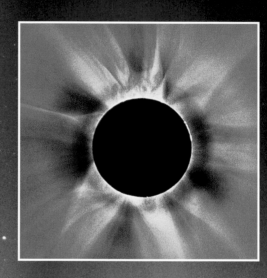

◀ *A total eclipse of the Sun. At this time, the solar corona becomes clearly visible.*

On August 11, 1999, on a trajectory reaching from the Atlantic Ocean to the Bay of Bengal, India, millions of people were gathered, all looking intently in the same direction – towards the sky. Wearing special glasses, they were waiting for the main event: the total eclipse of the Sun, a few moments when our nearest star would put on a fantastic live show. But the sky produces many other sights which are just as dramatic. Among the players are solar radiation and magnetism, and their effects on the magnetosphere and the atmosphere are an awe-inspiring spectacle.

The Sun, a superpowerful transmitter

The Sun is a wonderful machine for producing special effects. At its surface, the photosphere, made of hydrogen atoms, reaches a heat of 6 000°C. But, spreading out several million kilometres above it, is the solar corona which is much hotter, at 2 000 000°C. This temperature is high enough to extract electrons from the atoms that have escaped from the photosphere, which leads to the formation of a plasma, or ionised gas, some of which escapes the solar magnetic field and makes its way to Earth. This is the origin of the solar wind, which reaches the Earth at a speed of 400 kilometres a second. Some of the particles in this solar wind – protons and electrons – reach

◀ *The aurora borealis. The auroras spread around the magnetic poles of the Earth, along a ring-shaped area several hundred kilometres wide.*

the upper layers of the Earth's atmosphere where they produce a stunning sight: the polar auroras.

Curtains of heaven

Under the combined effect of the solar wind and the Earth's magnetic field (*see pp. 250-251*), the auroras spread out like curtains of light hanging in the sky. They occur simultaneously at the level of the magnetic poles in the north (the aurora borealis) and the south (the aurora australis), and are created in the following way.

As the protons and electrons from the solar wind enter the atmosphere they collide with oxygen and nitrogen atoms above the poles. These molecules of air are either ionised, in which they lose an electron, or become energised and go into an excited state. As the excited molecules become stable once more they emit radiation of a different colour according to type.

At a height of 400 kilometres, oxygen is dense enough for there to be collisions that produce green light. At an altitude of 90 kilometres, nitrogen predominates. Certain electrons, 1 000 times more powerful than those that create green auroras, may excite the nitrogen atoms. As these atoms stabilise they emit red light.

Light effects and illusions

Sometimes strange haloes, or coronas, can be seen round the Sun or the Moon. These large circles are due to a very common optical phenomenon:

the diffraction of light by drops of water or ice crystals in high-altitude clouds. You will see a sort of ring, which is bluish in the middle and redder towards the outside, rather like a rainbow. Other circles may appear around the first one, with the same spread of colours: blue in the middle and red at the edge. The size of these rings depends on the size of the drops of water suspended in the atmosphere.

The halo that appears round the Sun is seen most often in the cold of the polar regions. In addition you may see parhelia, or 'false suns', which are very bright points of colour situated on either side of the real Sun. But take care when observing these phenomena. It is very dangerous to look at the Sun with the naked eye.

In all its glory

Perhaps you have already seen, while on a plane journey, a shadow cast by the plane on the clouds below surrounded by a brightly coloured halo. This is called a glory. Always occurring opposite the Sun, it is, yet again, the result of the diffraction of sunlight by drops of water. The same phenomenon can be observed during a mountain walk, above a sea of clouds, when you

▲ *Areas of light sometimes appear on either side of the Sun. They are called parhelia, or false suns, and are images of the Sun caused by the diffraction of light.*

WHEN THE SUN IS HIDDEN...AND REVEALED

The eclipse of the Sun is one of the most moving natural phenomena that it is possible to observe from the Earth. But you need to be in the right place at the right time. The term 'eclipse' is actually inappropriate, and is more correctly referred to as 'the occlusion of the Sun by the Moon'. During the precious few minutes of this occlusion, you can see an immense halo round the Sun, the solar corona. Usually invisible, it reveals the existence of energy in surrounding space. Just before and after the stage when occlusion is complete, the final rays light up the irregular surface of the Moon. Very bright points called Baily's beads then appear, forming a ring of diamonds around the edge of the Moon. In addition, at the moment of total occlusion, it is possible to see the Sun's photosphere in the form of a thin pink border, from which emerge, here and there, incredibly long thread-like features, or prominences.

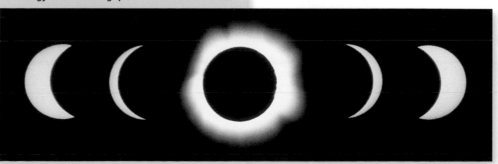

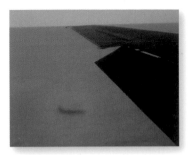

▲ *The shadow of the plane on the cloud layer below is often surrounded by a halo of colour, called the glory.*

are facing away from the Sun. Your silhouette is surrounded by a halo, which is also called the Brocken Spectre, after the mountain where it was first scientifically observed and described.

When the big blue turns flaming red

When you put a spoon into a glass of water, it does not look straight but seems to bend at the line where the air meets the surface of the water. In the same way, the light from the Sun travels straight through space because it is moving through a vacuum, but when it enters the atmosphere of the Earth, the drops of water and ice crystals in the atmosphere form obstacles which deflect and scatter the solar radiation. These behave like a prism that disperses light. But this does not mean that our sky takes on all the colours of the rainbow. During the day, when the Sun is overhead, the molecules that form the air scatter the blue part of the Sun's light much more than the red. That is why the sky looks blue.

At sunset, on the other hand, the Sun is low on the horizon and the layer of atmosphere between it and ourselves is thicker. Before reaching us, the light becomes so scattered that all the blue part is lost. Only the red colour reaches us and creates the magical hues of this time of day. The phenomenon is of course the same for the rising and setting of the Moon.

The green flash

Before disappearing completely over the horizon, the Sun has a final astonishing surprise for us. As

▼ *In the tenth of a second that precedes sunset, you may be lucky enough to see a fleeting green flash on the horizon.*

it moves lower, it is reduced to a strongly red-coloured hemisphere which is possible to look at without injuring your eyes. The disc continues to drop lower until only a point of light is left. In the final tenth of a second before the bright ray disappears, a distinct green light appears.

At this moment, it is as though the sunlight has come from a point source and the atmosphere is playing the part of a prism. For a brief instant, the light is dispersed. The remaining green is visible only at the level of the sea, while a deep blue might be observed by an observer at a higher altitude.

▲ *At sunset, only the red colour reaches us, because of a great scattering of the light in which most of the other colours are lost.*

PLANET UNDER CONSTRUCTION

Nature's building sites

*Lava flowing into the sea
(off the coast of Hawaii).*

Hidden faults

Together with volcanic activity, earthquakes are a major manifestation of the movement of the tectonic plates at the surface of the Earth. The most superficial earthquakes are caused by movements along faults, which are breaks in the Earth's crust.

In Algeria on October 10, 1980, a large earthquake (measuring 7.2 on the Richter scale) destroyed the surroundings of the town of El-Asnam (today Chlef), which had been destroyed once before, in 1954. Many cracks and breaks appeared in the ground, including an escarpment 40 kilometres long, which can still be seen. It divides the topography in quite a regular fashion, crossing hills and dry riverbeds, and is 50 centimetres high in the south-west and almost five metres high in the north-east. This is the surface manifestation of the fault that caused the tremors.

Earthquakes are an opportunity for geologists to observe sudden changes in the Earth's surface, and especially how faults and folds arise – which they can usually study only millions of years after these features have been formed.

The crust breaks

An earthquake, which propagates in every direction, often happens when there is a sudden movement along one of the large breaks in the Earth's crust. The rocks that make up the crust plates are subjected to force when these plates move, and they react differently, according to their depth. In the upper part of the Earth's crust, as long as the temperatures are not too high, the rocks exposed to excessive forces break, while lower down they become plastic and are pushed out of shape in a continuous process.

The movement often starts between 10 and 20 kilometres down. If it is slight, in the order

◀ The San Andreas Fault, in the Carizo Plain between San Francisco and Los Angeles, looks as if it has been carved into the ground with a knife.

THE DIFFERENT TYPES OF FAULT

▲ *Normal fault*　　▲ *Thrust, or reverse fault*　　▲ *Strike-slip fault*

of 10 centimetres or so, it does not reach the surface, but generates small seismic waves and is classified as a small earthquake (5 to 5.5 on the Richter scale). On the other hand, a movement of more than a metre affects a large area of the fault plane and may bring about a displacement of the topography that is called a surface slip, as at El-Asnam.

Vertically...

Just as the forces to which the plates are submitted are varied, so there are several categories of faults. In the case of El-Asnam, the convergence of the African plate and the Eurasian plate compresses this part of the Maghreb. The fault involved is called a thrust, or reverse fault: the part above the fault overlaps or straddles the other part. Geodesic measurements have shown that, in this particular case, the top part rose by nearly six metres, whereas the overlapped section sank down 70 centimetres at the base of the fault.

▲ A vertical surface slip created by the fault that appeared during the El-Asnam earthquake in Algeria in 1980.

Other faults, called normal faults, bring about a different result – the collapse of one of the two blocks of rock. These faults are caused by tension, which is common when the crustal plates diverge, and include oceanic ridges and troughs like the African or Icelandic rifts.

...or horizontally

Thrust and normal faults influence the formation of the landscape by a movement which is up or down. But there are other faults which move the ground horizontally. The most famous is the San Andreas Fault, in California. In 1906, it was responsible for the San Francisco earthquake. The fault affected an area of more than 100 kilometres, causing a slip of several metres in the roads, fences and houses along its route. In 1857 it had already seriously affected the area of California between San Francisco and Los Angeles. People who live there talk of 'waiting for the Big One', an earthquake measuring anything above 8.

From the bowels of the Earth

Not all earthquakes are the result of a sudden movement of the plates. Some are caused by volcanic activity, others even by human activity (for example, roof falls, when galleries in mines collapse). But these account for a tiny proportion. Some 95 percent of earthquakes are due to subduction phenomena which occur at depth, where slanted plates have to withstand major forces resulting from friction against the surfaces with which they are in contact.

Intermediate earthquakes (at a depth of up to 300 kilometres) and deep earthquakes (from 300 to 700 kilometres) are characterised by their great magnitude (8 and above). The most powerful earthquakes ever recorded – those of Chile in 1960 and Alaska in 1964, both higher than 9 – are examples. Fortunately, their epicentres were a long way from the surface, which lessened their impact.

It is thought that the deepest earthquakes are not caused by subduction. Mineralogical changes in the rocks of the mantle are believed to result in changes in their volume, which would trigger earthquakes.

▲ Many strike-slip faults run through Iceland where the landscape is marked with trenches called rifts, or grabens.

MEASURING EARTHQUAKES

Movement along a fault creates seismic waves. These vibrations move at a speed of several kilometres a second. The seismic vibrations in the Earth's surface are located by recordings from seismic stations, which since 1960 have formed a global network, and from local networks. The size of an earthquake is measured by seismographs, which record the size of the waves. As the seismic waves are of different kinds, there are several scales of measurement according to the type of wave (the Richter scale or the Gutenberg-Richter, and so on). They all use arabic numerals. The size of the waves increases by a factor of 10 between one degree of magnitude and the next, but the energy given off is 30 times greater. Thus, an earthquake measuring 8 is the equivalent of nearly 1 000 earthquakes measuring 6.

Work in progress

The way a volcano is built up and the shape it takes depend on the dynamics of its eruptions and the nature of the lava it spews forth.

At a depth of between 30 and 200 kilometres beneath our feet, magma, a mixture of molten rock and gas, is formed. Although most of it remains trapped underground where it crystallises, some (probably not more than 10 percent formed at depth) manages to rise to the surface by way of fissures in the Earth's crust. It is this that feeds into the volcanoes of this planet.

▲ *Sunset Point Volcano (Arizona, Unites States) is a prime example of a monogenic basalt volcano, which resulted from a single eruption. The lava flow can be seen below the crater, as an area of dark grey.*

A cone, a crater and a lava flow

At the end of its rise to the surface, magma (which at this point is known as lava) is expelled with a greater or lesser degree of violence. How long this goes on varies from a few hours to several months, and determines the shape of the volcano.

If the activity is limited to a single eruption, the resulting volcano, called a monogenic volcano, will be small. It is usually represented by a cone with projections of between a few dozen and a few hundred metres high surrounding a crater with a lava flow coming from it.

Growth spurts

But after a period of dormancy of varying duration, volcanic activity often resumes in the same place. The alternation of phases of dormancy and activity may recur many times. The volcanic construction, in this case called polygenic, may reach several thousands of metres in height, and take a more complex shape. The active phase of some polygenic volcanoes is measured in thousands of years, or even tens or hundreds of thousands. Stromboli, for example, a volcano 3 000 metres high (and 2 000 metres above sea level) has been active for at least 200,000 years in the Tyrrhenian Sea off the coast of Italy. In this type of volcano, activity rarely remains confined to the crater at the summit, and eruptions occur on the slopes.

In Sicily, Mount Etna is a good example. It is a generally conical construction, with craters on the top. But several dozen smaller cones have formed on the sides in the course of different, successive eruptions. These small cones are called

▲ *Fujiyama, so often depicted in Japanese art, is a magnificent stratovolcano (made of alternating pyroclastic material and lava flows of various sorts), with a perfectly symmetrical cone. 'Mount Fuji' last erupted in 1707.*

◀ *Mount Etna (Sicily) is an example of a polygenic volcano (resulting from a series of eruptions). Satellite cones formed on its flanks during subsequent eruptive phases.*

to spread. In the first case it produces thin but long and wide lava flows. In the second, thick layers of pumice spread from the outlet.

These types of volcano, known as shield volcanoes, often have a structure which stretches out very widely with gentle slopes. Hawaii's tallest active volcano, Mauna Loa (4 171 metres high), in the southern central part of the island, and its sister volcano Kilauea, the most active volcano in the world, belong to this category.

▲ Trölladyngja (1 438 m), or the 'castle of the trolls', is a shield volcano (very broad and shallow) in Iceland. At present dormant, some 3 000 years ago it emitted almost 15 km³ of basalt lava.

A pointed hat

When lava is viscous (as it is with andesitic and rhyolitic magmas), the eruptions are essentially explosive and result in nuées ardentes, or clouds of gas, ash and fragments which move fast along the ground. The result is a cone with steeply sloped sides formed by a heap of debris, full of ash. The debris accumulates on the slopes and at the base of these structures. A typical example is Fujiyama, the highest peak in Japan, reaching 3 776 metres. Short, thick, erratic lava flows sometimes escape from fissures opening on the slopes, but usually the lava remains confined to the crater, where it builds up a dome or a needle according to the degree of viscosity.

satellites. Mount Etna erupts regularly, creating havoc for those who live at the foot of her slopes.

A flat cap

The form of a volcano depends above all on the nature of the lava it ejects. When this lava is very liquid (like basalt magma) or when its gas content gives it a frothy texture at the time it erupts (in the case of ignimbrite magma), the eruptions tend

Very large volcanoes have a complex shape. They are called composite or stratovolcanoes because they are made up of alternate layers of solid fragments and lava flows that often change their nature over a period of time. In fact, the same volcano can emit liquid lava and viscous lava in turn, so that sometimes it can be spreading and sometimes explosive. Mount Lewotobi, in the Flores Islands of Indonesia, is an explosive stratovolcano.

A CAULDRON OVER THE FIRE

On the summit of many stratovolcanoes is found a vast, steep-sided basin several kilometres in diameter, with sheer walls and a flat bottom. These depressions, called calderas (shown right, the magnificent example of Crater Lake, Oregon, in the United States), are the result of the central part of these structures falling inwards. In fact, a few kilometres down, at the base of these large volcanoes are reservoirs that were once magma chambers. The magma may even stay there for several years. When there is a renewal of activity, the lava stored in these chambers can be expelled very fast. This rapid emptying leads to the collapse of the roof of the reservoir, which has the effect of forming a caldera at the surface.

Factories for oceanic crust

The Earth's surface is renewed in the middle of the sea, at the ocean ridges. Undersea volcanic activity is a discreet sign of the presence of huge factories where oceanic crust is manufactured.

▲ *Above, the Semail ophiolite of Oman; middle, a gigantic piece of oceanic crust; top, pillow lava identical to the lava observed at the base of the ocean ridges.*

In the mountains in the north of Oman, on the Arabian Peninsula, is an area of rock that fascinates geologists. Covering an area 400 kilometres long, 50 kilometres wide and 10 kilometres deep, it is the Semail ophiolite, a piece of oceanic crust cast up on the land when the Eurasian plate and the Arabian plate converged. This ophiolite, the finest in the world, is made of lava which flowed from the centre of an ocean ridge at the bottom of the sea 100 millions of years ago. So there is no need to dive to great depths to study oceanic crust. This ridge is accessible without getting your feet wet.

The egg and its shell

In the Earth's mantle, which is made of a very hard rock called peridotite, there is always some movement, in the order of a few centimetres a year. The rigid plates that make up the lithosphere entirely cover the mantle, like the shell of an egg. Riding passively according to the movements of the mantle, these plates move either towards each other or farther apart.

The oceanic plates, which carry the sea, are formed at the bottom of the oceans, in the ridges. These ridges extend for up to 75 000 kilometres and their centre is the site of continuous activity. Cracks are always opening, releasing lava that, as it cools, accumulates in the form of pillows characteristic of underwater eruptions, and creates the upper layer of the ocean floor.

Lava flows on the seabed

The ridge is under the lava that carpets the seabed. Twenty kilometres down from one of the flows that spreads over the ocean floor, in the middle of the peridotite of the mantle, at a temperature of more than 1 000°C, the solid rock is borne upwards by a few dozen centimetres each year. Coming from deep down where the temperature is higher, the rock takes a long time to cool because it is a poor heat conductor.

A small fraction of the mantle, between five and 20 percent, melts and forms a multitude of liquid droplets which gradually coalesce and move upwards by way of seams in the rocks. Basalt lava thus reaches the base of the crust in the form of a thick crystalline porridge and forms a gigantic cavity two kilometres high, the magma chamber, of which only the top is completely liquid.

On the walls of this vast cavity, the rock crystallises as it cools down and causes a layer of granular rock, or gabbro, to be formed two kilometres thick. Periodically, the roof of the chamber yields to the pressure of the lava that spurts violently into the cavity. A dyke forms above the chamber, which can reach a metre in width, one kilometre in height and 10 kilometres in

▶ *At the bottom of the Atlantic Ocean, the mid-oceanic ridge twists and turns for 15 000 km.*

25 km

THE RIDGE AT THE BOTTOM OF THE SEA

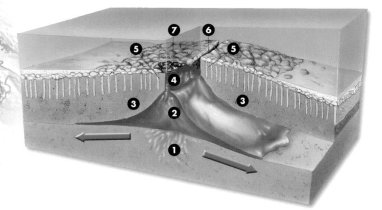

length, following the middle of the ridge. It thus forms a column of lava 10 cubic kilometres in volume. In 10 years, perhaps, a new dyke will form alongside it, and so on, until a continuous complex of vertical dykes is formed.

When one of these dykes reaches the surface, a basalt flow spreads over the seabed in the form of pillow lava. Each year, 50 cubic kilometres of oceanic crust are manufactured in the ridges. The whole of the Earth's seabed is therefore renewed every 200 million years.

The journey of a plate

Assuming that, on average, a dyke will empty the basalt magma chamber every 10 years, in a thousand years the ridge will have manufactured 100 metres of new crust, which means an amount of 100 kilometres in a million years and 10 000 kilometres in 100 million years.

But the production of oceanic crust by the ridges is actually quite variable, being as much as 15 centimetres a year in the Pacific Ocean but as little as five centimetres in the Atlantic Ocean, where the amount of basalt magma brought to the ridge is not enough to maintain a permanent magma chamber.

Created on both sides of a ridge, the young ocean crust is always moving away from it. Thin to begin with, it grows continually thicker as it cools during its lifetime of 200 million years. Finally, thick and heavy, it sinks into the middle of the mantle (by the process of subduction), where it will be used to create crust that will emerge again in a few hundred million years.

Thus, 'new crust' is always appearing at one end, and 'old crust' is disappearing at the other. This scenario is sometimes much more complicated because crustal plates necessarily interact with those surrounding them. An oceanic plate may encounter another plate, oceanic or continental, and straddle it, as in the case of the Semail ophiolite of Oman. We can be thankful that this happened in this example, because it is now exposed where the oceanic crust can be studied with ease.

▲ *The partial fusion of the mantle (1) feeds magma into a huge reservoir, or magma chamber (2), underneath the ocean ridge. The cooling of the walls of this chamber causes the formation of gabbro (3), while at the top of the chamber a new dyke (4) forms because of pressure from lava near older dykes (5). At the surface, on either side of the ridge (6), basalt spills have taken on a characteristic shape, given the name of pillow lava (7).*

When the sea takes a dive

Cordilleras (chains of mountains spiked with peaks and volcanoes) and island arcs (strings of low-lying volcanic islands) are caused by the same geological phenomenon: the titanic actions of oceanic plates.

▲ *The cordillera of the Andes (here in Bolivia). The peaks of this mountain chain, the only one of its kind, average about 3 900 m high, and the chain is more than 7 000 km long.*

On May 22, 1960, a giant wave swamped the shores of central Chile. Sixteen hours later, the archipelago of Hawaii was affected, and then Japan, which witnessed the worst tidal wave in its history. The cause of the series of disasters was an earthquake measuring 9.5 – amounting to a third of all the seismic energy let loose in the 20th century – recorded off South America. This double phenomenon, an earthquake followed by a tidal wave called a tsunami, is triggered by the sinking (or subduction) of an oceanic plate below another plate, usually a continental one.

The rise of the Andes

The size of this earthquake, which took place in the zone of interaction between two plates, demonstrates the power of the forces involved. The same forces were also behind the formation of the Andes. A product of the pressure exerted by young (40 to 60 million years old) oceanic crust in the process of cooling down, floating on the mantle without sinking under the South American continent, these forces built up this fascinating mountain range over millions of years. Its chain

of volcanic peaks, which at their highest rise to about 7 000 metres, attracts tourists in search of magnificent scenery, as well as geologists.

Here, geologists are able to study, in particular, another phenomenon connected to the subduction of the Pacific plate below the South American continent. Off the coast some 250 kilometres away from the impressive mountain scenery of the Andes is the Chile Trench, between 5 000 and 8 000 metres deep. The difference in altitude between this trench and its towering neighbours is the most spectacular on Earth.

Islands under tension

Oceanic trenches, which are evidence of intense geophysical activity, are usually found in the vicinity of island arcs. The Mariana Trench, the deepest in the world, is found off the archipelago of the same name. The arc of its 17 coral islands was born from the traction exerted towards the open sea by an old oceanic plate (between 120 and 160 million years in age), which was colder and

VOLCANOES UNDER THE SEA

Tens of thousands of volcanoes cover the ocean floor. With the exception of the most gigantic of them, like Hawaii, they are submerged. Thus, Mount O'Higgins (**1**), off the coast of Chile (**2**), rises more than three kilometres above the Nazca plate (**3**) and is ready to dive under the margin of Chile (**4**). Other volcanic structures preceded it, judging by the scars left on the edges of the cordillera.

denser than the surrounding mantle. This movement of separation from the plate bearing the archipelago is at the origin of the development of oceanic spaces that geologists call marginal seas.

These seas are of a more modest size than oceans and are always found along an area of subduction. Low stress between the two plates, or even the traction exerted by the oceanic plate, explains the low elevation of island arcs, which are dominated by volcanoes.

At the edge of the plates

The action of oceanic plates is also revealed by the extent to which the edge, or margin, of the overlapping plates is worn. The margin of the Mariana Islands, for example, has such a slope that rocks often fall off it down into the trench.

This phenomenon is thought to be caused mainly by the many volcanoes carried by the subducted plate under the archipelago. The upper plate is thus progressively nibbled away, and the products of this erosion are carried away, in this case by the subduction of the Pacific plate.

On the other hand, other trenches are hardly detectable, like those of the Lesser Antilles or on the edge of the Aegean arc. In these cases, the subducted plate is covered with sediment from the erosion of neighbouring massifs. This covering smoothes the interface between the plates and helps to build up a sedimentary strip on the edge of the arc.

▲ *In 1883, the eruption of Krakatoa (in Indonesia) caused a devastating tsunami.*

THE ARC OF THE LESSER ANTILLES

ISLAND ARCS AND CORDILLERAS

▲ **An island arc**
*The traction exerted by old oceanic crust (**1**), colder and denser than the mantle (**2**), on the upper plate (**3**) has caused a marginal sea (**4**) to open up. The presence of an oceanic trench (**5**) bears witness to this intense activity.*

▲ **A cordillera**
*Juvenile oceanic crust (**1**), floating on the mantle (**2**), resists sinking and exerts pressure on the plate above it (**3**).*

▲ **View from the north**
*1. Accretion prism of Barbados. **2.** Martinique.*
*3. Guadeloupe. **4.** Venezuela. **5.** Puerto Rico.*
*6. Caribbean Sea **7.** Atlantic Ocean **8.** Haiti.*

The Himalayas: a mega-collision

Large movements of plates on the surface of the globe have caused oceans to disappear and continents to converge. From the clash between India and Asia, 50 million years ago, a chain of gigantic mountains, the Himalayas, was born.

▶ *The Himalayas seen from above. In this photograph, taken by a satellite, the highest mountain peaks in the world can clearly be seen. Tibet is on the right and India is on the left.*

If you were to do this little experiment – jab a pointed tool onto an old plastic plate – you would find that a ridge of thicker material formed round the point of impact. Imagine that the tool is the Indian continent coming into contact with Asia, and that it goes on moving to the north, pushing the Asiatic crust out of shape. This formidable collision is in fact the origin of the Earth's tallest mountain range, the Himalayas.

It all began 50 million years ago, when the Tethys Sea disappeared by oceanic subduction under the continent of Asia. The plate bearing the oceanic crust was much denser than the continental crust and so sank down at an angle into the mantle.

▲ *The position of India 60 million years ago (1). Some 150 million years ago, the Tethys Sea (2) separated two continents, Laurasia (3) and Gondwanaland (4). A little later, the Indian plate separated from Gondwana, creating a new ocean and starting a slow movement in the direction of Asia at the rate of 15 cm a year.*

India sinks

The phenomenon of oceanic subduction has been known for a long time. On the other hand, it was not known until a few years ago that continental crust could also sink into the mantle to a depth of more than 100 kilometres.

As a result of this subduction, which is a major mechanism of plate tectonics involving enormous forces acting in conjunction with the internal dynamics of the Earth (convection movements of the mantle), the Indian continent continued its northward movement after the collision, all the time forcing its continental margin down into the subduction zone.

This process is highly unstable. Just as a cork will keep bobbing up in water because its density is less than that of its surrounding liquid, the light continental crust of India will resist being pulled under by subduction. When resistance to sinking (Archimedes' force) becomes greater than the mechanical resistance of the continental crust, the result is a fracture.

TRACES OF A VANISHED OCEAN

In the whole of the area immediately to the north of the Himalayan range, geologists today are finding traces of the Tethys Sea. They are able to pinpoint the suture zone, the geological point where India and Asia meet, by the presence of particular types of rocks. These are the remains of the vanished ocean. The rocks that are typical of oceanic crust are of magmatic origin, and include basalt pillow lava, seen on the right, gabbro and peridotite. The sedimentary rocks that covered them contain marine fossils that enable the ocean to be dated.

Stacks of crust

The crust of India then divided from the mantle in the subduction zone. The Himalayan chain began to form as stacks of gigantic 'flakes' of crust piled up, which caused the continental crust to grow thicker and the high peaks of the Himalayas to rise. At the same time, the lower plate continued to be subducted into the mantle.

Then, in a process that is still going on today, the deformation spread northwards into the interior of the Asian continent, forming the Plateau of Tibet, which has an average height of very nearly 5 000 metres. The effects of the convergence of India and Asia are perceptible even farther away and are responsible for folds and a set of large horizontal reverse faults (thrust faults) from Afghanistan to Mongolia and from China to south-east Asia.

Even now, the lowest flakes of the continental crust of this edifice are slowly moving on and deforming recent continental deposits in the Himalayan foothills (such as the Siwalik Range), where earthquakes frequently occur. The earthquake in Gujarat, India, on January 26, 2001, in which some 20 000 people died, occurred at the edge of the Himalayan front where the continental crust is beginning to be deformed but has still not separated from the mantle that moves it along.

Many other mountain chains were formed, and are still being formed today, in a similar way. This is the case of the European Alps, which were created farther west of the convergence of the African and Eurasian plates. The record height of the Himalayas may be explained by the greater speed of convergence of the plates, which is producing a thicker edifice of crust flakes.

▲ *The Annapurna peaks and the Manang Valley, in the Himalayas, seen from Nepal.*

THE FORMATION OF THE HIMALAYAS

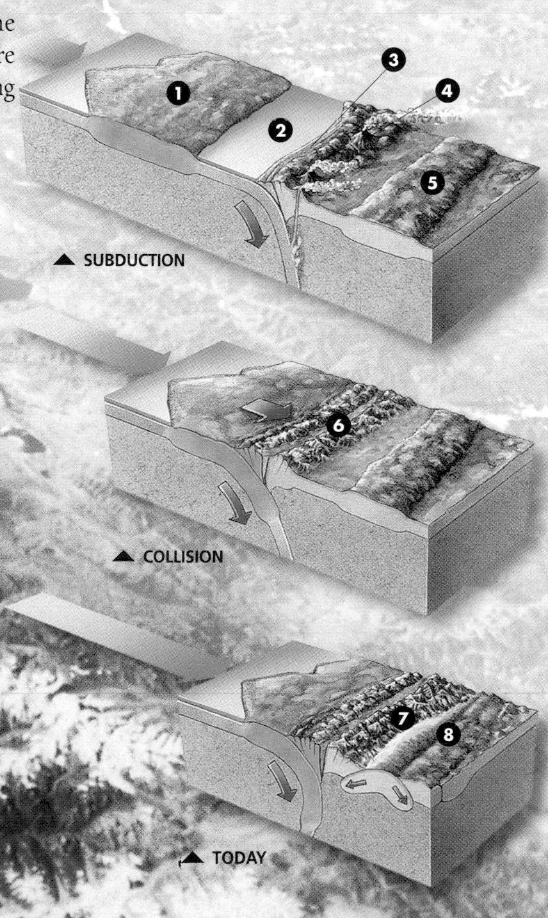

▲ SUBDUCTION

▲ COLLISION

▲ TODAY

◀ *Between 100 and 40 million years ago, because of subduction, India (1), moving northwards, met the Asian continent (5), bordered by a trench (3) and a chain of volcanoes (4). As a result of this collision, the ocean (2) closed up, creating a suture (6). Today the process continues, having formed the Himalayas (7) and the Plateau of Tibet (8).*

191

Earth, an active planet

Nothing on our planet is motionless. Its outermost solid casing, the continental crust, is divided into plates which move at speeds calculated to be, on average, centimetres a year or sometimes tens of centimetres. The whole surface of the Earth is thus constantly, but imperceptibly, moving. Certain plates grow larger or smaller while others meet and collide. Volcanoes and earthquakes are the most visible consequences of this dance of the land and sea. It was a visionary meteorologist named Alfred Wegener who developed the theory that landmasses moved during the course of geological time. Noticing the correspondence between the outlines of the South American and African continents, in 1912 he put forward the hypothesis that these two continents were once one. But how they drifted apart remains a mystery, as does the role played by the oceans, which he believed to be sunken continents.

THE YOUTHFUL SEABED

The global theory of plate tectonics was formulated in the 1960s, following new oceanographic studies. The sea floor is formed gradually at the ocean ridges, submarine elevations which rise above the abyssal plains that lie at a depth of over 2 000 metres, and extend for thousands of kilometres. In the middle of these ridges there is great volcanic activity. This means that the rocks emitted from them are older the farther away they are from the middle. Does this ongoing growth of the oceans mean that the Earth is expanding? The dating of these volcanic rocks and the sedimentary rocks deposited on them revealed a considerable surprise: the sea floor everywhere is very new compared with the continents that surround it. It is at most 165 million years old, whereas the oldest rocks on land may date back a thousand million years. This suggests that the creation of the ocean floor at the ridges is counterbalanced elsewhere. In fact the sea floor bends and submerges obliquely (by subduction) at the oceanic trenches, which causes, farther down from these zones, volcanic activity and very strong earthquakes.

PLATES AND MINIPLATES

Studies of plate tectonics soon proved that the whole of the Earth's surface is divided into plates, which may carry oceans as well as land. The boundaries of these plates are generally easy to find. Volcanism and the frequent seismic activity that occurs on the axis of the ocean ridges in the middle of the sea are signs of the points of separation between two plates. The boundaries are as clearly distinguishable in the region of oceanic trenches, where one plate bends and slides obliquely beneath another. In some instances, the plates fold horizontally along faults. This type of boundary, which is often an earthquake zone, crosses both the ocean floor and the land. The San Andreas Fault, for instance, which cuts across California from north to south, marks the boundary between the Pacific plate, which is almost exclusively oceanic, and the North American plate, which is continental. When the movement of the plates causes continents to collide, vast areas are subject to deformation accompanied by earthquakes, and give rise to the formation of mountain chains, such as the Alps, the Himalayas and the Andes. The breadth of these collision zones is such that it becomes difficult to trace the boundaries between the colliding plates, and this has led, in some cases, to scientists distinguishing multiple 'miniplates'.

THE LITHOSPHERIC PLATES AND THEIR MOVEMENT (IN CM PER YEAR)

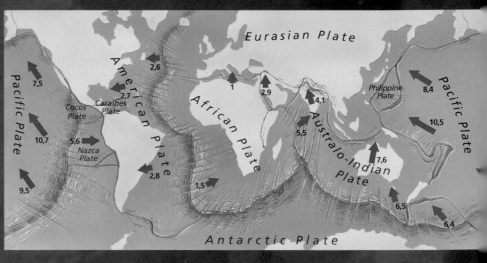

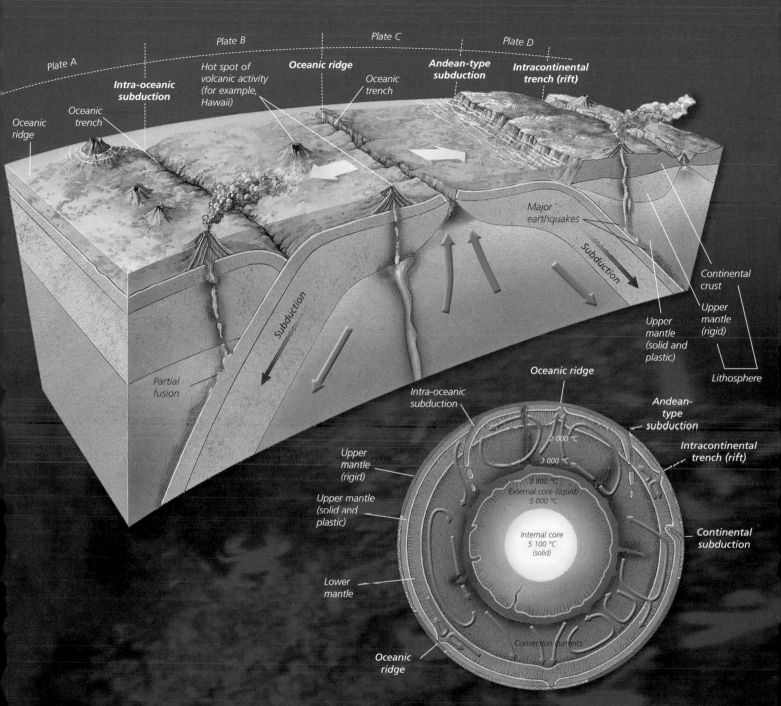

Plate A

Plate B

Plate C

Plate D

Intra-oceanic subduction

Hot spot of volcanic activity (for example, Hawaii)

Oceanic ridge

Andean-type subduction

Intracontinental trench (rift)

Oceanic ridge

Oceanic trench

Oceanic trench

Major earthquakes

Subduction

Subduction

Subduction

Partial fusion

Upper mantle (solid and plastic)

Upper mantle (rigid)

Continental crust

Lithosphere

Upper mantle (rigid)

Upper mantle (solid and plastic)

Lower mantle

Oceanic ridge

Oceanic ridge

Intra-oceanic subduction

2 000 °C

3 000 °C

3 800 °C

External core (liquid) 5 000 °C

Internal core 5 100 °C (solid)

Convection currents

Andean-type subduction

Intracontinental trench (rift)

Continental subduction

THE MOVEMENT OF THE PLATES

The Earth is made of several envelopes with varying chemical and physical characteristics. The outermost envelope, the lithosphere, is divided into plates. Its thickness varies, from 70 kilometres under the sea to nearly 150 kilometres under the surface of the land. Its composition also varies, being made up of different rocks according to whether the plates form continental or oceanic crust. Both types rest on the rigid outer layer of the upper mantle. The rocks of the upper mantle

(peridotites) may change their shape very slowly below the plates, which allows the plates, which are more rigid, to move. For, contrary to what is widely believed, the plates do not float on a liquid mantle. In spite of high temperatures in the mantle (between 1 000 and 3 000°C), the pressure is such that the rocks of which the mantle is made remain solid. Liquid rock, or magma, which rises and produces volcanoes, is present only when pressure decreases (where the plates diverge at

the ridges), when the presence of water lowers the point of fusion (where subduction occurs) or when a localised increase in heat (at the site of hot spots) causes partial fusion of the rocks. But, even though the mantle is solid, very slow convection currents move through it, nonetheless. It is these movements in the mantle, caused by differences in temperature and hence in density, that are thought to be one of the driving forces behind the movement of the plates.

Travelling volcanoes

Across the Earth's surface, there exist lines of volcanoes thousands of kilometres long. They are caused by the movement of a lithospheric plate above colossal upwellings of the deep mantle.

It is difficult to imagine any possible connection between the idyllic atolls of the Maldives, Mauritius or Chagos, and the volcanic peak called Piton de la Fournaise ('Furnace Peak') on the island of Réunion. Yet this little island with its enormous active volcano is the end of a line of volcanic mountains more than 5 000 kilometres long, which begins in the west of India, on the Deccan Plateau.

Similar alignments occur on both land and sea. Most of them contain active volcanoes at one end, but farther down the line the more dormant the volcanoes become, as if volcanic activity had moved over time from one end of the line to the other.

A garland of volcanoes

The most spectacular line of volcanoes on Earth is the one in the North Pacific Ocean that travels from the southern tip of the Kamchatka Peninsula on the eastern coast of Russia, to Hawaii in the South Pacific.

This garland of submarine volcanoes and volcanic islands is more than 6 000 kilometres long and is divided into two lengths. The first, the Emperor range, runs from north to south. The second, the Hawaiian range, extends from the north-east to the south-west. About 100 extinct volcanoes lie along this line. Only the island of Hawaii, at its southernmost tip, is now active. Its youngest volcano, Kilauea, is the most active volcano on Earth.

The volcanic activity began in the north, about 80 million years ago, and seems to have spread southwards. In fact, it was not the volcanic activity that travelled, but the Pacific plate that gradually moved (first northwards, then in a north-easterly direction) above a particular point of the Earth. This fixed point, deep in the middle of the Earth's mantle and continuously producing magma, is called a hot spot.

▲ The atoll of the Maldive Islands (top) is a link in the volcanic chain that stretches from the Deccan (India) to the island of Réunion. Of the old volcano of the Maldives there remain only shallows marked by rings of coral. Above, the Piton de la Fournaise, on Réunion, the active end of the chain.

A blowtorch in the Earth

Imagine a fixed blowtorch and a metal plate held over it horizontally. If the plate were moved slowly and smoothly over the flame, it would heat up over a certain area, the size of which would depend on the diameter and power of the blowtorch, the speed of the movement of the plate and the thermal conductivity of the metal. The position of the warm area would obviously depend on where the plate moved and this area would always be warmest directly above the blowtorch.

This is approximately what happens when a crustal plate moves above a hot spot. The heated area shows on the surface as a trail of volcanoes. In the case of the Emperor-Hawaii chain, the island of Hawaii is at present over the hot spot. But it won't be for long, because the Pacific plate is continuing its journey towards the north-east at a speed of about five centimetres a year.

Currents in the mantle

Between about 100 and 700 kilometres below the Earth's surface, the upper mantle, made of peridotite, is called the asthenosphere. Because of the pressure and temperature at this level, this layer acquires a certain plasticity which enables it to change shape without breaking and therefore to flow very slowly. The movement of material within it is then possible, either horizontally or vertically.

As this layer is much hotter at its base (about 2 300°C) than at the top (where it reaches about 1 300°C), convection currents develop within it. The

▼ A basalt flow on the slopes of Mauna Loa, Hawaii. With a temperature of more than 1 200°C and moving at a speed of more than 25 km an hour, this lava is the hottest and fastest on Earth.

THE HIGHEST, THE MOST ACTIVE, THE HOTTEST

The island of Hawaii, with its highest point of more than 4 205 metres (Mauna Kea, the highest island mountain in the world), rests on foundations 5 000 metres under the sea. In fact the island itself is only the protruding summit of a colossal underwater volcano measuring more than 250 kilometres in diameter at its base and nine kilometres high. Its activity is equally spectacularr – it erupts virtually every year. The lava, which is basaltic, reaches record temperatures of almost 1 300°C. The lava that flows down the sides of the two gigantic active volcanoes around which the island is formed, Mauna Loa and Kilauea, is the most liquid in the world.

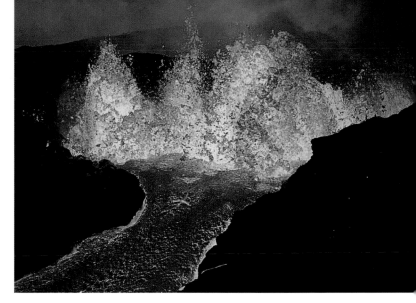

1. *Oceanic crust*
2. *Mid-ocean ridge*
3. *Continental crust*
4. *Hot spot and active volcanoes*
5. *Chain of extinct volcanoes*
6. *Lithosphere*
7. *Upper mantle*
8. *Lower mantle*
9. *Liquid core*
10. *Plume*
11. *Solid core*

hottest material, which makes up the lower part of the asthenosphere, tends to rise to the upper layers which are cooler, while the upper part, which is cooler, tends to sink.

Because of the viscosity of the material, the speed of these currents is not great (they move over a distance of a few centimetres to a few dozen centimetres a year).

Hot spots are one type of convection current running through the Earth's asthenosphere. Their lifespan is measured in tens of millions of years.

A HOT SPOT AND A CHAIN OF VOLCANOES

Gigantic thermal columns

The plumes that emerge from the Earth's core are like colossal cylinders of hot, rising matter, several hundred kilometres in diameter. Mostly, they come from the base of the asthenosphere. However, as the diagram on the left shows, the largest plumes are thought to begin still deeper down, on the edge of the Earth's core (at a depth of about 2 900 kilometres).

When they come into contact with the rigid lithosphere (at about 100 kilometres below the surface), the plumes flatten into the shape of a mushroom cap which can be more than 1 000 kilometres in diameter.

The rocks from the mantle that are drawn along in this movement undergo strong decompression, which causes their partial fusion. The liquid from this fusion, however, leaks from the residue of the rocks and rises to the surface again, where it will erupt in the form of lava through cracks in the crust. This is why hot spots can simultaneously feed volcanic activity at places several hundreds of kilometres apart.

The volcanically active area of Hawaii extends over a diameter of about 200 kilometres. It includes not only the great active volcanoes on the island, but also the undersea volcano of Loihi, situated to the south of Hawaii and active today.

▲ *The eruption of the volcano Mauna Loa (Hawaii) in 1984. Fountains of liquid basalt lava spurt from a crater in the side of the volcano and a lava flow is escaping down to the sea, in a scene typical of Hawaiian volcanic activity.*

DELTAS

Where the land conquers the sea

Deltas are fragile areas permanently under construction, relying simultaneously on the accumulation of sediment brought down by rivers and its removal by assaults of the sea. But the battle between land and sea always favours the land.

▲ *The huge Mississippi delta continues to advance into the Gulf of Mexico. The river divides into several channels and forms a shape like a crow's foot.*

The region of the Nile delta, in Egypt, an area of 24 000 square kilometres, is among the most densely populated in the world, like the delta of another legendary river, the Ganges, in India. If people chose to live in these low-lying areas between land and sea, it was because they were sure they would be able to sustain life. However, the boundary between the two elements is uncertain and very changeable. The branches of the river change course sometimes because of floods. Old watercourses are left high and dry, and new marshes and lakes are created. Dry land is always vulnerable to the invasion of water, even where the sea has already retreated.

Triangle or crow's foot

The Nile plain is the origin of the word 'delta'. At the mouth of the river, in the Mediterranean Sea, it has a shape so like the Greek letter delta, in its capital form and turned back to front, that

Herodotus gave it this name in the fifth century BC. Since then, the name 'delta' has been applied to any projection of land that is made of an accumulation of sediment at the mouth of a river or stream, either in the sea or a lake. The Rhône, River, for example, emerges into Lake Geneva where it has built up a delta much smaller than the one it makes in the Camargue as it arrives in the Mediterranean.

The shape of deltas may be very varied. Some, like that of the Mississippi, possess several branches and look like a crow's foot, whereas most others form an arch, like the deltas of the Ebro, in Spain, or the Pô, in Italy.

These various characteristics depend at the same time on the river that builds them and the resistance of the sea, where currents thwart the deposit of alluvial sediment or help build, by a slow process of erosion, sandy spits farther out, as in Venice.

Studies of the past positions of littoral zones show that changes in shorelines and in the course of rivers as well as their tributaries occur rapidly over the period of a few millennia. In the case of the Rhône and the Camargue, archives exist that enable us to trace the advancing or retreating movement, and measure it in kilometres.

THE BLACK GOLD OF DELTAS

Petroleum research is interested in deltas where several conditions for the formation of oil and its production are met. As the land extends into the sea there builds up a vertical accumulation of sediment which often began several million years ago and may be more than 1 000 metres high. Clay deposits found there are rich in organic matter, as there is considerable biological activity at the interface between the sea and the river waters. Rapid sedimentation allows this matter to sink and turn into oil, which is successfully trapped in the sandy material. Significant oil reservoirs have been found in the Mississippi delta (as the derrick on the right shows) and in the Mahakam delta, in Indonesia.

Repeated structures

A delta may also be the result of several deltas lying next to one another, as satellite pictures have revealed. The Mississippi is a good example of this sort of structure. After flowing over more than 6 000 kilometres, the river deposits its alluvial sediment in the Gulf of Mexico. This is estimated to be an amount of 20 tonnes a year. Transporting such a great solid load seems incredible, but the river drains more than a third of the territory of the United States, and has large tributaries like the Missouri, the Arkansas and the Red River. For 5 000 years, six deltas formed side by side until a final one, in the shape of a crow's foot, advanced out to sea on its own. In fact, a new delta is formed each time the watercourse changes direction, and the previous one is abandoned. In all, the front of the delta structure extends over 300 kilometres in length, with a width of up to 150 kilometres.

▲ *The Ganges delta is threatened by rising sea levels, in spite of the large amount of alluvium it deposits in the Bay of Bengal.*

The quality of materials

In order to build a delta, make a significant advance and in the end win the battle against the sea, the river must deposit a great deal of alluvium. But this is not always enough. The nature of the alluvium is just as important. This is why the Amazon flows into the Atlantic Ocean in a vast estuary and not through a delta. Although it carries a load estimated at 1.3 million tonnes a day, this is largely made up of fine elements that are carried by the currents along the coast towards the north. The Amazon and its tributaries in fact drain a region that produces little sand and gravel because chemical weathering predominates.

But the great forest-clearing operations in this region are removing plant cover from the loose soils, and will certainly encourage erosion. This may eventually modify the content of the load of the river, and then its course and rate of flow, and eventually transform the estuary into a delta.

In other regions, even though the present sediment load and type are adequate for the maintenance of deltas, the creation of dams and hydroelectric installations along the rivers and their tributaries may reduce the sediment and the sea may prove victorious.

▼ *The alluvium of the Pô, in Italy, carried by sea currents, has built up a coastal bar in the north of the river delta, which has cut off the lagoon from the sea at Venice.*

The battle between land and sea

Cliffs are visible proof of the undermining action of the waves,
and also of the variations in sea level throughout the Quaternary period.

After leaving the dark, rugged coasts of Norway, the Vikings entering the English Channel must have been surprised by the whiteness of the high chalk cliffs. On the south-east coast of England and on the Normandy coast opposite it, rise these impressive, often vertical walls. At their foot are beaches with pebbles of flint, a stone which is also found in surprisingly regular beds in the middle of the chalk.

The onslaught of the waves

At high tide, the waves throw out these pebbles which progressively hollow out a notch at the base of the cliff. This becomes top-heavy and panels of rock break away, falling onto the beach below and breaking up into many pieces. For some time the larger pieces of rock will protect the foot of the cliff. But in due course they will disappear, leaving only the flint they contained because it is not as easily worn down by the sea.

The erosion of the cliff occurs in fits and starts, but at low tide you can see how far it has been worn away. The chalk eroded by the sea will have formed a fairly level area just right for line fishing, called a marine abrasion platform. And then there are valleys, called valleuses in Normandy, perched at the edge of the cliff, opening out into thin air at the cliff face.

Fissures and cavities

Each type of rock resists the unceasing action of the sea to a varying extent. Chalk, with its cracks and fissures that enlarge as it dissolves, is a material that is ready to collapse in whole panels at a time. Chalk cliffs therefore retreat in a succession of vertical slices.

Other rocks will behave differently. The basalt organ pipes of the Giant's Causeway along the coast of Ireland, for example, are eroded along the geometric fissures that resulted from their cooling (*see pp. 144-145*). Granite rocks, a lot more massive, often form sheer coastlines, whereas schist will erode at a more sloping gradient.

Types of erosion that add their attack to that of the water also vary according to the nature of the rock, its disposition and the climate. Crystalline rocks like granite, for instance, are sensitive to salt spray. The cliffs of the granite coasts of Corsica and Sardinia have cavities, called taffonis, hollowed out of them, thought to have been caused by the crystallisation of salt in the minute pores in the rock, which gradually makes it crumble away. Elsewhere, in cold regions, it is water which, by freezing in the gaps in the bare

▲ *The cliffs of Tropea, in southern Italy, are gradually rising, as the region of Calabria itself rises in a movement begun in the Quaternary period, about one million years ago.*

▼ *The granite on the north coast of Sardinia (below, near the Capo d'Orso) is attacked by the sea, but also hollowed out in cavities, called taffonis, by salt-laden spray.*

THE RETREAT OF CHALK CLIFFS

▲ *The chalk cliffs of Etretat, in Normandy, France, are known for their beaches of flint pebbles and their arches, spared – for the time being – by erosion.*

rock on the cliffs, breaks up the rock and thus contributes to the retreat of the cliffs.

Triumph of the land

Since they gradually move back and break up because of erosion by the sea, cliffs should, logically, finally disappear. But there are other phenomena to counteract this development. The movement of the land-bearing plates – stable to a greater or lesser extent – causes land to rise or sink. And then there are variations in sea level dating from the Quaternary period, about one million years ago, or even before. The impressive

INLAND CLIFFS

The high coastlines of certain regions are rather like the recording reels of old seismographs. The elevated land documents the signs of sea erosion that were determined by significant changes in sea levels. During the ice ages these levels were between 120 and 150 metres higher than at present. The stepped appearance these coastlines have in profile is created by a succession of marine abrasion platforms with their old cliffs behind them. This is the case in the south of Calabria, where the oldest platforms, more than 1 000 metres high, are now several kilometres inland.

elevations on the coastline along the English Channel, for example, are simply due to the fact that the level of the sea is lower today than it was at the end of the Tertiary period, 1.8 million years ago. But the region is stable. On the other hand, in Calabria, in southern Italy, where a large part of the land was still under water 700 000 years ago, the region is rising by 1.5 to 2 millimetres a year. Near the straits of Messina, in spite of erosion by the sea, the land is actually creeping forward. Some cliffs, which used to be undersea escarpments, now have their heads above the water.

▲ The cliffs at low tide
1. Marine abrasion platform
2. Pebble beach
3. A valleuse
4. Chalk cliffs with seams of flint
5. Clay
6. Fissures in the chalk

Calcareous cathedrals

Coral reefs, enormous calcareous structures built by minute animals, grow in clear, warm waters. Without them, tropical seas, which are poor in nutritious matter, would be lifeless.

▲ *The atoll of Bora Bora (in the Society Islands, French Polynesia). The sea breaks in a fringe of foam against the coral reef. Behind the reef, a lagoon dotted with islands of vegetation surrounds the remains of a volcanic island. In places the lagoon is linked to the sea via channels. Here large predators lie in wait.*

In the Indian Ocean off North West Cape, Australia, a strange sight can be seen each year. Between the seventh and ninth night after the full moon in March, all the corals on the seabed in that area scatter their ovules and spermatozoa at the same time. The spawning is so dense that it looks as if it is snowing in the sea. This 'snow' is manna to all the marine life in a tropical ocean where there is not much food. Thirty-six hours afterwards, all the surviving eggs will have been fertilised. If they continue to survive, the larvae will attach themselves to a firm substratum to become polyps which, by division, will give rise to colonies and then to new reefs.

Polyps and algae in happy harmony

The Apollo astronauts were able to see the Great Barrier Reef from the Moon. It makes one's head spin to think that reefs – giant structures – are built by tiny polyps, a sort of miniature sea anemone. These animals, which belong to the group, or phylum, of the cnidarians (which includes jellyfish, anemones and gorgonians), share

▼ *The Great Barrier Reef of Australia, a giant coral reef.*

◀ *A coral polyp. Colonies of coral grow from these minute animals by means of budding.*

the peculiarity that each one encloses itself in a hard, calcareous casing for protection. All these casings together form a coral reef.

But where does this calcareous material come from? It is produced from calcium in the water. The chemical reaction by which it is synthesised is very unstable – no sooner has it formed than it dissociates. Fortunately, within their gastrodermal cells the corals host symbiotic algae, called zooxanthellae. The corals offer the algae not only sheltered lodging but also food in the form of nitrogenised material, a by-product of the digestion of microscopic animals they capture at night. These algae absorb the carbon dioxide and other waste products generated by the corals, which enables them to photosynthesise and survive. As a consequence of this photosynthesis, the algae release oxygen and nutrients which are absorbed by the corals and enable them to form their calcareous coating.

Near the coast: fringing reefs

Fringing reefs, which are widespread along tropical shores, are found on land that projects under the sea, between 50 and 1 000 metres offshore. A narrow stretch of water may separate them from the shore, in which the water is warmer and saltier than the seawater. The erosion of the inner wall of the reef (facing the shore) breaks off bits of coral which are reduced to white sand by waves and the stomachs of marine animals like sea cucumbers.

Fringing reefs grow outwards only. Their vertical development is limited by the average level of the water and their growth towards the shore is equally restricted. The water between the reef and the shore is trapped at low tide and is too hot, too full of sediment and far too saline. It is enough to kill off any polyp, a very fragile little creature.

However, these stretches of water, also known as reef flats, are not sterile places.

▶ *This soft coral is not a reef-builder, but it is essential to the coral ecosystem. Living in the current, it acts as a barrier to plankton, like the gorgonian with which it competes.*

You will often find there a collection of underwater plants, including cymodocea, turtle grass and fan-grass as well as calcareous algae that act as a support for filamentous algae, a favourite food of the herbivorous fish that come in from the open sea with each high tide.

Beyond the reef flats, towards the sea, the reef rises into a slight ridge or crest. Because it is the highest, most exposed point, subjected to the full force of the waves, only stumpy coral and calcareous algae can live there, being more resistant to the water.

The outermost part of a fringing reef, known as the fore-reef, is much richer. The diversity of corals increases from the inner to the outer side, because of the waves that disperse sediment and oxygenate the water. The force of the waves creates tunnels, holes and craters in this buttress zone in which a variety of marine creatures can live.

Finally the reef plunges into the sea. Primitive cnidarians flourish in the first few metres, such as fire corals, which are not true corals. These give way to branched corals which capture the light required by their house guests more efficiently.

▲ *On the surface of a reef (top), competition for light is intense. In the shallow reef flats, plants such as underwater grasses (above, cymodocea) often replace coral.*

THE THREE TYPES OF REEF

▶ **Fringing reef**
1. Shore
2. Sandy flats
3. Reef
4. Ocean

▶ **Ring reef or atoll**
1. Ocean
2. Volcanic island
3. Coral blocks
4. Lagoon
5. Reef
6. Channel between lagoon and sea

▶ **Barrier reef**
1. Shore
2. Sand
3. Coral blocks
4. Lagoon
5. Reef

▲ *Coral reefs provide a large number of habitats for animals. They are also an important source of energy for the warm-water ecosystem: dead polyps account for a great proportion of the nutrients in these waters.*

Out at sea: barrier reefs

Barrier reefs may be considered a development of fringing reefs. Several kilometres out from the coast they form a long linear structure resting on a rocky platform and capable of reaching an impressive length. The Australian Great Barrier Reef extends over 2 900 kilometres, for example.

A barrier reef may become a fringing reef if the lagoon separating it from the coast becomes smaller or shallower. But this is rare, and most frequently the opposite happens. Over time, the flats of a fringing reef become deeper, collapse or are flooded following a rise in sea level and become a lagoon. The reef then becomes a barrier. The two types are often found together, a barrier reef having on the shoreward side a more recent fringing reef.

Far from the coast, a barrier reef may extend towards the shore as well as towards the open sea. Then it takes on a ribbon or oval shape. Because the wind and the waves are stronger farther out than on the outer side of fringing reefs, barrier reefs suffer a lot more damage. Some appear ravaged, like a ragged line of scraps of ribbon, interspersed with channels or passages linking the sea to a lagoon on the inner side of the reef. Vast and deep, partly formed by wind-borne sand captured by the reef, this lagoon may be dotted with bits of reef and vegetation.

Between the inner and outer sides of barrier reefs, the size of the coral increases and it becomes more robust. The buttress that it forms is sometimes several hundred metres broad. The outer side, carved into jagged spurs by the sea, drops sharply to the bottom. It is in this buttress zone that the greatest number of species in the world is found.

The coral atoll

There is a third type of reef: the coral atoll, particularly common in Polynesia. It is often found where a fringing reef once surrounded a volcanic island that has since vanished under the waters, either because it sank or the sea level rose, and this reef has continued to grow upwards. A coral atoll may have two rings. And, in places, the accumulated coral detritus in the lagoon may form islands, with vegetation grown from wind-borne seed.

Coral reefs cover more than 60 000 square kilometres of our planet. In nutrient-poor tropical waters they play the same role as algae and plankton do in cold water. Their wellbeing is vital both to fishing and the coastal stability of many countries.

THE CORALS ARE LOOKING A BIT PALE

For 20 years, coral has been losing more and more of its colour. It is a fact that polyps are delicate. If the seawater is not clear enough to encourage photosynthesis, too salty (more than 35 grams per litre) or too hot (more than 20°C), their algae leave. Or else the polyp expels them because it is sick. Then the reefs grow pale, a phenomenon known as 'coral bleaching', and die within a few months. Another threat menaces this fragile ecosystem: the proliferation of a spiny starfish, *Acanthaster planci* (also known as the 'crown of thorns starfish', right), which unsheaths its stomach at night onto a coral colony to suck up the polyps.

BUILDERS IN THE ANIMAL WORLD

A roof over one's head

A fine net of silken thread, a spider's web is a miracle of strength and ingenuity.

The art of the cocoon

The nest plays a fundamental role in the reproduction of birds. But even if birds are masters in the art of building these 'homes', they are not necessarily their sole owners.

▲ *The nest of a pink flamingo consists of a mound of mud on the ground, with high enough edges to stop it being flooded if the water suddenly rises.*

In a garden of a Stockholm suburb, a pair of long-tailed tits has been busy since the fine weather began. The male and female take part in building one of the most elaborate nests in Europe. There is no time to lose, as it takes – on average – 18 days to create this masterpiece. Made against a tree trunk or in the fork of a branch, the nest, an egg-shaped ball, has a simple hole on the side as an entrance. The outside is constructed from twigs and lichen, with feathers, bits of string, hair, spider webs, scraps of wool and sometimes even scraps of material or flower petals all mixed in. Inside, it is cosily lined with feathers and all sorts of fur and hairs.

Appreciating creature comforts

A cousin of the long-tailed tit, the penduline tit, also builds an elaborate nest which often hangs from a branch over water. This species uses a lot of willow catkins or the woolly seeds of cat-tails or the common reed, which are especially soft, to line the inner wall of its shelter.

Only the male carries out this work. First winding wisps of straw round the chosen branch, it makes a sort of stirrup and then closes the sides to form a purse-shaped nest with a narrow entrance like a funnel. The young of these birds will have a taste of luxury from the moment they hatch from their eggs.

Minimalist builders

But there are also birds which are satisfied with the barest minimum. This is the case with small long-legged birds like lapwings or plovers, which nest on the ground and merely make a simple hollow in the ground where they lay their eggs. The guillemot lays its egg directly on a cliff ledge, often on a slight slope where it seems it may fall into space at any moment. The trick is that since the egg of the guillemot is pear-shaped it does not roll and so there is little risk. The record for simplicity is held by the white tern, from the tropics, which lays its single egg so that it balances on the branch of a tree.

◀ *A sophisticated and comfortable nest is built by the penduline tit, a passerine that lives in Europe and Asia. Shaped like a purse, the nest often hangs from a willow tree.*

◀ *The long-tailed tit, a small passerine of Europe and Asia, builds a woven, complex and very cosy nest, fit for raising numerous offspring.*

Nothing is wasted

Most geese and ducks build nests from plant matter, never far from water. To make the nest more cosy, the female uses some of the down feathers from its breast. The eider, a duck belonging to northern regions, is known for its particularly soft and warm down which is collected from nests to make high-quality eiderdowns, or duvets, and warm jackets.

Many raptors, and also storks, use the same nest year after year, to the extent that certain eyries may be several dozen years old and of an impressive size (as much as three metres high for a bald eagle). The large wheel shapes of branches in storks' nests attract small birds, like sparrows, which make their fairly rudimentary nests in the intertwined branches.

◀ *Young eider ducks live on the ground, on a soft carpet made of down from the breast of the female duck.*

Communal nests

All sorts of nests are found among species that nest in colonies. The emperor penguin, for instance, has a 'nest' made from the feet of the male which, during the incubation period, stands with its single egg placed on its feet and covered by the warm skin and feathers of its abdomen. The nesting males gather in groups to keep warm.

In sub-Saharan Africa, weaverbirds live in noisy colonies of nests. In some species, the nests are many individual 'hanging baskets' constructed on the branches of the same tree (*see p. 210*). Constructed from twigs and dry grass, the nests are very capacious and even very elaborate in some species. Weaverbirds, as their name suggests, are capable of interweaving fibres and making knots to hold them together. All this splicing results in a very strong structure, often anchored to its neighbour, as are the nests of the sociable

◀ *The nest of a northern lapwing. It is just a hollow in the ground, which the bird has lined with a few small branches and dry grass.*

▶ *Nesting on the wall of a cliff or on a rocky ledge, the guillemot lays its single egg on the bare rock. The egg is pointed so that it cannot roll off.*

weaverbirds. Their groups of nests look as though they are one huge structure with many entrances, one for each nest, and can weigh up to one tonne.

Masons' mansions

Some birds build their nests with wet soil or mud. You may have seen swallows' nests, lodged under the eaves of a house or under a beam in a barn. Using their beaks like a plasterer's trowel, they cement together scoops of mud to form a cup-shaped nest with an entrance that varies in size according to the species. Once dry, the mud is extremely strong.

Pink flamingoes construct a sort of mud mound that is slightly concave on top, where the female lays her egg. The nests are so close to each other that the incubating birds can hardly move. Gannets share the same habits, nesting on islands.

In tropical Asia, certain swifts, the sea-swallows, or salanganes, nest by the thousand in caves. Their nests, made of hardened strings of saliva, are sought after by gourmets and marketed as a dish in Asia that goes by the well-known name of 'bird's-nest soup'.

Another builder, the rock nuthatch, 'plasters' over very large holes or crevices in rock with mud to create a nest, or builds earth nests of a great size against a rock wall.

Better still is the rufous ovenbird (*Furnarius rufus*). This little bird, found on the pampas of Argentina, builds a spherical nest shaped like an old-fashioned oven, in which the first chamber is separated from the incubating chamber by a partition. Its structure weighs up to five kilograms, whereas the builder clocks in at only 75 grams.

Individual style

Other animals also build nests. The tiny harvest mouse makes itself a little cradle that swings from two ears of corn or tall grasses. The mouse uses the living leaves of the plants supporting it, which means that the material does not die off and the nest looks just like its background and escapes the eye of predators.

Many small rodents, and even rabbits inside their warrens, make a rough sort of nest from plant material and fur, which ensures a cosy environment for the babies.

Gorillas and chimpanzees also make nests, not for purposes of reproduction, but to rest in. As evening comes, they build large shelters of leaves and branches, on the ground or in the lower branches of a tree, where they can spend the night in a group. These primates never use the same nest twice.

Under the water, in a very different environment, some species of wrasse secret a large amount of mucus with which they surround themselves to make a sort of nest or bed where they, too, spend the night. Wrapped like this in their sleeping bags, these fish can hope to escape detection by their predators.

▲ *The barn swallow builds its nest by piling up blobs of wet mud mixed with wisps of straw. The inside is stuffed with dry grass so that the nestlings will be comfortable.*

▲ *Gorillas sleep in nests that they make each night before they go to bed. The nest is usually on the ground, or sometimes low down in a tree.*

THE TAILORBIRD

The tailorbird, a small warbler of the *Orthotomus* genus, from Africa and especially Asia, can really sew, hence its name. These birds make their nests in a large leaf by pulling the edges close to each other and sewing them together with their beaks, using a thread (a small liana or grass blade). In this way they make a sort of cone in which they pile up material in a cup shape to hold the eggs they lay. As the leaf remains on the tree, it stays green and the nest is completely invisible.

Low-cost housing for insects

The cells of worker bees are all made alike. This standardised, modular design saves time and space in a beehive.

▲ The wax produced by the abdominal glands of the bees is kneaded between their jaws. Then the workers stretch out the wax into a thin layer to make the cells.

If you were to look at a box of pencils you would see that between the cylinders of wood and lead there are gaps. Supposing the box were made of soft cardboard and you squeezed it in your hand, you would see the gaps between the pencils disappear and the pencils in the centre would arrange themselves into a hexagonal shape. The same thing happens in a honeycomb. The cells are initially round, one of the simplest shapes for nature to create, but they automatically become hexagonal when the bees build several more cells onto them.

Unrivalled geometricians

The six-sided cell is the bee's signature. And whether the cells have as their function nurseries in which to raise larvae or larders in which to store honey, they are all made to the same design. This explains the geometrical regularity of honeycombs that has long fascinated people, who have seen in the bee an exceptional architect working without ruler or compass in the darkness of the hive.

Apiculture made great progress in the 19th century when a few pioneers first understood the construction plan of a hive. The combs are always vertical and parallel to each other. The bees begin making a comb by fixing it onto the ceiling of their shelter and adding new cells onto the bottom. There is little variation from the average in the size and depth of the cells. The space between two combs is just sufficient for the workers and the queen to move around. All available space is occupied by the combs as the colony grows. Understanding these rules has enabled beekeepers to arrive at standard dimensions for the construction of artificial hives. Trays of embossed wax are used to start off a comb and help to speed up the work of the bees. And the combs are assembled on movable frames, making the harvesting of the honey and control of the colonies easier.

▼ To enlarge the combs, the workers cling onto each other and make what are called wax chains.

The queen in her palace

Adult bees capable of reproduction, that is fertile queens and the males, or drones, eat richer food than simple workers do. They are therefore larger and the standard workers' cell is too small for them. So the workers build them special, more spacious cells. But their dimensions do not fit into the regular plan of the basic comb. Instead, the royal cells take the form of large warts grafted rather haphazardly onto the bottom of the combs, where they lack any regularity. Our wonderful architects seem here to have lost their skill.

▲ The royal cell, grafted onto a comb looks like a large shapeless lump. Geometry does not seem to play a part in its construction.

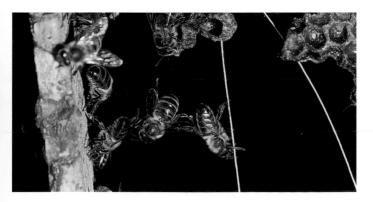

Skyscrapers in the savanna

In the African grasslands and the Australian desert there are termite mounds several metres high. They contain within them everything that is necessary for the life of a colony of several million inhabitants, from accommodation to air-conditioning, security systems and food.

A large pile of hard earth without any opening – this is what a termite mound looks like, a true fortress built to withstand the hazards of climate and predators. No termite is to be seen. These insects are blind and never appear in daylight. But there is certainly activity inside. To find food, to react to a threat, to repair damage or to enlarge the mound, millions of termites are in constant motion running around the extremely complex network of galleries inside the mound.

An underground city

Underground, this network is particularly extensive. It has horizontal passages leading into the surrounding area, so that supplies for the colony are available. Vertically, it goes deep down into the subsoil. The workers (the termites that build the mound) use clay mixed with saliva and sand to produce a very strong building material. When clay is not available nearby, they 'put on their miners' hats' and burrow down to extract it from a depth of 10 metres. And when there is a shortage of water, the efforts of the workers to find it

▼ The mushroom shape of this termite mound is the sign of a region where rain falls frequently (below, in the Central African Republic). The top serves to protect the mound from the water.

WHEN THE QUEEN SMELLS DIFFERENT

A particular pheromone given off by the termite queen sets off the process of rebuilding the royal cell. The variation in the size of the queen's abdomen means that her cell has to be frequently altered. If her odour is too diluted in the air, her workers are idle. If it becomes concentrated, the workers demolish the outgrown cell and rebuild it at the spot where the odour is strongest. Two arches begun at the two opposite ends by two groups working separately will end up meeting almost exactly in the middle. So it can be said that the perfect shape of the royal cell is due to a 'scent gauge'.

are amazing. During excavations for a well in Senegal, termite galleries were found going down to the water table some 52 metres below the surface.

Airy storeys

In the heart of the termite mound, a suite of vaulted chambers of various dimensions is built in two contiguous blocks over cellars, as a health measure. On the lower floors, the royal chamber, like a bunker, shelters the king and queen. The latter, with her enormous abdomen, is simply an

▼ Termite mounds in Australia, a field of 'apartment blocks' several metres high.

system for renewing the air supply in the termites' underground passages, the oxygen would run out in a few hours.

Building without an architect or a plan

The complexity of a termite mound, unequalled in the animal world, is truly amazing. It may rise eight metres above the ground, which, in relation to the size of the workers, is the equivalent of a building one kilometre high.

Scientists have tried for a long time to discover how such a complex design has come about. In the end, they have established that there is no design. Each worker builds on its own account, but a few simple features of their behaviour co-ordinate the work efficiently. If a worker has made a mud ball, it will go and put it on top of another ball if it can, or on the largest pile, attracted by its smell. Stimulating each other in this way, the workers build pillars that bend to form an arch.

An efficient security service

Reproduction falls to the lot of the king and queen, and the termite mound is run by the workers. This leaves the soldiers, whose duty is to defend the nest. Sterile, like the workers, but with well-developed heads and jaws, they make a strong army, always ready for war. If the mound is attacked by ants, dire enemies of termites, the workers scurry back to the middle of the nest while the soldiers hasten to the site of the attack. They 'go to the front', repaying their debt to the society that has always provided them with food.

▲ *The workers (white, top) build a wall of little mud balls. Soldiers with fearsome jaws (above) protect the site.*

▼ *The African aardvark, also called an anteater, feeds on termites which it catches with its long sticky tongue.*

**INSIDE A TERMITE MOUND
(AFRICAN TERMITES, IN NIGER)**

▲
1. *Ventilation shaft*
2. *Store of sawdust for making stacks for fungus cultivation*
3. *Stacks for growing fungus*
4. *Chamber for raising larvae*
5. *Royal chamber*
6. *Horizontal passage leading to the outside*
7. *Cellar under the papery floor*
8. *Large central pillar supporting the termite mound*
9. *Passage leading deep underground*

egg-producing machine. Round about are the chambers of the young, where the workers look after the larvae. Higher up, there are stacks of rotting organic matter where the workers grow the fungus on which the larvae feed. In the highest chambers spare food is stored.

One or several shafts open onto the top of the termite mound. These conduits communicate indirectly with the mound and come to an end in the depths of the soil. The fresh, damp air rising from the soil ventilates and cools the termite mound. Then, warm and no longer clean, it escapes by way of the shaft. Without an efficient

Master engineers

The prototypes of most human inventions existed long before in the natural world. To protect themselves and to ensure the survival of their species, animals are capable of amazing achievements and often solve problems with the skill of master craftsmen. In some cases, humans have made use of the same solutions found by animals to resolve technical problems as complex as the resistance of materials, for instance. But we are superior in one respect: we are constantly improving. Termites have been making the same cement for millions of years. Each animal in fact inherits its skill through its genes, and uses only the skills peculiar to its species. No bee knows how to weave grasses and no bird can make hexagonal cells.

◀ THE BEE, AN AERONAUTICAL ENGINEER

Wax is a fragile, malleable material that melts at 65°C. Yet the 20 grams of wax needed to build a honeycomb enables one kilogram of honey to be stored. The secret is the hexagonal cell, which owes its ability to hold its shape, and its shock resistance, to its form. These qualities are ideal for building walls that are both strong and light. Human aeronautical engineers therefore followed suit in the construction of aircraft. Between two corrugated iron sheets they insert a honeycomb structure made of hexagonal cells in a composite material, which means that the aircraft cabin is light and keeps its shape.

▲ WEAVERS, EXPERT BASKET-MAKERS

Hanging from flexible branches, the nests of weaverbirds are safe from predators. But these nests must also be strong enough to hold the weight of the chicks, light enough to stay in place and very flexible so that they can change shape without breaking. The highly elaborate building technique of the weavers meets the challenge. It actually has to be learned, and young male birds try it out on practice nests. The technique is somewhere between weaving and plaiting, and based mainly on the use of knots, rather like basketwork that uses blades of grass and green twigs instead of wicker.

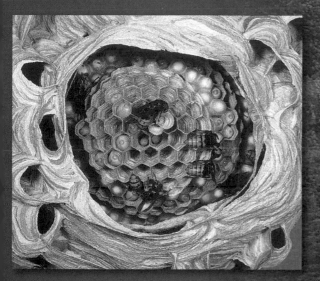

BEAVERS, HYDRAULIC ENGINEERS

In Montana, in the United States, a dam almost 700 metres long has for decades been controlling the flow of the Jefferson River. But humans have no hand in it. This structure is the work of generations of beavers, which regularly maintain it. Why do such small rodents make something so large? Because to ensure their survival, whatever the design of dwelling they choose, beavers need to make the entrance under the water. So their dams allow them to control the flow and create stretches of water with a constant level. They cut down small trunks and branches that they fix upright into the riverbed and then plug the gaps with mud, stones, reeds, branches and so on.

WASPS, THE INSULATION SPECIALISTS

To a community of wasps, nothing is more valuable than the larvae. And so they have made provision for their protection from cold spells at night as well as from the heat of summer. The drones wrap the nest (above) in a covering of 'papier-mâché' that contains alveoli, or cavities, full of air. This 'papier-mâché' is made of wood fibre that the drones have chewed between their jaws, and in it the alveoli overlap like scales. The trapped air acts as an insulator, protecting the middle of the nest from sudden temperature changes. The insulation of our houses is based on the same principle, in which layers of polystyrene or fibreglass containing pockets of air are laid on the floor of the roof space.

TERMITES, BRICKLAYERS
AT THE CUTTING EDGE

In the tropics, termites build structures that defy the laws of construction. To withstand the raging downpours that occur in these regions the termite builders use a hardening substance made out of their faeces with

which they impregnate the walls of the termite mound. Once dry, the mixture hardens to the point where it can withstand erosion and most attacks from predators. Humans have not really done any better by mixing cement, sand and gravel to make concrete.

A stitch in time

Weaver ants have an unusual technique for making their nests. They sew leaves together using a very peculiar tool: their own larvae.

In the tree chosen by the colony of weaver ants, there is intense activity. Leaves and twigs seem to be moving and twisting about by themselves. The community is hard at work. They must make a nest of leaves that is firmly sewn together with smooth seams. It will be the nest for the larvae. It is difficult to accept that this meticulous work, genuinely 'hand sewn', has been carried out by little insects.

Weaver ants form part of a limited group of animals that use tools – not manufactured objects, but objects used for a purpose other than their original one. And in this area they show unusual initiative because they use their own larvae, which thus participate in the building of the shelter where they will eventually hatch.

Teamwork

To protect themselves both against bad weather and predators, the weaver ants of Africa and Asia build a strong, waterproof nest in the trees.

With a remarkable sense of teamwork, the workers line up side by side along a leaf. They cling firmly to the edge with their rear legs and catch hold of the edge of another leaf with their jaws and their front legs. Then they pull to bring the two sides closer. If the leaves are too far away, they form chains as long as 10 ants or more. Gradually, with the combined effort of many individuals, the two leaves eventually touch. Now they have to be fastened. It is here that the larvae come in, temporarily promoted to sewing machines.

▼ In the second stage, the larvae are brought in to fasten the two edges together by means of a thick silk seam.

Baby stitches

There is no question of giving this work to the youngest of the larvae. The workers need larvae that are approaching the stage of metamorphosis, because only then have they developed special salivary glands in their heads that are able to spin the silk for the protective cocoon necessary for their transformation.

When the two leaves are next to each other, the worker ants go and fetch some larvae by taking hold of them in their jaws. The larvae make themselves stiff and almost motionless, a passive tool in the service of the adults. The workers place the larvae head down onto a leaf and the larvae then emit a drop of silk. The workers tap the larvae with their feelers, which makes them begin to put out a thread, and they quickly move the larvae onto the opposite leaf. The manoeuvre lasts hardly a second, but repeated dozens of times by many ants it produces a strong silk seam that would be a credit to any professional tailor.

▲ In the first stage of making the nest, the weaver ants join forces to pull the edges of two leaves together.

Webmasters

Strong, elastic and tough, the silk thread produced by spiders has many excellent qualities. And spiders have plenty of ideas for using this wonderful material: to make trapdoor-webs, nursery-webs, or bell-webs.

A common resident of houses, the *Tegenaria* or house spider likes to spin her web there in peace. Simply visiting any attic or cellar will allow you to see how very strong this delicate mesh of silk threads can be. Out of the wind and rain, it collects dust for decades without coming to harm. Stronger, weight for weight, than the best steel wire, coated in an antibiotic substance that prevents its destruction by micro-organisms, a spider web is able to withstand almost anything – except deliberate destruction by human hands. It is not surprising that this wonder fabric has such varied uses.

A very special cradle

Despite its renown, due probably to the fascinated horror many people feel about spiders, the trapdoor form of web is not used by most spiders. All species, on the other hand, spin a protective cocoon for their offspring. Each chooses its pattern according to its way of life and its environment. Some are thick to keep out parasites, others are covered with debris for camouflage, and still others are perched on a sort of stem to make them less vulnerable to predators.

Some very prudent mothers prefer not to leave their offspring

▲▶ *After carrying its cocoon full of eggs everywhere with it (right), the nursery-web spider has spun a nursery (above) in which to protect its newly hatched young.*

to their own devices, like the large *Pisaurus mirabilis*, a spider which hunts in low vegetation. She carries the cocoon in her jaws or under her body. When the eggs hatch, she spins a second web round the cocoon, called a nursery because its sole function is to protect the babies' first outings under the watchful eye of their mother. But as soon as they have moulted for the first time, all the little ones rapidly disperse, because spiders have no objection to eating their siblings.

A diving bell of silk

The prize for originality in the area of spun homes must go to the argyronete, the European water spider that lives in ponds. But how can it breathe under the water and survive without any diving equipment?

The water spider has found the solution by building a proper diving bell. It begins by making a first web that it attaches to water plants, then rises to the surface for air, which it keeps next to its abdomen, protected by the hairs on its body so that it cannot escape. Then it releases the air under its web by brushing the bubbles off the hairs. The spider repeats this process until the air bubble under the first web is large enough, and then uses the web as a mould for spinning round it a second web in the shape of a bell.

An opening at the bottom is provided for hunting or renewing the store of fresh air. The spider will spend its whole life under the bell, quite safe to eat its prey and reproduce.

▲ *This water spider, or argyronete, lives in water inside the cocoon it has spun. But as it has lungs, it needs oxygen to breathe. So it brings a supply of air bubbles to its home, and travels regularly to the surface to restock this vital supply.*

The pleasures of life in a hotel

Some animals move into homes they have not built themselves. But they sometimes have to reach a compromise with others.

Excreta falling on your shoulder or your head may betray the presence of a nest of sparrows or swallows under a beam or a high windowsill of your house. Apart from this slight inconvenience, this is an example of an accepted cohabitation between species. It is often the same in nature.

The hermit crab has a soft underbelly to protect

The hermit crab is a crustacean closely related to crabs and langoustines. Almost all the 500 or so species of hermit crab, found everywhere on Earth, have one common feature – a soft abdomen. And unlike crabs or lobsters, they have no exoskeleton with which to shield themselves from danger. They therefore cannot survive without a solid home to live in.

The hermit finds a home in the form of an empty shell, usually that of a gastropod, such as a snail. When it comes across a home that suits it, the hermit crab walks off with it. But as it needs a home from the time it is very young, and it

moults and enlarges several times during its life, it has to move house regularly.

The hermit crab is a decapod, which means it has 10 legs. The last two pairs of appendages on its abdomen enable it to cling onto its protective shells one after the other. The second and third pairs are used for walking, and the first pair in the front provides the tools for eating and defence. Ironically, the hermit crab's home is sometimes shared with other squatters. Sea anemones may attach themselves to the shell of a hermit crab and travel about with it. This arrangement benefits both. The anemones have tentacles that are armed with stinging cells which prove a good defence for the hermit crab and

▼ *A crustacean without a 'crust', the hermit crab finds refuge in the old empty shells of gastropods such as snails and whelks.*

▲ *This adult long-eared owl feeding its young has most probably used an old crow's nest in which to raise its young.*

frighten away possible predators. In a fair exchange, the hermit crab provides the anemone with an effective means of locomotion and access to a variety of food.

Nest thieves

Bird squatters may use a nest abandoned by its owners. The long-eared owl, for example, often uses the old nest of a crow in which to lay its eggs. This way of doing things saves it time and, once it has been patched up a bit, the nest makes an excellent place in which to raise its offspring.

Some species choose holes that have already been made by other animals in which to nest. The red-rumped wheatear of north Africa tends to favour the old burrows of small rodents. The shelduck, a large, brightly coloured duck that lives on European coasts, lays its eggs in old rabbit warrens, usually on a dune. If it cannot find a hole, it will make its home in the den of a badger or fox.

At the other end of the world, the little blue penguin of Australia and New Zealand sometimes borrows the nest of a petrel. What is more, it sometimes happens that the two species live side by side during the breeding period without the house-sharing giving any problem.

Other birds are much more invasive. This is the case with the cuckoos of Eurasia or the cowbirds of North America, which do not build their own nests and are happy to be parasites on other species by laying their eggs in their nests. Each hen cuckoo has a favourite 'target species' (at least 15 have been counted) that has eggs of

about the same colour as hers, so that the extra egg is not noticed. Moreover, because the young cuckoo hatches first, it throws out the remaining eggs in order to be the only baby bird fed by its adoptive parents.

Every floor a hotel

Houses and apartments harbour all kinds of animal squatters. Invertebrates, reptiles, birds and mammals all make use of humans and will try to live as discreetly and as well as possible under their roof.

The best examples are surely rats and mice, which take advantage of the lodging and table of *Homo sapiens*. They are not well tolerated, as they are in direct competition with their host, especially as far as food is concerned.

The dormouse and the stone marten are less likely to be noticed. As they are strictly nocturnal, they are seen only rarely. The first seeks out the warmth of the house in which to hibernate, and might even be found asleep in a large book that it has nibbled out to make a cosy bed for itself.

The stone marten, which lives in Eurasia, likes attics and lofts, and has the annoying habit of depositing the remains of its food there as well as its droppings. The smell of the decomposing food eventually spreads over the whole of the upper floors and betrays the animal's presence.

▲ *The cuckoo lays its egg in the nest of another species a lot smaller. It is a burden for the adults whose nest has been taken over to feed the young giant.*

▲ *The stone marten shamelessly takes over old attics and lofts, which does not always please the owners.*

SQUIRREL SQUATTERS

The red squirrel makes nests that it uses for breeding or to rest in. However, if the opportunity arises, it does not hesitate to take up residence elsewhere. So an old crow's nest will do. But because you can only do things properly yourself, it will improve on this new home by using its own materials. This 'revamped' nest may in turn be occupied by other squatters, like blue tits or wrens, which do not consider it beneath them to spend the night there.

A happy life is a hidden life

Some animals dig under the soil to make dens or burrows in which they will spend a few hours, a few weeks or the whole of their lives. Others choose to move into a lair that is ready and waiting for them.

A field studded with conical mounds of brown earth betrays the presence of a mole. This solitary insect-eater spends its life underground, digging a complex network of passages, some of which may be 50 centimetres deep.

The mole regularly sends the excavated soil up to the surface, which produces the molehills. Under the largest of these, the animal will have dug out its bedroom, which will be lined with hay and dry leaves.

The tunnels of the females often surround those of a male, but encounters occur only during the breeding season.

Digging for happiness

Other mammals are also inveterate diggers. Animals like the water vole or mole rat are rivals of the mole for burrowing. They too throw up mounds, but these differ from molehills in that they have fine-textured soil and a side opening. Unlike the mole, they do not hibernate, but often excavate a network of passages that they use in bad weather to shelter from the cold.

◀ *The black woodpecker is able to make a deep hole in a tree without harming it. It is here that it will lay its eggs.*

▼ *The king of underground passages, the vole can multiply rapidly. The more voles, the more tunnels. They can thus dig over a field in record time.*

Rabbits also dig underground passages, but unlike moles they live there only occasionally. In the breeding season, the female isolates herself in a special nest in the burrow, where her young will be born, naked and blind.

In North America, prairie dogs too live in a labyrinth of chambers, tunnels, passages and blind alleys. At the first sign of alarm, individuals that have gone out to feed in the open air race to take refuge in their shelter.

The badger is another expert at burrows. It digs a huge lodging place, with several entrances, many chambers and passages, in which the female will give birth to her young. This animal, which likes to be clean, has the habit of depositing its excrement in small holes near its living quarters. It sometimes happens that it shares its home with a family of foxes, which is usually a peaceful arrangement. Foxes also often choose to live in an old rabbit warren, which they will dig out to make larger. Less often, they dig a home from scratch.

Living in a hole

Birds too can show a talent for digging. Some species choose to nest in dark holes. Woodpeckers, for instance, lay their eggs in holes in trees. They may make a previous hollow larger and wider with their strong beaks, but most often they drill out for themselves a new space intended for their eggs. The black woodpecker, the largest of the European species, thus makes holes as large as 40 or 60 centimetres deep and between 20 and 25 centimetres across.

The nuthatch, a bird of more modest size, also nests in a hole in a tree, usually a natural hollow or a hole abandoned by a woodpecker. But it fills up part of it with mud so that it has an opening that suits its size. Once dry, its work will be as hard as cement.

Among the hornbills of Africa and Asia, this system is taken to the extreme. The female establishes herself in her chosen hollow, and then the male cements up the opening so that its mate is totally enclosed. He will feed her through a tiny hole for the whole of the incubation period. If he dies, however, the female will not be able to get out of this prison alone.

Sometimes fish use the rough surface of rocks in which to live or hide. Moray eels hide at the bottom of a deep hole, with their heads at the opening of the passage. When prey goes past, they jump out at it. Some gobies, which live mainly around estuaries, also hide in holes in the rocks, either to sleep or to lie in wait for their prey.

Bird burrows

Puffins like burrows too. They most often build them in the soft earth near the coast and dig a passage which gives them access to the place where the eggs are laid.

Other seabirds, like petrels and shearwaters, share this custom. But digging a burrow is a lot of work, so some shearwaters and Humboldt's penguins prefer to nest in natural rock cavities along the edge of the sea.

Skilled nest-diggers are also found on the banks of rivers and streams. The sand martin makes its nest at the end of a short tunnel dug into a sandy cliff. The kingfisher digs passages in a bank over a stream. The main tunnel, which may be as long as one metre, leads to a chamber where the female lays her eggs. In drier environments, the magnificent European bee-eater does the same, digging into a sandy cliff or soft soil.

◀ *This burrowing wasp digs a tunnel in sandy soil, in which it buries its prey. It then lays its eggs there so that the young will have a supply of food when they hatch.*

▼ *Lurking in a hole in a rock on the seabed, the honeycomb moray from the Red Sea patiently waits for its prey to pass within reach so that it can jump out after it.*

Microtunnels

Invertebrates also are excellent at drilling. Bees of the genus *Colletes*, for example, live near sea cliffs where they drill deep passages into the earth and make cells in them where they put nectar and pollen before laying eggs. Those of the genus *Hylaeus* do the same, but they set up their nests in existing passages.

Other solitary bees, such as *Andrena thoracica*, dig passages of a complicated design and make small rooms for storing food and laying eggs. Small bees of the genus *Heriades* nest in minute cavities that have already been carved by worms in places like dead wood, and they back into them to deposit their pollen.

▲ *The puffin digs a hole in the soft soil near the shore to lay its eggs.*

Megachilids, or leaf-cutting bees, take up residence in hollow, dry stems or other little cavities and line the cells of their nest with pieces of leaf they have cut out.

Some species of weevil tunnel into wood to create their nests. Their skill is not always looked on favourably, however, because they can cause great damage.

Timber-boring beetles of the Scolytinae subfamily, insects that are related to weevils, are able to reduce an old wooden wardrobe in which they have decided to live to a pile of dust.

The bark beetle, which is another kind of timber-borer, builds networks of tunnels for its eggs. It attacks living wood and also causes considerable damage. The smaller European elm bark beetle, for example, which lays its larvae under the bark of elm trees, has been responsible for the destruction of millions of elm trees across Europe and North America. In spring, when the larvae have developed into adults, they feed on the new twigs and so infect the trees with the Dutch elm disease fungus that they carry. Weakened, diseased and stressed trees are particularly vulnerable to this disease.

▲ *This graceful pattern on a tree trunk was made by a bark beetle. This insect is a great builder of passages and likes wood, doing great damage to trees.*

COMMUNAL DENS

Meerkats, or suricates, which live in southern Africa, dig tunnels intended as homes for several families. This digging species has hind feet with strong claws. Meerkats are often seen standing straight up on their hind legs inspecting their surroundings. If danger threatens, the alarm is given and the community takes refuge in the deepest tunnel. It is unusual for carnivores to have recourse to an underground shelter.

TASTE, SMELL, TOUCH, SIGHT, HEARING

The empire of the senses

The multifaceted eyes of a fly are highly sensitive to movement.

The gift of perception

One closes its leaves to make a trap, another folds up when touched. Sensitive plants are able to make voluntary movements, visible to the naked eye.

In the marshlands of a small region of the eastern United States, at the boundary between South and North Carolina, there is an extraordinary plant. Its leaves, consisting of two valves round a central vein, and covered in fine, very sensitive hairs, have long pointed teeth along the edges. They close suddenly as soon as an insect touches them, and the teeth then meet like a grid to prevent escape.

▶▼ Drawn to the bait of the sugary nectar secreted by this Venus flytrap, a fly has settled on it. But insect beware! It has only to brush against the plant's hair-triggers and the two halves of the leaf will suddenly close over it. The slow digestive process can then begin.

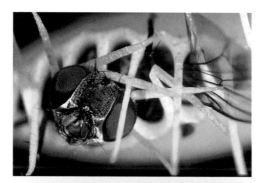

This sprung trap, like many old traps used to catch animals, is unique in the plant world.

The Venus flytrap

The Venus flytrap has developed this sophisticated trap for the purposes of its survival. In the poor, acid soils where it lives, there are not sufficient nutrients for the plant to grow. So it has sought them elsewhere, and has become carnivorous.

The leaves of this plant are covered with red glands with different functions. In the middle of the leaf they are digestive glands. On the outside they are nectar-secreting glands. For insects that are attracted by the colour and the sugar bait, it is a very appealing flower. For the greedy ones who approach and touch one of the hair-triggers in the middle of the leaf, it is certain death.

The leaf will open again about 10 days later, when digestion is complete. The skeleton will be carried away by the first breeze and so make room for the next victim.

Involuntary movements

Plants do not remain motionless, because outside agents act on them. These movements, which may be called passive because the plant does not use any energy to effect them, are indispensable in reproduction, for male and female reproductive cells to meet and for seed dispersal. Many of us are aware of these movements indirectly,

when we experience the familiar allergy called hay fever, which is caused by wind-borne pollen that fills the air in certain seasons. Among plants, active movement that means expending energy is extremely rare.

Many unicellular algae, very simple organisms that are plant-like, are flagellate, which means they have whip-like appendages that they use to propel them through water. As more complex plants and multicellular organisms evolved, this ability to move disappeared. Plants adapted to living on land and the emergence of roots literally anchored them to the ground. But when difficult living conditions make radical adaptation necessary (a situation scientists call 'selective stress'), some plant groups, not closely related, can be particularly inventive.

The modesty of the sensitive plant

The Europeans who discovered the sensitive plant in Brazil in the 17th century found it very modest in its behaviour – it withdrew as soon as it was touched. This behaviour earned it its Latin name of *Mimosa pudica*.

This herbaceous plant has pretty pink flowers with petals arranged in a spherical shape, or inflorescence. Valued by gardeners, especially in hot regions, it is widely cultivated. Its leaves are composite, which means they are divided into leaflets. If one of these leaflets is touched, it folds

▲ A climbing bryony moves imperceptibly to twine round its support. This movement is connected with its growth and is too slow for us to see.

A PLANT THAT GETS TIRED

Each leaf of the Venus flytrap can digest a maximum of three insects. As each bulb grows six leaves, it catches fewer than about 20 insects a year. Misled by a falling blade of grass, the leaf sometimes closes on nothing and quickly reopens. But closing and opening the trap uses a lot of energy. If it happens too often, the Venus flytrap risks wearing itself out.

---

inwards onto a vein, and the movement is transmitted from one leaflet to the next until the whole leaf has folded. In the case of a major shock, the whole plant reacts by shrinking back. It then appears to fade, its leaves fold up and they droop from the stems as though they were dead. This reaction can even spread to neighbouring plants of the same species.

It seems that this sensitivity to touch is the result of an adaptation of the plant to protect itself from herbivores. By almost immediately appearing dead, it can take away the appetite of a mammal preparing to eat it.

Almost an animal

So is the sensitive plant only a plant? There may be some doubt. Studies have shown that this organism can be truly anaesthetised, and for several hours it loses its ability to close its leaves when ether or chloroform are administered to it. And an even more remarkable fact is that it is capable, if not of learning, at least of forming a habit.

If a sensitive plant is regularly submitted to the same stimulus, it will gradually react less and less strongly. After a few hours, it will even fail to react at all. If the stimulation is stopped for a time, and then resumed, the plant will again react, but less strongly. It somehow retains a partial memory of its experience.

According to Japanese scientists, the sensitive plant owes this extraordinary ability to a protein called actin, also present in animals, which produces a chemical reaction each time the plant's leaves are touched.

▲ The sensitive plant is wide open, but at the slightest touch the lobes of its leaves will fold up along the veins.

▲ Sainfoin, or holy clover. The leaves are divided into three leaflets. The two outer leaflets go up and down as fast as 60 times a minute, which makes the plant seem to be beating its wings.

221

Look before you eat

Among animals it is only vertebrates that taste their food. This is not in order to find out if they will like it, but to make sure the food is not toxic to them. Tasting is a means of being on the alert, a safety precaution before ingestion.

With loud cries, a chimpanzee calls the fellow members of its troop. It has just discovered a fallen branch laden with juicy fruits. The troop gathers round. They all watch the finder attentively. It begins by sniffing one of the fruits. This passes the test. Then it raises the fruit to its lips, tastes it, then violently throws it down, pulling a face of disgust.

None of the other chimpanzees will repeat the experiment. The taster has performed a valuable job and warned its companions that the food might be poisonous. Other primates will taste and then invite their fellows to join the feast or, on the contrary, will give loud cries to keep the group away if the food is considered inedible.

▲ *Fruit-eating primates, like this chimpanzee, always taste before they eat.*

Taste buds

Fish have taste buds on the whole of their bodies to test the water around them constantly. These taste buds are small barrel-shaped cells with sense cilia, or hairs. As they come into contact with molecules dissolved in the water, the cilia generate electrical impulses that are conducted to the brain by specialised neurones. The taste buds of sharks can detect one drop of blood in a volume of water equal to that in a municipal swimming pool.

The taste buds of land vertebrates are situated in the upper part of the digestive system (the mouth and larynx) and on the tongue, where saliva acts as a solvent. This obliges these animals to place what they want to taste in direct contact with their tongue or lips. In mammals, the taste buds are grouped in papillae, almost all of which are on the tongue. They are particularly numerous in primates, with the exception of humans. Unlike our tree-dwelling cousins, we have an insufficient number of papillae to ensure that we can tell whether our food is perfectly safe or not.

▲ *Vibrations first tell the shark that there is prey around. As the vibrations grow stronger, the shark locates the prey by its sense of taste (at less than 100 m), vision (at about 15 m) and its electromagnetic field (at less than 1 m). Finally it bites, but it will kill deliberately only if the prey tastes right.*

TASTING FROM INSIDE

In mammals, the carotid sinus (at the bifurcation of the aorta from which the carotid arteries go up to the brain) is also equipped with taste buds, which measure the pH and the levels of carbon dioxide and oxygen in the blood and then inform the brain. During apnoea in a mammal, for instance, the carbon dioxide level rises. As soon as it passes the threshold tolerated by the organism, the brain triggers the reflex to breathe in.

Sensitivity to vibrations

Hearing is a key element in communication among and between species. In particular, it gives information about the presence of prey or predators and allows passionate declarations of love to be received.

The cricket 'hears with its feet', by means of complex listening equipment. Tympanic organs on its legs enable it to capture even ultrasound. It also perceives low frequencies by means of other sensory receptors in the same place. Chordotonal organs, a sort of stretched guitar string at the joints of its legs, wings and especially antennae, play a similar role.

Vibration receptors

The cricket is not the only creature to hear in this way. All insects are sensitive to vibrations. And, in the way of almost all insects that emit sounds, the cricket's vibration receivers have become proper sound organs.

In order to hear it is often necessary to be able to generate sound. Thus crickets attract their mates by chirring. Migratory locusts also chirp in order to assemble their large groups. They are all capable of discerning the low frequencies emitted by the sound of a predator passing and sometimes can also hear the ultrasound produced by the 'radar' of bats.

Amplified vibrations

In vertebrate animals, sensitivity to sound is concentrated in a single phonoreceptive organ. In fish, for example, vibrations are transmitted to the inner ear through the bones of the skull. Species with a swim bladder hear better than those without one, because this organ is up against the bones of the head. The swim bladder of the herring, for example, transmits sounds much more accurately than its skull and the inner ear.

This structure is further improved in white river fish, like the char. Between their swim bladder and their two internal ears is a system of small bones called Weber's ossicles, which greatly amplify the sound vibrations.

The hearing organ of land vertebrates becomes ever more sophisticated from amphibians to mammals. The latter are probably the most sensitive to sound, since they possess a complex system of bones in the ear (the hammer, anvil and stirrup).

Feathers to hear with

The night vision of owls is excellent. But for these birds, hearing is even more important. They have discs of feathers round their eyes that absorb sound and in fact enable them to locate prey (like a mouse) that is a long way away and sometimes even under the snow.

An owl like the snowy owl can therefore catch an animal without being able to see it. Perched on a post, it listens and then swoops down on its prey which may be moving about 10 or 20 centimetres beneath the snow. Owls have a further advantage in the fluffiness of their feathers, which softens the sound of their beating wings and enables them to fly through the air without being heard.

▲ *The snowy owl lives in arctic regions and its main food is lemmings. Its advantage in hunting is that it has such acute hearing that it can capture its prey without seeing it.*

▶ *Ears 15 cm long enable the bat-eared fox to perceive the movement of its prey on the sand several dozen metres away from it in any direction. Its prey includes invertebrates, lizards and small mammals as well as birds.*

▲ *The 'ears', or 'tympani' of the cricket are situated on either side of its tibiae, one on the outer side and another smaller one on the inner side. By rubbing its elytra (wing-shields) against each other it produces a loud chirring.*

Extrasensory perception

Animals can hear, see and smell better than humans can. They are also sensitive to information that we do not even notice. According to their environment, their sense organs have developed to the extent that they contain receptors unknown in human beings. They have acquired the capacity for an electrical, magnetic, aural, thermal or chemical representation of their world. These 'extrasensory' perceptions, invisible to us, fascinate zoologists and engineers alike. The first wonder whether we have them, and the second aim to copy them.

▲ FOLLOWING THEIR NOSE

Pheromones are chemical substances produced by organisms, which play a very important role in communication. The sense of smell in humans is not able to identify pheromones, although they may play a part in our behaviour without our being aware of it (especially sex pheromones). On the other hand, many animal species are able to 'smell' a pheromone and to adapt their behaviour according to this chemical signal. For example, in certain moths (above, a Spanish moon moth), the females emit a sex pheromone, bombicol, which males perceive through receptors situated on their antennae. Attracted by this 'smell' even from a great distance (sometimes as much as one kilometre away), the male or males will soon locate the female they desire.

◀ A WARM FRONT

Some animals are able to perceive infrared light (corresponding to an emission of heat), which humans cannot detect. Snakes find their prey in this manner, by the heat they give off. They 'see' the infrared of their victims and, guided by this source of heat, they make their way towards it. Polar bears on the ice also perceive the slightest emission of heat. For example, if a person were inside a tent set up in the middle of the snow, a bear would find him or her just by the infrared emitted, even though it could not see the person.

◄ HUNTING BY ECHO

Bats (left, a long-eared bat), like dolphins, emit ultrasound and hunt with sonar equipment. The ultrasound waves travel through space, and when they encounter an obstacle or prey they are reflected back to the animal, which judges the distance of and identifies the object responsible for the reflection. In bats, these return waves are captured by the ears and in the region of the nostrils. In dolphins they are received in the lower jaw. Infrasounds (low-frequency sounds) are also a mode of communication totally inaudible to the human ear. Whales and elephants use it to communicate among themselves, sometimes over distances of several dozen kilometres in the case of whales, and several kilometres in the case of elephants.

▼ NAVIGATION BY COMPASS

The navigation of night migrants is thought to be guided in the first place by variations in the Earth's magnetic field. In the laboratory, artificial modification of the direction of magnetic north leads birds like pigeons astray. For the first time in the animal kingdom, crystals of magnetite (an iron oxide) have been identified in pigeons. Between the brain and the skull, these small particles behave like the needle of a compass. They orient themselves to the north and record the value and direction of the local magnetic field at the departure and arrival points. These crystals are also found in the bodies of bees and in certain bacteria. There is no reason to think that there is not a built-in compass like this in other animals, to help them migrate or simply find their way – in sea turtles, for example.

▼ ELECTRIC RELATIONSHIPS

Some fish can detect the most minute changes in the electric field of the water. Others surround themselves with a weak magnetic field that acts like a security system for their protection, and which enables them to find prey, predators or obstacles that their other

sense organs would not reveal to them. This is the case with some African and South American freshwater species and especially with sharks (above, a whale shark), and rays. The electric eel, a freshwater fish, is able in this way to distinguish between two objects of different colours in the dark, because the electrical conductivity of each is slightly different. The Nile perch, which emits 300 electrical impulses of five volts a second, has a whole sonar at its disposal. Sharks are sensitive to electrical fields of less than 0.005 millivolts per square centimetre. Some electroreceptive species, like the electric eel, are able to use the electricity as a weapon.

IMITATING NATURE

We copy nature, when this is economically profitable and technologically possible. But in order to imitate nature, it is necessary to have a perfect knowledge of the mechanisms at work. The organs of extrasensory perception in animals have not been reproduced because we don't fully understand how they work. However, we use them as models for trying to improve what we have manufactured. The thermoreceptive organ of rattlesnakes (below), for instance, is less sensitive than the infrared receptors on missiles, but has a far higher resolution, which is guiding scientists in their attempts to improve alarm systems. Observation of tropical freshwater fish equipped with sonar equipment assists engineers working on new guiding and detection systems. And engineers dream of imitating the mysterious navigation systems of migrating birds.

Catching the light

All animals can perceive light, but it appears that very few can see properly. Sight organs have the same structure, whatever the species. Only the level of complexity varies.

▲ *The large eye of the bigeye fish can capture more light than others that are smaller. It tells us about the lifestyle of its owner: the bigeye is a nocturnal animal. Its big eyes are a great asset when it has to leave its shelter after dark to go hunting.*

Light reaches these half-eyes from all directions, so these animals have global vision. But they can only perceive light levels, contrast, vague shapes and some movement, and cannot locate a light source with any degree of precision.

Eyes like telescopes

Most arthropods (spiders, scorpions, crustaceans, mites and insects), on the other hand, can see better. How do they do this? Look at a dragonfly, for example. Like all insects, it has true eyes. They are called composite eyes, because each consists of a large bunch of 2 000 geometrical facets. Beneath each facet lies a photoreceptive structure, the ommatidium, which is a combination of light-sensitive and sense cells. Pointing in a very slightly different direction from its neighbour, each ommatidium functions like an astronomical telescope that covers only a small part of the field of vision.

In daytime, bright light fills the eyes of the dragonfly. Their ommatidia all react in the same way, each giving an image. The dragonfly therefore sees the world in colours like those in a pointillist painting, and, in particular, is able to see the slightest movement.

At night, however, the low levels of light taken in by all the ommatidia enable them to produce only one image, not in colour, but with enough

▲ *A bee's eye. The composite eye of insects enables them to see in almost every direction, but not in detail and only in two dimensions. Very sensitive to contrast, the composite eye is an excellent motion detector.*

The earthworm cannot see much – just black and white. Its body has a scattering of light-sensitive cells. When they are excited, this tells the worm that there is light, and where it is coming from. In the absence of light, these cells do not react. That is all. The earthworm cannot see colour, contrast, movement or shapes. It can tell only whether there is or is not light.

From skin to ocellus

The earthworm's sense of vision is particularly primitive. Its scarcely specialised cells react rather like our skin does to the sun. They represent the basic unit from which the sight organs in other more complex species are developed.

Together with sense cells, these photosensitive (light-sensitive) cells constitute photoreceptors, or ocelli. As the species becomes more complex, the ocelli sink under the skin until they form an eye with a crystalline lens over it. Most invertebrates see their world only through ocelli.

LIGHT-SENSITIVE PIGMENTS

Many animals can see because of light-sensitive pigments in the retina. Some mammals have trichromatic vision: their eyes contain three types of cones, each with a different pigment, which reacts to blue, green or red. A pigment is a combination of two molecules. One, retinol, is a derivative of vitamin A. The other, opsin, includes a family of complex molecules, each of which reacts to particular wavelengths. When light strikes the pigment, the two molecules dissociate. This movement generates electrons that, guided by a sense cell, form a nerve impulse that travels to the brain giving it information about colour.

▶ *The seagull is thought to see in four colours, whereas humans see only in three.*

▶ *Bees see in three colours: ultraviolet, blue and green. They can also find the planes of polarisation of light.*

▶ *Frogs can see neither shapes nor outlines, only colours and movement.*

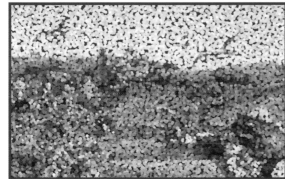

contrast to at least allow the dragonfly to avoid flying into obstacles.

Eyes like video cameras

Vertebrates see through eyes still more complex, eyes with a chamber, like the inside of a camera. The film, or retina, is lined with two types of photoreceptive cells – cones (which are responsible for seeing colours) and rods (which are sensitive to contrast, and are particularly efficient in low light). The cornea, which is curved, acts as the front lens.

The crystalline lens, just behind the cornea, makes focusing possible because muscles change the shape of the lens according to how much light falls on it (except in fish and amphibians). Between the retina and the crystalline lens, saline ensures a good convergence of the light rays.

The eyelid and the iris act like the diaphragm in a camera, which means that they control the amount of light entering the eye.

Most mammals do not see colours, because their retina contains almost only rods. Bulls, therefore, cannot see the red of the bullfighter's cape, but when the cape is waved in front of it the movement is transmitted to the bull's eyes together with the light it reflects.

In contrast, tree-dwelling mammals (like the primates, and therefore human beings) and fruit-eating mammals all appear to see colours. Herbivores, which have eyes with relatively limited movement, have less acute vision than predators. Predators have eyes that are relatively close together, which allows them to see in three dimensions. Each eye produces the same image with a very slight shift, and the brain superimposes the two images. It is the difference between them that gives the image depth and allows the predator, for example, to determine how far away an object is.

The eyes of cuttlefish, squid and octopus are very different from those of vertebrates. Yet, apart from a few details, their sight is the same as that in humans. They probably even see colours better than we do, and in particular they have three-dimensional vision. Because of their enormous eyes, they have become far more successful predators than fish.

Bird's eye view

Birds are known to see colours, but in some species (particularly seagulls and diurnal passerines), this colourful world is quite different from ours. It is thought that they might have more acute vision by increasing certain contrasts. Migrants, like pigeons, are probably sensitive to ultraviolet light and are able to detect the planes of polarisation of light.

The cylindrical eyes of raptors act like telescopes. Their retina contains special zones called fovea where the image is enlarged.

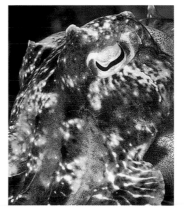

▲ *This snake is particularly active at night and lives in trees. These two factors explain why it has such large eyes: they capture more light and give more detailed vision.*

▲ *The cuttlefish sees in colour and in three dimensions, just as we do.*

A good nose is not good enough

▲ Smelling with its feet – this is what a fly does, each of its feet (above, seen through an electron microscope) having sense cilia and sensitive lobes called pulvilli.

The sense of smell is receptive to molecules in a gaseous state. In marine animals, it goes with their sense of taste. In land animals it offers a far richer range of perceptions than even sight does.

▲ Snakes (above, a boa) use their tongues to analyse the surrounding air and so detect the presence of possible prey.

Throughout the world, stories celebrate lost or abandoned pet dogs that have found their way home, even when it is several hundred kilometres away.

These are not just myths. Dogs possess an awe-inspiring sense of smell, 10 000 times stronger than ours. Impregnated with the smell of its owners, a dog's sense receptors can bring it home if it follows a route, no matter how long, that has been traced by the scents its nose has already received.

This is not a particularly extraordinary feat, because in most vertebrates the sense of smell, together with sight, is the sense that gives them the most information. Pigs and dogs are trained to seek out truffles underground, which they can do because of their strong sense of smell. For the same reason, dogs are trained to sniff out contraband substances. In humans, however, this sense has greatly regressed.

Many carnivores also make use of their fellows' sense of smell to make their presence known. Male cats and dogs, for instance, urinate at specific places to mark their territory. The pheromones they leave behind (chemical substances that we cannot smell) warn off others of their kind.

Smelling with feelers and feet

In the world of the invertebrates, insects have the most highly developed sense of smell. The basic anatomical structures that allow them to capture smells are modified sensory hairs that form small excrescences on their hard carapace.

These organs, called sensilla, play a major role in the life of the insects. Reproductive behaviour, the formation of colonies and the flight reflex in many species are under the command of the smells emitted by the animals themselves. The sensilla of male moths are thus able to find females within a range of several kilometres in any direction. Those of cockroaches give the flight command as soon

CAN BIRDS SMELL?

It is sometimes said that birds have no sense of smell. This is not strictly true. Black vultures (right), which are New World vultures, have a highly developed sense of smell that allows them to find a dead animal from a long way off. Many seabirds, like petrels, are also able to smell their food – the remains of fish or even microscopic animals – from kilometres away in the open sea, whereas the human nose can detect only the salty smell of the sea. On the other hand, whales and dolphins do not have a sense of smell.

◄ A cheetah marks its territory by spraying urine. It is relying on the sense of smell of its peers and is a way of telling them that the area is occupied.

as a message from one of their number is received. Ants invariably follow the chemical routes traced out by scouts.

Flies also have sensilla, but essentially 'sniff' with their feet, which have sensory hairs at the end that are capable of distinguishing between sugars, salts and other chemical substances.

More than one smell up their sleeve

The sense of smell detects only molecules that are in a gaseous state, but vertebrates can nevertheless analyse them only when these molecules are in solution. In the world of these animals, smelling is therefore very close to tasting, which also analyses substances in solution. Volatile molecules, when they reach the organs allotted to the sense of smell, are made soluble in mucus. The sense cells can then analyse them and pigeonhole them as salty, sweet, acid or bitter.

◄ The albatross is able to smell the remains of fish on the sea from a great distance, and then to find its way unerringly to the food.

These olfactory organs constitute a sort of bag where smells taken in by a nostril accumulate. Within this 'bag', a membrane analyses the molecules. Its surface is permanently covered with a film of mucus produced mainly in the lachrymal glands. This olfactory sac, or nasal cavity, in humans, is very often connected to the back passage, or pharynx, that leads to the throat, by an internal nostril, also called a choana.

From reptiles to birds to mammals, this choana opens progressively deeper in the pharynx. And in humans it opens above the soft palate.

In reptiles, Jacobson's organ, which lies between the olfactory sac and the mouth, enables them to smell what they intend to eat while keeping their mouth closed. To do this, they put out their tongue, which collects molecules from the air that will then be analysed by their receptors. Information obtained in this way lets them know that there may be prey nearby.

▶ The particularly impressive muzzle of a brown bear shows that the sense of smell is of major importance to this animal. Its small ears give away the limitations of its hearing. And as for its vision, this is poor. In fact, many bears are shortsighted.

Sensitive skin

All animals are endowed with touch-sensitive organs that allow them to identify prey or enemies. The way they function is based on the presence of nerve-endings in contact with the skin.

The raccoon, known to North American urbanites as having a talent for scavenging, also has an outstanding talent for fishing. It has a highly developed sense of touch, especially in the region of its nose and front paws, which enables it to catch a large amount of small water animals.

Touch is not a game

From the sponge to the bear, animal organisms are sensitive to the slightest touch. The touch may be interpreted as an attack or, on the contrary, as a gesture of affection. Immobile species like sponges defend themselves when they are touched by pulling back or emitting repellent or even poisonous substances (like sea anemones do). Mobile animals flee or attack.

The sense of touch may, besides, be required for finding food. The implanted hairs round the mouth of the manatee, for example, give it the ability to touch, even before seeing them, the underwater plants it feeds on.

▼ The manatee is very shortsighted, but its nose has hairs that enable it to detect its food even before it has seen it.

The sense of touch is common to all animals and needs little to function, just a few free nerve-endings in direct contact with the skin. The tactile corpuscles of vertebrates and the hair-like (trichoid) sensilla of arthropods (crustaceans, mites, spiders, scorpions and insects), equally have this structure.

Very efficient nerve-endings

In fish, the touch organ consists of a line of receptors running along each side of the body. This lateral line, sensitive to vibrations, allows them to 'touch' without entering into direct contact.

In most land vertebrates, the touch organ is located in the region of the mouth. The tactile receptors of snakes, monitor lizards and chameleons, for example, are on their tongues. In primates, there are additional touch organs, at the tips of the fingers and on the palms of the hands and the soles of the feet.

Touch organs become more complex when combined with the hairs of mammals. Some sensory hairs, stiffened and elongated, transformed into feelers or whiskers, act as a tuning-fork, amplifying the slightest contact.

The unexpected function of the ear

In all vertebrates, especially mammals, the inner ear has touch organs which serve as balance organs. These are the semicircular canals. When the head is turned or the body moves in any direction, the movement of the liquid inside these canals induces the displacement of a membrane, which is measured by sensory hair cells. The utricle and the saccule, two chambers in the labyrinth of the inner ear, contain calcite crystals, or otoliths, and their displacement is also monitored. The brain is thus kept constantly informed of the position of the head in three dimensions, information necessary for keeping the head straight, situating the body in space and moving the body without losing balance.

▲ A herring, above, like all other fish, has a lateral line which functions as a sense organ, evaluating the direction and speed of flowing water.

▲ The whiskers of cats, like the jaguar shown above, are actually tactile organs, having many nerve endings at their base.

LANGUAGE AND COMMUNICATION

Cries, murmurs and whispers

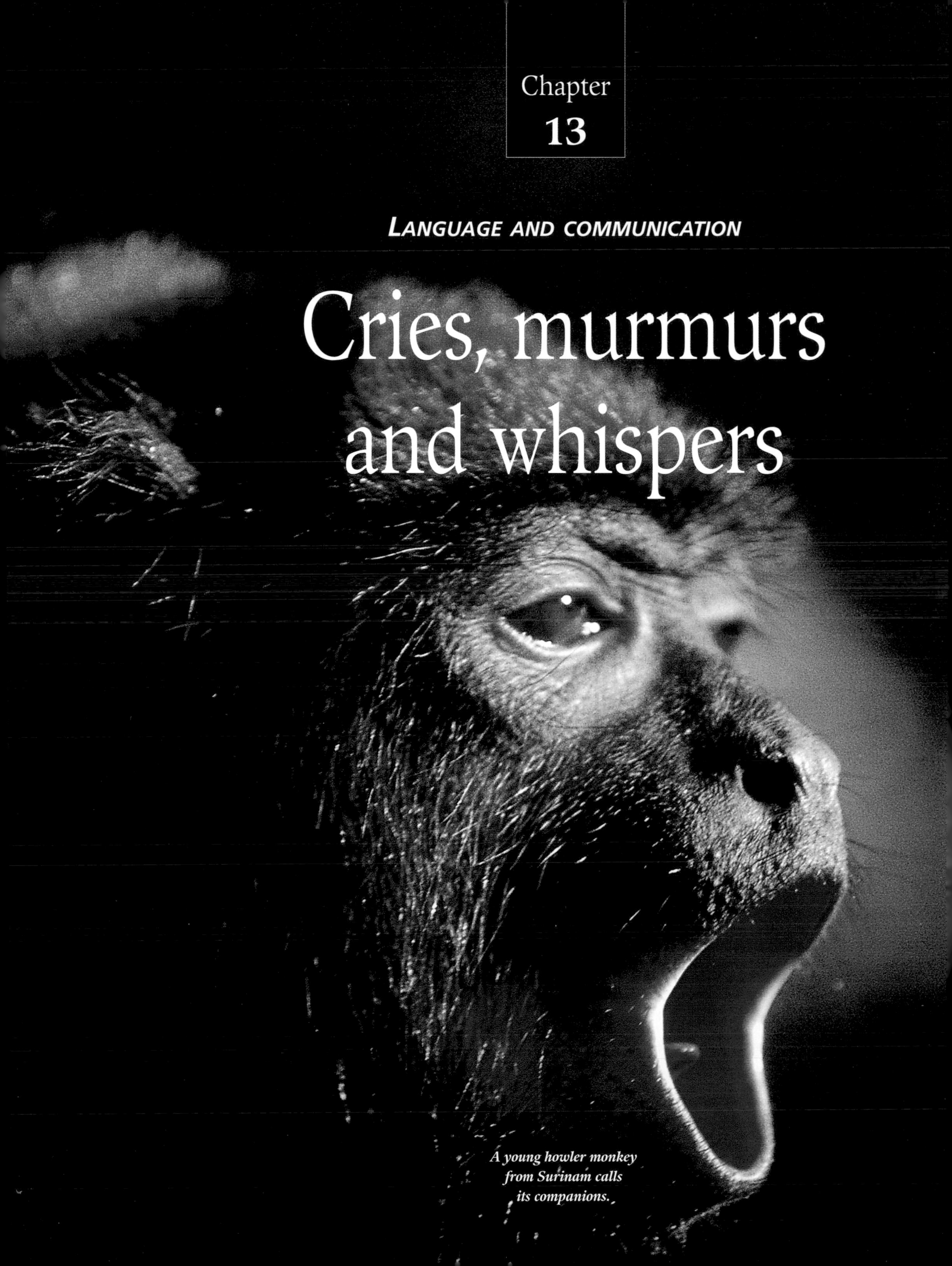

*A young howler monkey
from Surinam calls
its companions.*

Body language

Communication is extremely important for the survival of animals, particularly those living in groups. To pass on information, insects dance or touch antennae and monkeys make sounds, gesture or touch each other.

Inside a hive, a bee twirls round and flies about frantically. At least, that is what it looks like at first sight. But if you were to look closer, you would see that the bee is flying in a figure-of-eight pattern and its abdomen is quivering in a certain rhythm. The bee is doing a dance and repeating figures in the air according to a set choreography.

Bees dance in the hive to pass on information to their community about the place where they have found a new supply of flowers, or pollen. Their movements observe a code. Their 'dancing' is a language.

The language of bees

When the bee describes a figure of eight on a ray of sunlight, the angle between the central line of the figure and the vertical axis of the light ray indicates the direction of the food source from the nest. The faster its abdomen quivers, the nearer the supply of flowers. And the faster the dance, the more plentiful the food.

This amazing means of communication has regional variations, like the dialects in human languages. Thus, to bees in the south of Europe, the same quivering speed indicates a shorter distance than it does to bees in the north. The nectar-seekers of the south need not travel as far as those in the north, where there are generally fewer flowers.

Communicating with antennae

Insects feel and taste with their antennae. Their tapping movements are done in order to detect molecules. It is in this way that ants, for example, decode the scents emitted by one or other of their

◀ The grimace of a male orang-utan is quite frightening to a human observer. But it is the orang-utan that is scared. This expression denotes fear rather than aggression.

TALKING MONKEYS?

Between 1966 and 1971, a husband-and-wife team, the Gardners, taught a chimpanzee called Washoe nearly 150 words in sign language. So it is possible to have a simple conversation with these animals. It has also been noticed that monkeys that had assimilated a certain number of sounds foreign to their language were capable of transmitting them to their fellows.

number in case of danger – an alarm pheromone in the atmosphere – or to indicate a food supply – by making a scent trail on the ground where they regularly deposit a pheromone with their abdomen. So it is possible to speak of an 'antennary language'. An ant deprived of its antennae is incapable of communicating with its fellow ants.

Insects also use their antennae to ask to be fed and to exchange food. It is through these incessant exchanges that many chemical substances circulate in the group, which ensure the order of the society, notably by regulating the formation of the different social castes. The secretion by the queen bee of a certain molecule, for instance, which is diffused throughout the nest by food, stops the workers from laying eggs and prevents new queens from being bred. When the queen dies, the source of this molecule dries up, and the behaviour of the workers changes as they prepare to breed new queens. Similarly, when soldier termites become too numerous, they secrete a very concentrated substance that prevents new soldiers from being born. Conversely, if this same substance is present in a weak concentration, after significant losses during an ant attack, for instance, this favours the birth of more soldier termites in compensation.

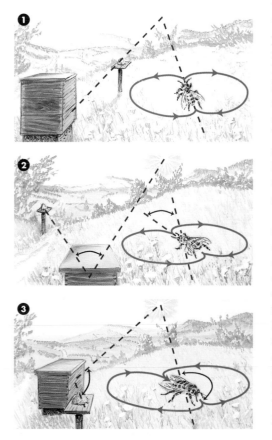

Passing regurgitated food from mouth to mouth, called trophallaxis, is common among social insects such as red ants.

Chimpanzees shout and pull faces

Nobody today would be very surprised to hear that the behaviour of chimpanzees is similar to that of humans. From the time that it is very young, our nearest relative in the animal world faithfully follows its parents, especially its mother. It expresses its joy, fear and needs like a young human, attracting the mother's attention by tapping her or with cries, to which the female replies. When she refuses to comply with a request and the offspring becomes too insistent, the mother begins to shout and does not hesitate to issue a slap.

Monkeys, notably the great apes (chimpanzees, gorillas and orang-utans) thus have at their disposal many different means of communication. Take facial expression, for example. Their faces become very eloquent and each expression has a meaning. When it is preparing to attack another chimpanzee, or in any other situation causing an aggressive state, a chimpanzee remains silent and keeps its lips together. When it smiles broadly, with its mouth open, it is afraid or excited. This smile is often accompanied by piercing cries. But a smile with the mouth closed shows a less

If the food source detected is between the hive and the Sun, the bee will dance upwards. If it is on the opposite side, it dances downwards.

► **The dance of the bees.**

1. The food source lies on the axis of the hive and the Sun: the dancing bee makes a vertical figure of eight by flying up towards the Sun.

2. The food source is 30° to the left of the axis of the hive and the Sun. The bee inclines the plane of its dance to the same angle in the direction of the source.

3. The source is 120° to the right in relation to the Sun. The bee draws the middle line of the 8 at the same angle.

➊

➋

➌

▲ *Calling*

▲ *Fear, excitement*

▲ *Aggression*

frightened animal. Chimpanzees do not utter cries indiscriminately. They make sounds according to set codes.

For example, faced with a new supply of food, a chimpanzee informs the others of its discovery and invites them to come and join in. Its mouth forms a closed O to make a succession of panting 'hoo-hoo' sounds. Then it opens its mouth wider and makes a sort of 'ooa' sound that is the last phrase of the call. This is called a panting hoot.

On other occasions, for instance at night or when they are on the move, chimpanzees use voice communication for the purpose of maintaining contact with their fellows.

Bonobos and free love

Communication between individuals may take a form that is much freer. In another species of chimpanzee, the bonobo, sex plays a major role in social relationships. These animals are polygamous (one male for several females) and polyandrous (one female for several males). Masturbation, fellatio, incest and homosexuality are all common practice among this species. Their function is to calm aggressive tendencies and

tighten the links between individuals and cement the group. For the young, it is doubtless also a question of initiation into adult life.

In a large number of species of polygamous apes, the heat, or oestrus, of a female lasts several weeks with very visible manifestations of red ano-genital swellings. In the males of certain species, such as macaques, the testicles are large and bright blue in colour, and the penis is red. These very pronounced visual messages, in addition to great sexual dimorphism (males are larger in size), exacerbate sexuality.

The rather ostentatious ovulation period of the females incites the males to unusual boldness and even to strategic alliances. A male may rope in two companions to keep off a rival who is trying to establish himself as a favourite, while he gives himself over to the joys of reproduction with the female he desires.

◀ *The great apes (top and bottom: baboon; centre: chimpanzee) possess a large range of facial expressions, some of which are like those of humans, although they have very different meanings.*

▼ *The adult and the young bonobo below have recourse to pacifying gestures that allow them to communicate without violent clashes.*

Chemical messages

Animals and plants emit and perceive strange chemical substances that determine a large part of the communication within their species. There is a whole world of relationships out there that cannot be detected by the nose of humans.

In a large game reserve in South Africa, it is the rutting season among the rhinoceros. The males are quite excited and rather aggressive. The females are on heat. There is no question of disturbing them. But what can possibly set off this turmoil of the senses? It is not the return of spring or the sudden growth of the grass. It is simply a sort of internal clock that these animals – and others – possess, which leads them to secrete chemical substances that are called pheromones.

Stimuli vital to life

Animals receive information from external chemical stimuli, through the intermediary of their own chemical receptors. These 'detectors' respond specifically to chemical molecules in their environment.

Of course, each detector can cause different reactions according to the situation. We therefore salivate when we smell the delicious aroma of chocolate wafting from the kitchen and retch when we scent the odour of decomposing matter. Animals react in the same way.

The chemical signals known as pheromones are a form of communication between individuals of the same species, triggering actions to do with relationships. Pheromones are in fact small molecules – peptides – the production of which requires only a very little energy. Just a few of these molecules, which means very small doses, are sufficient to trigger specific behaviour immediately (the best known have a mainly sexual function), or a series of physiological changes (often serving a social function, and still not properly understood).

They are all around us

Pheromones are never alone: they act in bunches made of several molecules. Chemists have only recently been able to describe them, for they are produced in very weak concentrations. They have been found to be released by every organism, from bacteria to the rhesus monkey, fungi, algae, trees, fish and especially insects. Most of these pheromones facilitate the coming together of male and female sex cells, notably in fungi and algae, and perhaps also in animals.

Pheromones also play a major part in the physical act of mating between males and females. This role may be direct (allowing a female insect to attract a male insect from a distance, for

▲ *At the approach of a female, a stallion first sniffs the air, then assumes a specific attitude, the flehmen. This 'grimace' is thought to be a way of activating the pheromone receptors located in its nose.*

▼ *Like most of the so-called higher animals, the rhinoceros mates under the influence of hormones, but it also does so partly under that of pheromones.*

▲ *Cockroaches group together when they release a 'scent beacon'.*

example), or indirect (it is thought that in many cases mating rituals are precipitated when the males receive pheromones emitted by the females.

A tree that scents danger

Pheromones also play a role in marking (territory or paths), indicating social status, aggregation (by allowing the gathering of individuals) or alarm. This last function has recently been shown to exist in trees and shrubs. It has been demonstrated that an acacia suffering from the 'stress' of a kudu's tongue excreted a pheromone that, when it was received by other acacias nearby, could trigger in them the synthesis of molecules that would making their leaves inedible.

▲ *By emitting a pheromone the queen bee dominates her workers.*

In other plants, defence molecules produced in this way allow them to withstand, for a time, attack by insects.

Are insects giving industry a lead?

The agricultural industry is attempting, although without much success so far, to synthesise defence molecules produced by plants so that farmers may limit the use of insecticides. However, better results have

▲ *Ants exchange pheromonal information from one antenna to another.*

been obtained with biological control. By putting the female sex pheromones of those insects responsible for damage in orchards inside a trap, it is now possible to capture the males, and thus partially to control the reproduction of these unwelome creatures.

Another method is to saturate the sense receptors of these insects with a very large quantity of sex pheromones, with the aim of confusing their senses and thus preventing them from mating and reproducing.

Aggregation pheromones

Pheromones play a major role in the cohesion of animal societies, especially in insects. The prime factor in the gathering together of the same species is a chemical stimulus, a specific scent that allows an individual to recognise another of the same group.

Among cockroaches it has been shown that the extract of a pheromone has as great a power of attraction as a live individual, and that, in a maze,

◄ *In the female bonobo, sexual arousal does not go unnoticed. Her vulva becomes enormous and she emits quantities of sex pheromones that will lead the males by the nose.*

a cockroach will find its way with the help of a 'scent' emitted by the pheromone, which has obviously triggered a behavioural response in the cockroach receiving the scent.

▲ The antennae of moths are like chemical sponges. They capture the smallest molecule emitted hundreds of metres away and are also sensitive to the slightest vibration.

A redoubtable remedy for competition

But in other cases, pheromones are used to reduce competition. In bees, the queen secretes a pheromone that aims to prevent the development of the ovaries in the workers, to stop them breeding new queens, which also stops them from building new royal cells in the hive. This pheromone, produced in the region of the mandibular glands, will later allow the workers to surround the queen as a team when they leave the hive and swarm. This shows the multiple functions that the same chemical substance may have.

Ants also use pheromones, either to trace out a path that leads to a food source, or as a call to

◀ The queen mole-rat can inhibit the production of reproductive hormones in her subjects by using pheromones she gives off in her urine. This feature is rare in mammals.

◀ In the inky depths of the sea, female angler fish attract the males by pheromones.

arms when an individual is attacked. In addition, pheromones serve to recruit new workers.

The secret weapon of the male

In the world of mammals, urine is of capital importance as a signal because it contains pheromones.

The urine of male mice contains molecules that hasten puberty in juvenile females. Other male pheromones synchronise ovulation in sexually mature females. But as soon as the dominant male is absent or has been chased away, his pheromones no longer act on the females in the group. The females he has impregnated may even miscarry as soon as they detect the presence of the urine of another male. Still under the influence of pheromones, they will immediately ovulate, in order to be ready to produce descendants of the new dominant male.

LOVE AT FIRST SCENT?

Humans can see very well but have a poor sense of smell, especially as the various fragrances with which we spray ourselves mask our natural odours. It is therefore difficult to believe that we emit pheromones. But at the University of Philadelphia, about 30 years ago, scientists amused themselves by impregnating the seat of a chair in a waiting-room with a man's sweat. Some 80 percent of the women they then placed in the waiting-room chose to sit on the impregnated chair or the chairs next to it, whereas 80 percent of the men sat a long way away from it. Yet the smell of sweat was apparently undetectable. Pheromones – sex pheromones in this case – probably exist in humans. But they have not yet been isolated, as they are certainly produced in very small amounts. A growing number of biochemists are nevertheless advancing the hypothesis that love at first sight is partly due to pheromones. Others even think that infidelity may be due to reduced receptivity to the sex pheromones of the spouse.

Noise can be a calling card

Sound is the simplest means of communication, and is therefore the most widespread language in the animal world. To communicate, it is enough just to create vibrations in the air, which is possible even without a mouth or vocal cords.

The wolves are attacking. Under the authority of their leader, they harass an old reindeer. Wary of the hoofs and horns of their victim, they try to tire it out. The scramble will come when the exhausted reindeer stops fighting.

From the start of the hunt to the finish (throughout the chase, the kill and finally the division of the spoils), the members of the pack will have co-ordinated their actions by howling. They also will have kept other wolves in the group, who may be prowling nearby, informed of the progress of the hunt by modulating their cries, in the same way that hunters sound the horn to let all the other participants in the hunt know what is happening.

▲ By howling, wolves communicate within their pack, but also inform other packs of their activities.

Alarm signals

Warning is the primary aim of communication by sound. Like most mammals, when it is frightened the Burchell's zebra emits alarm cries. As soon as one of the herd has picked up the scent of a predator and then located it by sight, it whinnies or snorts noisily. The whole herd gathers and flees from the danger.

Faced with a similar situation, wild horses neigh in a fashion their fellows understand. In the same way, marmots are known for their squeaking. Each clan has its watchman. As soon as an intruder comes nearer to the group than is considered safe, the sentinel utters a very sharp cry that makes the whole group scramble for the protection of their burrows.

In some animals there are different alarm cries corresponding to different kinds of danger. In vervet monkeys, young babies may give voice as soon as they see a harmless warthog, for instance. After the age of a year, the vervet also cries out when there are other mammals about, or snakes and birds that worry it. It is only when it reaches adulthood that the vervet develops a howl that indicates to its fellows the presence of a leopard, its chief predator.

▲ Among birds like this male bluethroat, singing serves to claim territory and to attract one or more partners.

Master singers

In the animal world, birds have no doubt pushed the language of sound to the peak of achievement. Indeed, there is nothing more beautiful than the voice of a song thrush or a nightingale in the middle of the night. The rich phrasing and the pure notes have inspired writers and musicians through the ages. Yet the song, besides its beauty, has a very simple aim: to announce the presence of the bird.

In most birds, it is generally enough for the female to utter the cries particular to the species and it is only the male that sings. This is especially true during the breeding season, and the song sends two messages. The first message is that the male is searching for a future mate. The male shows off the entire tuneful repertoire at his disposal. This is because there are good and bad singers within the same species,

◀ The male elephant seal utters low-pitched cries in order to gather his females round him and to prevent any rival intruding into his harem.

▲ *A monkey (above, a howler monkey), a camel and a zebra: all mammals communicate through sounds. The breeding season is notably the time for songs or calls particular to each species.*

Although the males are generally the only ones to sing, there are some bird species in which the females also give voice. The result is a duet. This occurs among birds such as the European eagle owl, the African mousebird (a bird which is related to the cuckoo) and also the barbets that are found in Asia and tropical Africa.

Cries from the heart

Sound language goes together with mating displays, and not only in birds. Among insects and amphibians, males also make themselves heard at mating time. As soon as night falls, the concert of frogs and toads is heard in all the ponds and wetlands across the world. Their harsh and seemingly tuneless croaking is just a love song.

Among mammals it is the same. In the forests of northern Europe during September and October, the autumn months, the male elk can be heard uttering low groans. The female he desires replies by neighing back. He will then follow her the whole night, bellowing.

During the same season stags are just as loud. They bell, or roar, to impress each other and attract females. These tactics are deployed if possible on cold, clear nights, when their powerful cries will resound over long distances.

and the better he sings, the greater the chances of his success.

The second message is that this is the male bird's territory, which means saying to the neighbouring males: 'This is my turf, so keep away!' This song is the most sophisticated form of communication in birds during the all-important breeding period.

Birds' words

Outside the breeding season, calls replace songs. These are generally simple, but they are also diverse. They may be sounded to warn others when there is danger (as in alarm calls) or at the time of migration, to maintain contact and cohesion among the flock (as in contact calls). Furthermore, young birds possess their own cries, notably when they call for food.

▲ *The cicada makes its song by rubbing its elytra, or wing shields, together.*

▼ *In spring, the male frog inflates its vocal sacs in order to make the well-known sound that comes from the edge of ponds.*

that opens the concert with a strident cry. Her excited suitor will howl interminably until the affair is over.

During the rutting display, the male camel too performs its own sort of song. It makes a gargling sound by exhaling forcefully and thus inflating its soft palate, which it then pushes outside its mouth. This is not a particularly elegant performance, but it seems to indicate to other male camels who is who.

Whale song...

It is well known that whales sing under the water. The human ear is able to hear snatches of the song, but special equipment (called a hydrophone) is needed to hear the whole message.

As in birds, whale song is used by males at the time of the mating display, but also within the group to keep it together. And, as in birds, groups or pods of whales may have their own dialect.

▲ *This pair of jackals bonds by nibbling each other, rubbing against each other and making 'contact calls'.*

...and monkey cries

In primates the use of territory is proclaimed by vocalisations. Each social group announces the occupation of its territory in this manner.

In the tropical forests, monkeys utter cries from quite high in the trees in the morning. This is true of the gibbons of Asia, for example. Their voices sound repeatedly for a long time at a frequency rarely exceeding one kilohertz, so that they are heard through the foliage and their sound cuts through the continuous chirring of insects in the forest.

But it is the howler monkeys that utter the loudest cries. Because of a sort of echo chamber in their throats, their voices carry for more than a kilometre. Savanna-dwelling monkeys like baboons also utter alarm cries that are at once very stereotyped in structure and very specific. In most species, it is only the males that are vocal in this way.

▲ *The male gibbon has a pouch that fills with air and enables him to produce a sound which carries a long way and is typical of Asia's tropical forests.*

Similarly, but earlier in the year (in July and August), during its rutting season, the roebuck announces himself with a bark like that of a dog. Normally, however, both the stag and the roebuck are rather quiet creatures.

Unlike them, the black-backed jackal is one of the noisiest animals in Africa. It howls, barks, cackles and snorts at any time, according to how it feels. During the mating season it is the female

ULTRA-ACUTE HEARING

When an adult emperor penguin, which nests on the ice in Antarctica, returns to feed its chick among the colony, where there are thousands of others, it only has to utter a cry and the young bird replies at once, despite the surrounding hubbub. This means that each individual bird is capable of uttering a distinctive cry that the young will recognise. To human ears, the cries are identical, but penguins are able to distinguish minute shifts in frequency, which makes all the difference.

Lights, please

For reasons which are not fully understood, many animals have the capacity to make themselves luminous by a chemical reaction.

▶ *At night, squid communicate among themselves by means of light signals.*

▶ *Most animals that live deep under the sea emit light spasmodically (or react to the slightest photon), whether it is to hunt, frighten off a predator or to communicate.*

The male *Photinus puralis*, a firefly, sits on a branch. His abdomen is turned to the horizon, and it glows with a green light for a moment. Exactly three seconds later, a similar flash is seen in the distance. He flies off at once in the direction indicated by this exciting signal, surely emitted by a receptive female. He soon settles again, contracts his abdomen again and sends another signal. And the game of lights goes on until the eager couple find each other and possibly engage in a sexual union.

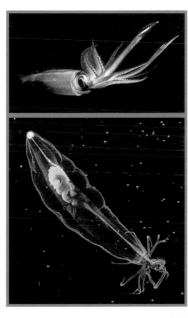

Each according to his light

A surprising number of living organisms give off light. Among them are bacteria, fungi and many animals, like protozoa, dinoflagellates (the micro-organisms that look like shining dots in the sea at night as the water moves), cnidarians (jellyfish, for example), mud worms, deep-sea squid, krill (small crustaceans filtered from the water by baleen whales), glow-worms and many fish.

The manufacture of light in animals, called bioluminescence, is due to a chemical reaction between two proteins, luciferine and luciferase. When oxygen is present they emit energy in the form of photons, the little particles that make up light. The synthesis of the two proteins is carried out by individual animals themselves or by bacteria living symbiotically with them (in lantern fish, for example).

▶ *Fireflies give off light to attract a sexual partner. They have alarm systems which interrupt the light emission as soon as a predator approaches.*

The most surprising thing is that luminous crustaceans, such as krill, cephalopods and phosphorescent fish, have developed very similar light-emitting structures and the way these function is not totally unlike that of the eyes of vertebrates.

Adapting to oxygen?

Photogenesis, or the capacity to emit light, occurs so widely in animals and plants that it must have some kind of use. But this facility does not seem indispensable to their survival.

The signalling function of the phenomenon is obvious in fireflies and some cephalopods. Light emission in some squid serves to help them get away. And a sort of fishing lure glows above the head of deep-sea angler fish, perhaps to attract prey.

But almost all other phosphorescent species emit light for reasons unknown. This feature might be left over from the past. When oxygen appeared on Earth, bacteria, which were at the time the only living organisms on the planet, had to adapt to this molecule, which may have been dangerous to their DNA. Some may have developed chemical reactions that enabled them to use the oxygen and at the same time stop it from entering their cells. Photogenesis could have been one such reaction, which evolution has retained because it was harmless.

Without communication, there is no social life

All animals are social beings. And sociability only exists because information, be it visual, auditory, tactile or chemical, is transmitted among the members of a group. It is the continual interaction of this varied information that ensures the cohesion of an animal society and the continuing existence of a population.

◀ COMMUNICATING CATERPILLARS

Communication is not the sole prerogative of the most evolved creatures. The pine processionary caterpillars have a sort of Ariadne's thread that enables individuals to communicate with each other. When, for reasons so far unexplained, a group of caterpillars decides to leave the top of the tree where they normally live, each of them secretes a silken thread, so that in the end there is a long string that unfolds. The move takes place in single file, led by a single caterpillar at the head of the line. Besides maintaining constant tactile contact (through the antennae) and chemical contact (by pheromones), processionary caterpillars thus also have a visual form of communication: the silk thread. Once the walk is over, the first in line has the further responsibility of leading the colony back to where it came from, by following the thread it wove on the way out. When it loses the thread, the lead caterpillar is sometimes obliged to make large circles to find the way. If, as an experiment, you were to remove the lead caterpillar and deprive the group of its guide, you would find that the caterpillar next in line would take over.

▼ VISUAL GRAMMAR?

The 'coat' of a cuttlefish is a stylist's dream. Its skin is studded with papillae that model it according to its needs. In a fraction of a second it can go from being as rough as rock to as silky as seaweed. In certain species, during the mating displays the males make their bodies look larger by adding stripes. This colour magician intrigues scientists, who wonder whether the skin does not serve as a means of communicating according to a subtle visual grammar. But nobody has yet found an explanation. The cuttlefish, capable of memorising and learning, has very complex behaviour. It is one of the most efficient hunters in the underwater world, and its eyes and brain are in some ways like ours. So why should it not be able to communicate? The sudden colour changes of the octopus and the squid, near relatives of the cuttlefish, could be a similarly interesting field of study.

▲ COLOUR SCHEMES

The fish that live in coral reefs are known for their bright colours. These are displayed essentially by small-sized species, which are hunted rather than being hunters, and function primarily as a deterrent or as camouflage. Predators often associate vividly contrasting colours with the presence of toxins, so are more likely to leave the bright little fish alone. Stripes allow the small fish to blend in with the corals and seaweed and so hide from hunters. And false eyes help to deceive the enemy. As predators always attack where the eyes are (they are easy to see), it is in the interest of the prey to have false eyes near the tail. Once the attacker has realised its mistake the prey has already had time to escape. Colours also allow members of the same species to identify one another. And, during the breeding season, competition between males is expressed in an orgy of colour meant to impress the females.

▼ THE INFORMATION CENTRE

Starlings are gregarious birds. At nightfall in winter they gather in large noisy dormitories, often in trees or on lampposts or telephone wires in a built-up area (there are fewer predators in towns). Scientists have shown that their continuous chattering serves chiefly as a means of sharing news and information. It is in these great gatherings that birds which have found an abundant food source tell the others about it. So the next morning they will all go to the place where the discovery was made.

Kittiwakes, which live in vast colonies on cliff faces, do the same in the breeding season, when they need to find extra food for their chicks.

▲ WHALE DIALECTS

Not only do humpback whales communicate by songs and calls, but they do not all speak the same language. The humpback whale, also called a megaptera, is a whale known for its high leaps. Only the males sing, during the winter breeding season. The song is so powerful that it sows havoc when it reaches the listening devices of nuclear submarines, which perceive it from more than 180 kilometres away. This low-frequency, far-carrying song identifies each individual by sound. It is partly innate and partly acquired, and may change from one week to the next. It is composed of six distinct basic themes made of sentences which can be broken down into syllables (each sentence is a series of *chip* and *yup* sounds). Individuals belonging to the same population are thought to speak a common dialect, different from that of other groups.

▼ WHAT ABOUT TALKING WITH OUR HANDS?

The great apes (orang-utans, chimpanzees and gorillas) have developed a complex system of communication. Among chimpanzees, this communication is not only by sounds but also with gestures. To make peace between themselves or to reassure each other, two chimpanzees will embrace and become calm again. In the same way, the grooming ritual is preceded by gestures that allow each individual to identify itself. A dominated animal will present its head in order to receive a tap in greeting which guarantees it against any aggressive behaviour from its grooming partner.

A tree which talks to others

An inability to move would seem to prevent plants from communicating among themselves. And yet they do exchange information, by way of chemical messages.

▲ *Kudu are antelope that live in small herds in wooded areas; the head of the male bears fine twisted horns*

In the grasslands of north-east South Africa, near the Kruger Park, many game ranches shelter a diversity of animal life, including kudus, which are large antelopes. When grass is scarce in winter, the kudus graze on acacia leaves… and risk their lives, for this change of diet can cause their death. The acacia tree, to protect itself, synthesises a molecule which is highly toxic to the antelopes and destroys their livers. And the astonishing thing is that trees on which the kudus have not yet set their sights also begin to produce the molecule. Somehow they have been warned of the danger and are taking preventive action.

The attack of the kudus

Anchored to the ground, a plant has few means of defence. Thorns are its best protection and poison is the only thing that enables it to fight back. In this domain, plant life shows unequalled inventiveness. The game-reserve acacia trees use tannin to poison the kudus. The antiseptic properties of this compound are well known. Also extracted from the bark of oak trees, tannin has been used since time immemorial for the preservation of skins, a process to which it has lent the name 'tanning'.

Acacias produce tannin when herbivores come to graze on them, but also when they receive a chemical message from one of their neighbours in distress. This message is transmitted through the air. By studying the tannin content of the trees surrounding a grazed acacia, botanists have observed that trees that are downwind and therefore directly exposed to the action of molecules in the air react more quickly and produce a greater concentration of tannin than the others do.

The response of the acacias

Normally, the foliage of the acacia contains little or no tannin, which is a very complex compound that requires resources of energy to be produced. The action of a herbivore browsing on an acacia triggers the production of the poison, but it takes some time for this to reach an effective level of concentration.

However, this poison is of use to the plant only if the attack can be anticipated and sufficient quantities of tannin produced just before the browsing begins. The antelope would not be so eager to eat leaves that taste different and would be poisoned sooner. By picking up the chemical message sent by another tree that is under attack the acacia gains an advantage over the browsers.

A tree only 'warns' its neighbours when it is attacked repeatedly, so that tannin, costly to produce, is not used needlessly in false alarms. But as soon as the message is received, only between five and 10 minutes are needed for the concentration of tannin in the neighbouring trees to rise to 15 percent. No animal organism can withstand this dose, and will die.

ETHYLENE AND FRUIT

Ethylene is well known for its role in making fruit ripen. Just one overripe or rotten banana triggers the ripening process of a whole bunch by the ethylene it emits. So a piece of fruit that has gone bad should always be removed from the fruit bowl so that the rest do not ripen too quickly.

▲ *The giraffe has no competition for resources. Its very long neck makes it the only herbivore able to reach the high branches of trees.*

The acacias warn each other of the danger by emitting ethylene. This is a fairly simple hydrocarbon gas (containing one carbon molecule and two hydrogen molecules) familiar to chemists and easy for plants to synthesise with water, carbon dioxide and light.

Plant communication (either within a species or between species), and even communication between plants and animals, is a new field of scientific research only feasible with very complex equipment. Today there is proof that several species of trees, and also plants like maize, emit molecules when they are under threat.

Communicating freely

Kudu which are raised in ranches, contained by fences, cannot move about freely, and a scarcity of food means that they will rely too heavily on acacias and end up being poisoned. In this situation, it is easy for the acacias to defend themselves.

The kudu in the neighbouring Kruger Park, on the other hand, live relatively freely in this much larger area, are often on the move and therefore do not have to feed on several trees in the same place. They will not graze a tree downwind from the one off which they have just eaten the leaves. So they avoid the risk of tannin poisoning.

In natural conditions, communication between neighbouring trees appears not to entail poisoning the herbivores, but to encourage them to graze over a wider area. Food supplies are distributed better, to the greater good of both trees and animals. It is captivity that upsets the balance and, by preventing the kudu from moving away from areas already grazed, brings about their death. Ranchers have since learned to feed alfalfa to their kudu herds to reduce the grazing of acacias.

▶ *The bushveld is a vast grassland with scattered shrubs and acacia trees, on which antelope feed during the long dry season.*

Private codes

Certain animals emit and receive very low or very high sounds that are imperceptible to the human ear. They use this ability mainly to move about safely and to obtain food.

◀ *Elephants communicate by emitting low-frequency sounds. In this way they can exchange information over long distances without being able to either see or smell one another.*

A group of elephants is moving around on the banks of a pool. The young are drinking and the adults are on the watch. Suddenly, one of them raises its head and freezes. With its ears laid back, it moves first one front leg and then the other. At each of its movements it changes position slightly. Others soon imitate it. They are doing this because another herd of elephants has just started to move, about 20 kilometres away. The elephants at the pond have heard them approaching, through the vibrations they are transmitting to the ground.

Infrasound communication on land

So an elephant, by walking, is able to send infrasound messages that, according to recent studies, may travel unaltered for more than 30 kilometres or so.

By shifting their weight alternately onto each of their front legs, the elephants at the pond are acting like we do when we turn our heads first one way and then the other in order to determine where an indistinct sound is coming from. The vibrations (infrasounds) in the ground are captured by the elephants' feet, travel up through the bones, are amplified by bones in the head and are finally captured by the ossicles in the middle ear, which are highly developed in the elephant.

Elephants thus perceive infrasounds, which have very low frequencies that are inaudible to humans. For about 20 years we have also known that elephants emit these infrasounds through their nasal fossae. Of very low frequency, between 14 and 35 hertz, these waves have low energy but carry a long way. They enable elephants to inform their fellows that there is water, or that danger is near. Through them, also, females in oestrus notify solitary males who may be roaming long distances away. The sexual message is furthermore confirmed by the emission of pheromones as the males approach.

This sensitivity to infrasounds, at the same time telluric (through the ground) and aerial (by the nasal fossae), would explain how elephants can make their way unhesitatingly towards a stormy area, since thunder generates waves in the atmosphere. Once they are facing in the right direction, their trunks tell them whether or not there is pollen emitted by plants on which rain has just fallen. The area they are aiming for is sometimes more than 150 kilometres away from their starting point.

Infrasound communication in water

Insects, spiders, scorpions, amphibians, reptiles, some rodents and other mammals of a much more respectable size send out and perceive infrasound. By contracting the muscles of its gigantic throat, the hippopotamus is able to make its skin vibrate like a drum. In this way it emits low-frequency messages that will very quickly be

▲ *The alligator emits infrasounds by contracting its throat muscles.*

◀ *The hippopotamus seems to be laughing or yawning when it opens its mouth wide. Under the water, it sends out infrasounds by means of its nostrils and by making the skin of its throat vibrate.*

▲ Dolphins can knock out small prey by attacking them with ultrasound. This 'sound laser' even sometimes bursts the swim bladder of certain fish.

melon, a sort of fatty lump on the front of their skull. The ultrasound emitted by the melon propagates in the water over a long distance and allows the dolphin to locate prey and obstacles. These send back an echo to the dolphin, which receives it in its lower jaw, the back part of which is in contact with the inner ear. The captured echo is then immediately analysed by the brain.

▼ Bats started using radar a long time before humans did. Like whales, they find their way by emitting very high-frequency sounds and analysing the direction of the echoes.

transmitted a long distance by the water of the pool it is in.

Crocodiles are thought to be capable of the same feat, and they are certainly very sensitive to infrasound. Florida alligators – especially rutting males – are known to be very perturbed by the infrasound produced when the space shuttles take off from Cape Canaveral.

Whales, too, emit infrasound, which is not particularly surprising, since low-frequency waves are transmitted much better by water than by air. Pods of whales are able to communicate perfectly even over several hundred kilometres.

Ultrasounds that kill

Sea mammals also send out high-frequency sounds, known as ultrasound. Dolphins, for instance, generally produce two types of high-frequency sounds. The first are half-second whistles, on wave frequencies between 7 and 15 kilohertz. These emissions of sound are audible to man and therefore are not strictly ultrasonic. They allow conversations at any time so they are like a sort of language.

The second type consists of clicks, and is made by all cetaceans. This is true ultrasound, since the waves occur from five to several hundred times a second on frequencies between 20 and 250 kilohertz. Cetaceans use them to find prey and obstacles by echo sounding, and also to kill or knock out their victims. This is the treatment meted out to shoals of cod on which orca whales feed in winter.

A clever melon

To emit these killer ultrasounds, dolphins have an organ with the inappropriately gentle name of

I hear, therefore I see

Bats also practise echolocation. Their noses direct the ultrasound they emit wherever they choose, and the sound waves propagate in the air until they meet an obstacle or prey. The echo then sent back is captured by the bats' ears which have a well-developed flap. From the changes that the waves undergo following their encounter with a foreign body, bats can work out the distance that separates them from it, and also identify it.

Bats generally emit a series of sound waves, the number varying from 20 to 80 a second. Inaudible to the human ear, they are on the other hand very powerful compared with the weak echo sent back by the insects on which these winged mammals feed in the air. This technique is so effective that a bat is able to find its way in complete darkness through a room criss-crossed with wires without touching a single one.

MYSTERIOUS BEACHINGS

Cases of the beaching of numbers of whales along coasts are frequently reported in the press. The globicephala, or pilot whale, a good-sized gregarious whale, is particularly susceptible to this sort of accident. How is it that these animals with such a highly developed system of echo sounding can go astray in this fashion? According to some hypotheses, schools may follow their leader who is sick and unable to navigate properly underwater. Others suggest that electrical interference on the coast may disturb the sonar system of the whales. The truth of the matter is that we do not yet know the cause of these mysterious cases of beaching.

NATURE THAT PROTECTS AND HEALS

The healing power of nature

The leaves of the ginkgo, golden yellow in autumn, are known for their anti-inflammatory properties.

The fresh air on Earth

The Earth is protected by a fine layer of gases spread evenly around the globe. This air that we breathe, itself produced by organisms living on the planet, is essential to our existence.

Confined inside their space capsule, astronauts know that they will only be able to leave their ship with protective equipment. Without a space suit and a good supply of air they would face certain death. Human beings have become so used to the Earth's atmosphere that today we could not survive without it. Yet immediately after the formation of our planet, the atmosphere was not a healthy one.

The right formula

The primitive atmosphere was a mixture, among other things, of carbon dioxide, nitrogen, water vapour and a little methane, and it was not really pleasant. It was also lacking an ingredient that eventually was going to reshuffle the cards: oxygen.

Oxygen came onto the scene a short time after the beginnings of the Earth, when various upheavals caused the disappearance of a large proportion of the carbon dioxide and the first life forms appeared, in the safe shelter of the oceans. Oxygen, such a determining factor in our survival, was in fact generated by these first living organisms. At first it was toxic to organic matter, and living things had to learn to isolate it before using it for their own development. Today the atmosphere is made up largely of nitrogen (77 percent), oxygen (21 percent) and water (1 percent).

Ozone, the protector of life

Shortly after oxygen appeared, there was another newcomer: ozone. This oxygen compound plays a major role in protecting life. In fact it absorbs some of the ultraviolet radiation emitted by the Sun. These extremely harmful rays would prevent the development of any living organisms if they were allowed to strike Earth at full force.

Ozone is concentrated in a layer of atmosphere 25 kilometres above the Earth's surface. But human industrial activity and the gases emitted by cars and aeroplanes have tended to diminish the

quantities of ozone in the atmosphere, in 'holes' over the polar regions. It is more than ever necessary to keep an eye on this precious protective layer.

Between the ground and interplanetary space

Only 80 kilometres thick, the atmosphere seems very thin compared with the 6 300 kilometres that separate the surface of the Earth from its core. Interplanetary space is continuous and has no real borders. The aurora borealis, for example, which is mostly a phenomenon of the atmosphere (the interaction between charged particles from the solar wind and certain atoms from our atmosphere, like nitrogen and oxygen), takes place at an altitude of up to 400 kilometres.

In the atmosphere, several layers can be distinguished, differentiated by their temperature. These are, from the closest to the Earth to the farthest away, the troposphere, the stratosphere, the mesosphere and the thermosphere. The temperature is lower in the first two layers than in the last two. Large variations in temperature are the cause of disturbances that continuously stir

▼ *Weather balloons (the one shown below is about to take off) enable information to be gathered in real time about the fluctuations in the atmosphere.*

▼ *When the Earth rumbles, the atmosphere shows signs of it. This false-colour map reveals the effects on the atmosphere of the eruption of Mount Pinatubo, in the Philippines. Below: just after the eruption; bottom: a few months later.*

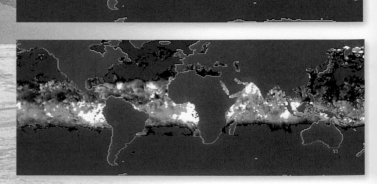

▲ *Remote sensing makes it possible to observe atmospheric phenomena from space. On this false-colour globe the decreased level of ozone above Antarctica is clearly visible.*

up the Earth's atmosphere, thereby favouring heat exchanges between the polar and equatorial regions. These phenomena act particularly on the base of the atmosphere, at low altitude.

What is the weather on Venus?

The atmosphere of Spaceship Earth certainly seems unique in the Solar System. There is one obvious distinction between the four planets nearest the Sun and those farther away. The first have a dense enough surface on which it is possible to land, the others (the giant planets) are made of gaseous matter of varying density.

But another great distinction is that they have undergone a very different evolution and do not all possess an atmosphere. If you were to compare Mercury, Venus, Mars and Earth, you would see considerable differences in temperature. On Mercury, for example, which has no atmosphere, the temperature varies between 350°C in the areas facing the Sun and -150°C in the shade. The pressure – 90 times higher than that of Earth – is such that we could not land on this planet without being immediately crushed.

(CAST) IRON PROOF

The oxygen released by the first living organisms 200 000 years ago first oxidised a large part of the rocks protruding from the Earth's surface, before it could accumulate in the atmosphere. These rocks reflect part of the gradual evolution of the atmosphere. Little by little, less oxidisable soils covered the Earth, and as it had nothing with which to react, the oxygen accumulated in the atmosphere.

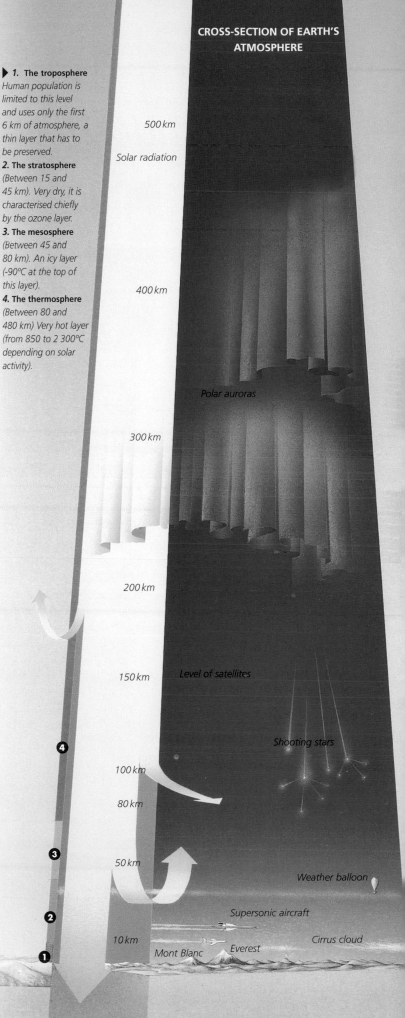

CROSS-SECTION OF EARTH'S ATMOSPHERE

▶ **1. The troposphere**
Human population is limited to this level and uses only the first 6 km of atmosphere, a thin layer that has to be preserved.
2. The stratosphere
(Between 15 and 45 km). Very dry, it is characterised chiefly by the ozone layer.
3. The mesosphere
(Between 45 and 80 km). An icy layer (-90°C at the top of this layer).
4. The thermosphere
(Between 80 and 480 km) Very hot layer (from 850 to 2 300°C depending on solar activity).

500 km

Solar radiation

400 km

Polar auroras

300 km

200 km

150 km — Level of satellites

Shooting stars

100 km

80 km

50 km — Weather balloon

Supersonic aircraft

10 km — Cirrus cloud

Mont Blanc — Everest

A tremendous sun filter

Some 150 million kilometres away from the Sun, the orbiting Earth is in the path of the full blast of this star's high-energy radiation. Luckily, the magnetosphere repels it.

▶ *Seen from a satellite, a superb auroral circle glows red above the Earth. On the left is the side of the Earth lit by the Sun; on the right is the side in darkness.*

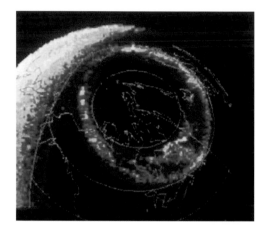

Place a compass on a flat surface, and the needle will settle immediately pointing in the direction of magnetic north, demonstrating the existence of terrestrial magnetism. From somewhat higher up, in space, satellites reveal that the Earth itself is like a magnetised bar, with a north and south pole. However, this magnetic north-south axis is not exactly in line with the Earth's rotational axis.

Above a height of 1 000 kilometres, an immense feature surrounds Earth's magnetic axis, which is called the magnetosphere. This covering, which protects us from the fieriness of the Sun, is not spherical. On the side facing the Sun, it is flattened and its width is up to five times the radius of the Earth, while on the other side it stretches out like a sort of tail over several thousand times the Earth's radius.

▼ *An aurora unfolds in the upper atmosphere, seen from a space shuttle.*

The raging heat of the Sun

The Sun functions like a colossal energy factory. The combined effect of its extremely high temperatures and very strong magnetic field is to cause portions of hydrogen gas to be dissociated as elementary particles and to be ejected at nearly 400 kilometres a second.

Other solar events, such as those in the photosphere, may aggravate this phenomenon by producing even more violent and erratic winds. Thus, depressions in the corona let matter escape at a speed of 800 kilometres a second, the brief solar flares cause a wind of 1 000 kilometres a second, while the rare but furious eruptions in the corona (prominences) blow winds towards the Earth that have a speed of 2 000 kilometres a second (*see pp. 166-167*).

Saving the planets

These outbursts could be extremely dangerous to our planet, but the magnetosphere protects us from them. However, the Earth is not the only planet at risk. The interaction of the solar wind with a planetary magnetic field is frequent in the Universe, and particularly in the Solar System. The planet Mercury, for example, has a smaller magnetosphere than Earth does. The giants Jupiter, Saturn, Uranus and Neptune also have magnetic fields, each very different from the others.

winds penetrate it with ease. Electrons and also protons are trapped there and move periodically round the field lines of the magnetosphere in well defined zones called the Van Allen belts. Some of these particles rush to the upper layers of the Earth's atmosphere in a movement that becomes almost circular.

▲ *The bubble that is the Earth's magnetosphere is flattened on the side facing the Sun (1) and misshapen on the opposite side (2). It protects the Earth from the solar winds (3). But it has weak points (4) in the region of the poles.*

Faults in the system

The misshapen bubble of the Earth's magnetic field usually deflects the solar winds, but certain occurrences can allow particles to enter the magnetosphere. In addition, there are weaknesses at the boundary between the side facing the Sun and the side facing away from it. Here the magnetic field is almost zero, and particles in the solar

A strange glow

Between 1 000 and 100 kilometres above the Earth's surface, electrons and protons interact with the elements that are present, essentially nitrogen and oxygen, to create a light phenomenon called ionisation. This produces an immense oval glow similar to an aurora, which is visible from space. On the ground, it takes the form of an arc easily visible to the inhabitants of the most northern or southern regions of the Earth. This phenomenon occurs usually at a height of 100 kilometres, with a depth of one kilometre. Seen from Earth, it presents a superb spectacle in which green and red predominate.

POWER FAILURES ON EARTH

From the beginning of its life in the Universe, the Sun has had regular bursts of furious activity. These are sometimes so violent that it has been possible for the effect of these energy surges to be felt on certain equipment on Earth. Our technology is based to an increasing extent on electricity, and electronic components are fragile objects that are sensitive to energy fields. Consequently, on August 2, 1972, a 230 000-volt transformer exploded in British Columbia, in Canada. And on March 13, 1989, at a time of maximum solar activity, the whole network of the Hydroquébec company went down, leaving a large area of the province of Quebec without power for nine hours. Solar activity can also cause sudden breakdowns in telecommunication satellites or, even worse, cause manned space flights to become high-risk expeditions.

BAD RAGAZ
Ragaz Spa-Switzerland + Ragaz les Bains-Suisse

Taking the waters

Water, which may remain in contact with deep geological layers for millions of years, is able to dissolve numerous mineral salts and other trace elements, which give it medicinal properties that have been used for thousands of years.

▲ *This poster of 1930 proclaims the merits of the thermal waters of the Swiss spa of Ragaz-les-Bains. Filtered by alpine rocks, the spa's water has a reputation for being particularly pure.*

There is no such thing as pure water on Earth. This is not a disillusioned comment on pollution, but rather a physiochemical reality. The H_2O element – the chemical formula for a molecule of water – has such a natural propensity for linking with other chemical substances that it is not likely to be found in its original pure state unless it is in a laboratory.

From the depths of the Earth
All water is therefore in the company of a multitude of molecules. Leaving aside the undesirable ones, like nitrates, heavy metals, pesticides and organic waste, many of the chemical elements of some underground water are beneficial to human beings.

While it remains in contact with various geological layers, water dissolves mineral salts and trace elements. It resides in deep fossil beds, formed during past periods of volcanic activity when molten rocks released the water they contained, or alternatively as a result of slow infiltration seepage. It rises to the surface after remaining underground for periods of between several hundred and several million years.

This mineral water becomes hot while deep underground and gushes out in the form of thermal springs when its surface temperature rises above 20°C. The term 'thermal springs' is, however, often used to describe any water with medicinal properties.

More than 40 elements in solution
The use of thermal waters as a healing treatment is a form of therapy that goes back at least as far as Roman times, when it was valued highly. Nevertheless, if just the effects of the chemical elements dissolved in the water are considered – apart from the thermal or mechanical effects like Archimedes' upthrust, viscosity and pressure – the results of these treatments are often empirical and the way they work is not well understood.

In France alone there are almost 1 000 mineral water springs, some of which contain more than 40 different elements. About a hundred of them are used for medicinal purposes – chiefly in the regions of the Auvergne, Rhône-Alpes and Midi-Pyrénées – by nearly 700 000 people each year. The number of people throughout the world who 'take the waters' annually for medical purposes is estimated to be about 10 million.

Bicarbonates, chlorides and sulphates
Mineral water can be classified according to its chief mineral constituents and their effects on the human body.

▼ *The tradition of public baths has been widespread for centuries in many countries, with the exception of those where the Church condemned their use on moral grounds. They continue to be popular throughout the world. Below, from left to right: hot springs and mud baths in Japan, and the Szecheny baths at Budapest, in Hungary.*

THE BENEFITS OF SEAWATER

S eawater is the prime medicinal water. It is considered to be the richest and the most complete with regard to its range of dissolved mineral salts, and its composition remains almost constant, regardless of the variations in concentration according to the ocean or latitude. On average, one litre of seawater contains 35 grams of mineral salts: 27.3 grams of sodium chloride (cooking salt), 3.4 grams of magnesium chloride, 2 grams of magnesium sulphate, 1.3 grams of calcium chloride, 0.6 grams of potassium chloride and 0.1 gram of calcium carbonate. Seawater treatment, or thalassotherapy, is generally recommended for joint or motor problems and rehabilitation. It produces results in cases of skin infections and certain respiratory allergies, and also in cases of developmental problems in children.

▶ *Diana bathing (fourth-century mosaic). The Romans took bathing to the point where it became an institution. Some establishments could accommodate thousands of people.*

– Water with bicarbonate releases gases, in particular when it emerges at the surface. Sometimes it is naturally sparkling and is sold as a drink. It facilitates digestion and water elimination, lowers blood pressure and has beneficial effects on the skin.

– Water containing chlorides has acquired its high salt content by dissolving deposits of rock salt left by ancient seas. Similar to seawater in composition, it is used for respiratory, neurological and gynaecological infections. Because of its density, this type of water is recommended for motor rehabilitation.

– Sulphuric water, which contains sulphur in a chemical form that can be directly assimilated by the body, stimulates blood circulation and the nervous system. Sometimes it emerges in hot springs. Its use is recommended in dermatology and rheumatology, and in the treatment of locomotive problems.

– Water containing sulphates, a compound of sulphur (which is not very soluble and therefore only present in small amounts, generally combined with other chemical elements), acts on the joints and skin. Water with calcium sulphate in it is used to complement the treatment of urinary and metabolic disorders. Water containing sodium sulphate acts on the liver and gall bladder. This is taken cold as a drink in treatments that are sometimes complemented by baths and local applications.

– Water containing trace elements has hardly any mineral salts, but on the other hand it contains trace amounts of elements such as iron, copper, arsenic, etc. Water with a high iron content has an effect on metabolic iron absorption and intestinal transit time.

In addition, all mineral waters are more or less radioactive at their source. Use is made of this property in some treatments, but with no health risks, of course, because the level of radioactivity is very low.

Nature's pharmacy

The term 'folk remedies' is sometimes used as an expression of scorn when describing traditional medical practices, because modern medicine has a tendency to disregard any knowledge that has not issued officially from scientific experiments. This is rather hastily writing off thousands of years of observation and use of nature's resources in healing. Common and essential medicines in today's pharmacy are derived from plants or animals over five continents. Empirical uses of these natural resources going back to the remote past anticipated certain surgical techniques and therapies of today. The recipes of our ancestors were certainly not always effective, nonetheless we still benefit today from a number of their discoveries.

▲ **FROM AUSTRALIA: EUCALYPTUS FOR COUGHS AND COLDS**
A large, evergreen tree with a grey-blue trunk, the eucalyptus – or gum tree – from the dry regions of Australia is a giant in the plant world. It can grow to a height of 100 metres. The Aborigines traditionally used the highly aromatic leaves to treat fevers and infections. The leaves contain an essential oil, 80 percent of which is made of cineol, which is an antiseptic and a bronchodilator. Because of these powerful properties, it is a medicine widely used in the treatment of colds, coughs and sore throats. Who can say they have never sucked a eucalyptus sweet?

▼ **RADIOTHERAPY IN HOT SPRINGS**
In volcanic regions (below, Sakurajima, in Japan), there are numerous thermal or mineral springs which are known to possess therapeutic properties. Many of them, venerated from primitive times, were subsequently used by great civilisations, like those of Japan or the Roman Empire, to treat certain diseases. It is generally the minerals dissolved in the water that are the active ingredient. However, some springs that are low in minerals are undoubtedly effective. It is known today that as it goes through hot rocks, water collects rare gases and traces of radioactive elements, which give it this therapeutic effect.

▲ FROM CHINA: EPHEDRA FOR ASTHMA

Growing in the desert regions of northern China is a small evergreen shrub with many branched, long stems. It is known locally as ma-huang, and to the Western world as ephedra. This plant was used as long ago as 5000 BC by the Chinese in the treatment of chills, fever and respiratory problems. The active compound it contains, ephedrine, has hypertensive, anti-asthmatic and psycho-stimulating properties. Three families of synthetic medications inspired by the natural model have been derived from it, in particular some of the most effective substances for the treatment of asthma, as well as amphetamines. These drugs have a stimulating effect on the brain that has sometimes been used for strictly non-medical purposes.

◀ FROM CANADA: GOLDEN SEAL, AN ANTI-INFLAMMATORY

A discreet little perennial not more than 30 centimetres high, golden seal bears a single, red, inedible fruit at the end of a stalk. Originating in the wooded mountain regions of North America, it became rare in nature but is now widely cultivated. The native Indians used it in a lotion to heal ulcers, sore eyes, swellings and wounds. An antibacterial agent and a tissue toner, golden seal is prescribed as an anti-inflammatory and to treat the mucous membranes. In particular, it is an effective stimulant of the uterus in difficult pregnancies.

▲ FROM THE AMAZON: THE ARMY ANT MAKES A GOOD NURSE

The army ants of South America march through the jungle devouring everything in their path. These voracious insects are equipped with extremely powerful jaws. The Amazonian Indians use them to stitch up an open wound. They pull the two edges of the wound together and place on it an army ant, which closes its fearsome jaws over the area to be stitched. Then they cut off the head and it remains fixed in this position. Healing takes place in the best conditions and the ant's head eventually falls off when the wound has healed.

▼ THE ANTICOAGULANT BENEFITS OF THE MEDICINAL LEECH

The leech is a large worm that lives in water and feeds on the blood of mammals. Because of several products that this blood-sucker injects into flesh, its bite is totally painless and the blood does not coagulate. Doctors in ancient Rome were already making use of its properties. Today, living leeches are used in microsurgery to reestablish good blood circulation after a transplant. But science is particularly interested in the reservoir of enzymes the leech represents: it synthesises about 20 substances, among them hirudin, the most powerful anticoagulant known.

ANIMAL MEDICINE

A pharmacy for the beasts

People are not the only living beings able to look after their health. Animals sometimes show great ingenuity in this field. Moreover, in some cases, humans have watched them and followed their example.

A cat sunning itself on a warm patch may look harmless, but it is a predator, a carnivore, which means that it essentially feeds on meat, even if it does not refuse the occasional little saucer of milk. It may get up, stretch, scratch and lick itself, and then make its way to a clump of grass – which it begins to eat. We have cows – which are strict herbivores – that are fed on animal protein, and now we see a cat grazing. Has the world turned upside down? Not in the case of the cat.

Like dogs – which are also carnivores – cats regularly need to eat greens, especially grass. It allows them to clear their digestive tract, just like eating salad improves the intestinal transit time of food being digested in humans. In fact, eating grass causes a strong itching sensation in the cat's (and dog's) oesophagus which induces a vomiting reflex, a cleansing purge.

The salt route

Most animals need food supplements, whether these are proteins or mineral salts. This is why you may see large blocks of salt in fields in which cattle are grazing. This salt supplement is of prime importance to domestic animals, as it is to herbivores in general. Salt licks are also placed in forests, for deer. But what do animals in the wild do for their salt supply?

A FALSE TEETH STORY

Some stories about the animal world are very persistent. Crocodiles are often shown with their mouths wide open, 'having their teeth cleaned' by a small wader, the Egyptian plover. This bird is credited with the ability to clean the reptile's teeth to rid it of food debris and any parasites it may have. Scientific observation has shown that this is not the case and if plovers are sometimes seen near or even on crocodiles, the birds are probably serving to warn them of danger rather than looking after their teeth.

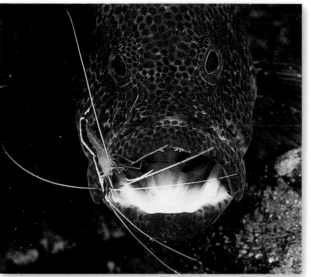

▶*The red shrimp of the* Stenopus *genus is skilful in the art of cleaning. Here it cleans a grouper of its parasites.*

In the case of African elephants, they will sometimes make long journeys in their search for the salt they need to survive. For this reason, in Kenya, the elephants go so far as to venture into deep caves in order to lick the walls that are covered in salt deposits left behind from infiltrating water. In particular, they go into the passages of old volcanoes, and pursue their search for salt more than 150 metres underground in the mountain caverns.

Forest elephants, which live chiefly in central Africa, dig into the ground in order to extract the mineral salts they need.

Free pest control

Paying attention to health and hygiene is a golden rule for the survival of all living beings. For some, their skin, fur or feathers unfortunately become a refuge for parasitic animals. These bite and cause itching, and the infernal creatures are often quite inaccessible. Because nature thinks of everything, some species function as parasite-removers.

In the sea, the master at this business is undoubtedly the cleaner wrasse. This fish does an excellent job, ridding other fish of parasites or moulds at the same time as it feeds itself. In the first stage, it performs a sort of dance in front of the other fish, which is intended to make it passive. Then it sets to work and cleans the skin of its 'customer'. Whole queues of fish, moreover, have been observed waiting their turn to be cleaned by a wrasse.

The same habit is found among some cleaning shrimps of the genus *Paraclimenes* and *Stenopus,* which are peerless cleaners of parasites attached to the skin of fish. They attract the fish by waving their long antennae from inside crevices on reefs and then climb on to perform their work.

Dry-cleaning too

But it is not only sea creatures that make a clean sweep to avoid parasites and the diseases they sometimes carry. Among land-based vertebrates pest control also occurs. Oxpeckers are past masters at this practice. In Africa these birds find their food almost exclusively on the backs of large wild herbivores (such as buffalo, elephant, hippopotamus, etc.) and domestic cattle, which tolerate them unprotestingly because these birds rid their hides of irritating parasites, such as ticks, and flies.

Monkey doctors

Because monkeys are at the top of the evolutionary ladder, they follow quite remarkable procedures to care for their health. Take the example of the red colobus monkey that lives in Zanzibar. This little monkey often eats the leaves of almond or mango trees. These taste good, but can cause bad indigestion. However, in the forest where this monkey lives, there are men at work making charcoal by leaving logs of wood to burn slowly – and wood charcoal has been known through the ages as a treatment for digestive ailments as it absorbs toxins.

▼ *The Zanzibar red colobus has become expert at pilfering charcoal, which helps it to digest certain delicious but toxic leaves.*

▼ *A little cleaner wrasse is busy with a giant moray eel, showing no fear of its client's sharp teeth.*

▶ *A dust bath. The practice enables this hen sparrow to rid herself of the parasites deep in her plumage that sometimes carry diseases.*

Over the years, the monkeys discovered that the charcoal was an effective remedy. So today, as soon as a wood-burner has his back turned, one or more monkeys nimbly go up to the fire to steal a piece of wood. When the wood has cooled down, the monkey, sitting in a tree, will slowly suck this excellent medicine for stomachache. Thanks to this 'medicine', the colobuses can eat to their hearts' content the delicious leaves that used to be bad for them.

Gorillas too eat certain leaves, but as a cure – and an effective one – for diarrhoea. In the same way, chimpanzees have been observed practising phytotherapy by using plants to rid themselves of parasitic infections. The plants all have in common a degree of abrasiveness, and when swallowed they scrape parasites from the chimpanzees' gut lining as they make their way down the digestive tract.

Female baboons even eat certain plants, such as the leaves of some species of *Sansevieria*, a genus to which mother-in-law's tongue belongs, to suppress menstrual cramps. Finally, monkeys cheerfully practise delousing, to rid each other of the parasites in their coats. This is done hierarchically by animals that are lower in the order of the group than others, and so also serves to maintain the social harmony of the troupe.

Dusting and fumigation

Certain birds, such as sparrows and pigeons, take dust baths to rid themselves of parasites. Others, like those of the crow family, and particularly the jackdaw, resort to 'ant baths'. The bird settles down next to an anthill and spreads out its wings so that the ants can become embedded in its plumage. But ants secrete a particular substance – formic acid – which stings. So how can we explain this masochism on the part of the jackdaws? It is simply that this acid kills bacteria and parasites.

Similarly, jackdaws and other crows have been observed fumigating themselves round an outdoor fire. Once again, the smoke enables the birds to get rid of their parasites.

Boars, warthogs and buffalo all have the usual mud baths, which allow them to rid themselves of parasites. Moreover, the crust of mud that remains on them protects them, at least for a time, against new infestations.

▼ *To African warthogs, there is nothing like a good mud bath for getting rid of the flies and parasites on their skin for a while, especially when the weather is very hot.*

Pillars of modern medicine

Whether they have been used since time immemorial or discovered recently, the willow, quinine tree, poppy, digitalis, periwinkle and various moulds are the basic ingredients of a large proportion of the medicines in the pharmacopoeia of today.

The willow grows with its feet in water and has very supple branches. It is therefore recommended for treating chills and rheumatism. This theory, which establishes a relationship between the characteristics of a plant and its curative properties, was put forward in the 16th century by a doctor, a colourful character and a charlatan in the eyes of some, by the name of Paracelsus.

As Professor of Medicine at Basel, in Switzerland, Paracelsus scandalised his contemporaries by abandoning Latin for German as a teaching language and claimed to reform medicine by throwing Hippocrates, Galen and Avicenna out of the window. His theory and explanation – called the doctrine of signatures – was in actual fact far-fetched, and represented an attempt to justify, after the event, examples of ancient usage: the willow, for instance, had been used since antiquity in the treatment of fever and as a painkiller.

The fever of knowledge

In Mesopotamia at the time of the Assyrians, around the 6th century BC, willow bark was being used for fevers, which were particularly frequent in this marshy country. This folk remedy continued to be used through the ages. In 1829, the French pharmacist Leroux isolated an extract from willow bark, which he called salicin from the Latin name of the plant, *Salix*. Shortly afterwards, his Swiss colleague Pagenstecher, by distilling flowers of meadowsweet, another marshland plant used for treating chills, isolated a closely related substance: salicylic aldehyde. In 1853, the Frenchman Gerhardt obtained acetylsalicylic acid. In 1899, it was Hoffmann, a chemist in the German laboratory of Beyer, who succeeded in preparing this acid in a pure and stable form. From

▼ *The white willow, growing in Europe in damp woodland and on river banks, was used from ancient times for its fever-reducing properties.*

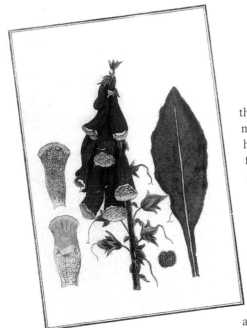

▲ *The large flowers of the purple digitalis, which easily fit onto the fingers, earned it the popular name of foxglove.*

the Latin name of the meadowsweet, *Spiraea*, he called his discovery Aspirin. The foremost modern drug, and still the most widely used medicine in the world, had been born. More than a century later, it is still a bestseller in pharmacies.

An affair of the heart

Authors of detective stories at the beginning of the 20th century were fond of inventing murders by digitalin. This substance in high doses causes a heart attack which appears entirely natural. Only a tissue analysis enables the culprit to be identified, a fact that has inspired many pages devoted to medical law.

A tall plant with long spikes of large purple tubular flowers, the foxglove, or digitalis, flourishes in the acid soils of western Europe. Its use as a medicinal plant has been attested from the Middle Ages, even though it is not known exactly what properties were attributed to it at this time. An English doctor at the end of the 18th century showed how effective it was for stimulating a weak heart. In 1828, the active ingredient of digitalis, digitalin, was isolated. It remains to this day one of the most effective medicines of its kind.

Nothing has yet been discovered in cardiology or pharmacology that calls into question the use of digitalin. But, even though it is used in medicine, the effect of an overdose is fatal; this means that the plant is classified as a dangerous poison and the drug is to be taken only on prescription and under medical supervision.

The other sort of Inca gold

The greatest treasure discovered in the Andes by the conquistadores was not the gold of the Incas, but the bark of the quinine tree. Known to the local Indians for a long time, its properties for treating fever earned it its native name quina-quina – 'bark of all barks'. It was used to treat the Spanish conquerors suffering from malaria.

In Europe, the new drug became a popular remedy only at the end of the 17th century. It was taken macerated in wine. Numerous recipes for quinine drinks date from this time, some of which are still drunk today as aperitifs. In 1820, the active ingredient, quinine, was isolated. The European armies engaged in the colonisation of Africa and Asia used it widely in the struggle against malaria, which often killed more soldiers than the enemy. Today, quinine remains one of the best drugs for treating influenza, and is still widely used for malaria.

Opium, the smoke of dreams

The opium poppy is closely related to the field poppy. The latex that flows from its cut seed head is dried and becomes opium, which is burned and the smoke inhaled. This plant was first used at least 3 000 years ago, initially in a religious context for its stupefying effects, and later for its therapeutic properties. The ancient Hindus and Assyrians knew of its remarkable value as a painkiller. So did the Chinese, and its use spread quickly to Europe.

◀ *The opium poppy (*Papaver somniferum*). It is from this poppy that opium is extracted. The plant is illegally cultivated for international drug smuggling.*

◀ *The quinine tree is an evergreen that originated in the tropical forests of South America. The bark contains the active substance, quinine.*

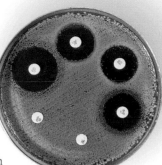

▲ *A bacterium culture test. The resistance of bacteria to antibiotics has become a major medical problem. Above, the bacterium* Escherichia coli *has been submitted to six different antibiotics. Four of them effectively stop it increasing (dark rings).*

Magical mould

During the slaughter of the First World War, medical teams were sorely tested. The poor nursing conditions caused many deaths from infection. Alexander Fleming, a young British doctor who served in this war, came away from it with the determination to find a drug that could combat infectious bacteria. At the end of the 1920s, his research resulted in the discovery of the bactericidal properties of a mould belonging to the genus *Penicillium*. Fleming called the active ingredient extracted from it penicillin: it was the first antibiotic on the market.

Less than a century ago, country folk in the south of France still used to apply a little Roquefort cheese to wounds. Conventional medicine fought against this practice, even if this meant the infection getting worse, thinking it ineffectual without considering its beneficial effects. This went on until Fleming's discovery, which explained everything, for the mould in Roquefort is *Penicillium roquefortii*. Today, antibiotics remain indispensable to modern medicine in the fight against infectious diseases, even if bacteria show an unfortunate tendency to become resistant to them.

For a long time opium was used by apothecaries as one of the ingredients that made up tranquilliser drugs. Now that the active ingredients of opium have been isolated, it is these that are used, such as alkaloids like codeine, for fever, and morphine, which relieves pain. Although morphine, and especially its derivative, heroin, is an illegal drug with tragically addictive effects, it remains indispensable for pain relief in the seriously ill, especially for terminal-phase cancer patients.

Pretty flower and cancer-fighter

On the plateaus in the south of Madagascar grows a pretty pink flower with a purple centre, a cousin of the delicate garden periwinkle. Used in the traditional medicine of that country, the Madagascar periwinkle (*Catharanthus roseus*) attracted the attention of scientists in the 1950s. Several of its alkaloids have the ability to halt cell division. As cancerous tumours are caused by the anarchic division of diseased cells, the significance of these substances is clear. Today, drugs derived from this periwinkle or synthesised from substances it contains are some of the most effective for certain cancers.

The star of today's botanical pharmacology, the Madagascar periwinkle did not grow naturally in sufficient abundance to meet the needs of medicine. Today it is cultivated on a large scale, notably in Hungary.

▲ *The Madagascar periwinkle is intensively cultivated exclusively for medical purposes.*

A CATHOLIC REMEDY

The Jesuits in Peru were quick to see the value of quinine and obtained the monopoly for its use. They introduced it into Europe at the beginning of the 17th century. In those troubled times of religious and political conflict, quinine thus acquired a bad reputation. It is said that the Protestant Oliver Cromwell refused this Jesuit remedy while suffering from the disease from which he later died. But after some important social personages at the court of Charles II of England and Louis XIV of France were cured, its virtues were definitively recognised by both sides.

Therapies of the future

Less than one percent of the plants growing on this planet have been exploited by the pharmaceutical industry to date. They nonetheless constitute a store of possible medicines, as do insects and marine animals.

▶ *The common germander, widely used in traditional medicine, is found in European woods and on dry hillsides. It flowers from the middle of spring to the end of summer.*

The common germander is a humble plant from dry regions in Europe that was used for a long time in folk medicine, especially for inflammation and pain. Its use is so harmless that it is one of the traditional ingredients of the liqueur Chartreuse, and of wine aperitifs.

But the unrestricted sale of capsules containing germander powder caused acute attacks of hepatitis in people taking it on a regular basis. Studies have since shown that the incriminating substance is not soluble either in water or alcohol, which explains why teas and alcoholic drinks based on germander carry no health risks.

Plants, the raw material of the pharmaceutical industry

Plants considered ordinary may still hold many secrets: each of the chemical substances they contain possesses particular properties. To develop new drugs, scientists at present have about 400 000 plant species available to them, each containing several dozen different substances. The number of possible combinations is therefore mind-boggling.

Sophisticated equipment involving highly developed automation and cutting-edge information technology enables plant molecules to be extracted and analysed. Batteries of tests performed on precise biological targets (enzymes and cell cultures, for instance) are continuously

▼ *The pharmaceutical industry shows a close interest in folk herb remedies, and is ready to invest considerable sums of money in the production of new drugs.*

carried out to find the most promising substances. The selected candidates are then submitted to further more precise studies to investigate their therapeutic effects, their secondary effects and possible derivatives. In these tests there are many failures and few passes. But only a rigorous selection process will enable new 'parent compounds' to be found.

This term is used by scientists to designate the chemical products that are extracted from plants that, with slight modifications, can produce a family of drugs with different actions. Cocaine – extracted from the South American coca plant, traditionally used to allay hunger or to fight fatigue – is, for example, one of the best known parent compounds. Its numerous derivatives constitute valuable local anaesthetics, particularly widely used in dental surgery.

An anti-cancer league

The Madagascar periwinkle provided the basis for the first anti-cancer drugs in the middle of the 20th century. In the 1970s, scientists became interested in the properties of many other plants: tropical trees, like the jacaranda and the ochrosia; species from temperate countries, such as yew and mistletoe, and cultivated plants like the castor-oil plant. None of the substances discovered acts on all tumours. On the

▲ *From left to right: yew, castor-oil plant, Japanese white pine and St John's wort. The first two are used to combat certain types of cancer. The antiviral properties of the last two are still being studied.*

other hand, each of them allows specific cancers to be treated, such as leukaemia, Hodgkin's disease, skin cancer and breast cancer.

The research goes on, needless to say, for the disease is far from conquered. In addition, the drugs used today, which have meant real therapeutic progress, cause very significant side effects. The future belongs to derivatives of the present substances, which will provide the same efficacy but with far fewer side effects.

An antiviral armoury

Antibiotics enabled infectious diseases caused by bacteria to be conquered. But viruses, which are at the root of many infections, from the common cold to Aids, are immune to antibiotics.

To fight the plague of Aids, scientists are studying the properties of the Australian chestnut, formerly used by the Aborigines to make the poison for their arrows, and also those of the Japanese white pine. Blue-green algae in the ocean off Hawaii have been found to contain a class of anti-HIV chemicals, called sulpholipids, and are the subject of further research. Similarly, new anti-HIV agents have been found in the seeds of the bitter melon, the root tuber of the snake gourd (*Trichosanthes kirilowii*), and the leaves of the carnation (*Dianthus* spp.), and these too are being investigated and submitted to trials.

In Europe, St John's wort, a beautiful yellow flower, is traditionally used to heal wounds. But it is also of interest to science on account of its remarkable antiviral properties and its effectiveness in treating cases of mild depression and fatigue of

nervous origin, conditions which are increasingly common in modern societies.

Hormones for life

Numerous functions of the human body are coordinated by chemical messengers, known as hormones. If some of them are missing, or if, on the contrary, there are too many of them, there will be a malfunction. Many specialists think that there is a hormonal disturbance associated with ageing. Future anti-ageing drugs will doubtless make a major use of hormones.

That is why scientists are very interested in plants containing substances related to human

▼ *Yams (here* Dioscorea villosa) *are an important food source in tropical regions, but also an irreplaceable source of hormones for the laboratory.*

▲ *Two plants with promising properties: above left, the agave, the traditional source of sisal, and above right, the Australian chestnut.*

▲ *Sponges, primitive animals living in colonies, are today used in the pharmaceutical industry. A Mediterranean species is famous for its soft skeleton, which absorbs water and has been used since ancient times as a bath sponge.*

The secretions, venom, enzymes and other hormones produced by their glands afford a variety of mechanisms from which we may learn, and hint at an immense range of possible products.

Neither traditional nor modern medicine has much recourse to these animals. The Spanish fly was used at one time to make remedies to help heal skin lesions, whereas bee venom and preparations based on maggots were known as remedies for rheumatism in the case of the former and for some kinds of gangrene in the case of the lattter. Both, moreover, still have some supporters today.

The sea is another good hunting ground for laboratories. Although flora in the form of seaweed are an extremely rich resource, at present it is especially marine fauna that interest scientists because it is actually more probable that molecules related to those synthesised in the human body (like hormones and enzymes) will be found in animals rather than in plants.

Species like sponges and corals already supply drugs used in leukaemia and nervous disorders. The review of active ingredients goes on and holds out much hope for the future.

hormones. The one most sought after today is the wild yam (*Dioscorea* spp.) of Mexico, which is related to the sweet potato. It provided the chemical basis of the hormone progesterone, which facilitated the manufacture of contraceptive pills. It subsequently enabled cortisone, a hormone able to relieve rheumatism as well as certain skin conditions, to be mass-produced.

Nowadays, the wild yam is used to make DHEA, the anti-ageing hormone, the merits of which are contested among the medical community. The agave plant and certain periwinkles have now also been shown to be promising sources of substances with chemical formulas related to human hormones.

Our animal friends

Two even more promising avenues are open to science. Firstly, the study of insects, of which there are reckoned today to be more than 750 000 species, which constitute as many chemical factories on legs.

▲ *In the 17th century, Spanish fly powder, from the cantharis, or blister beetle, was fashionable as an aphrodisiac.*

THE AMAZING DIVERSITY OF THE LIVING WORLD

Today, about 1 600 000 living species have been recorded, but scientists think that there are perhaps 10 or 20 times as many. There are 5 000 bacteria known to science, but there are thought to be 100 times more. Insects, with about 750 000 known species, are easily the largest group of living things. It has even been suggested that there may be 20 million species. But this diversity is concentrated especially in the tropics and is threatened by deforestation, marine pollution and the intensive exploitation of natural resources. How many sources of possible drugs have already been lost in this way?

HUNTERS AND THE HUNTED

Eat or be eaten

*A water snake seizes
its prey.*

All the better to eat you with!

Not all predators fall on their victims as brutally as the lion on the antelope. The animal world has at its disposal thousands of less spectacular strategies which are just as effective.

In the forests of Madagascar, miniature dramas are being played out. A chameleon, perched on a branch, looks as though it is asleep. Yet its eyes are turning round in all directions very slowly. The animal gradually raises one foot and then seems to hesitate as it puts it down again, while its body ripples strangely from front to back. How is it possible that such a slow creature feeds on insects?

On the tip of the tongue

The next moment the question is answered. Suddenly, the chameleon's very long tongue shoots out and just as quickly flashes back again into its mouth – with an insect stuck to the end of it, which is immediately swallowed. The whole operation has taken just a fraction of a second. Then the animal continues its slow journey along the branch. Chameleons possess a tongue which is as long as their body and which they keep neatly

◀ The chameleon's tongue, as long as its body, is able to catch an insect sitting some distance away. This ability allows the animal to merge into the background in order to hunt.

rolled up at the back of their mouth. Their eyes move in all directions and independently of each other, allowing them to judge very precisely the distance between themselves and their prey. The larger species, like the Oustalet chameleon, a giant from Madagascar which can grow to 60 centimetres long, can even catch birds with their sticky tongues.

Fatal fangs

Some snakes have fearsome, even deadly, weapons, in the form of fangs. Planted in the upper jaw, these teeth have a channel running through their whole length, which is linked to a poison gland. As soon as the prey is caught, the venom spurts from the gland through the fangs, which inject it, and the captured animal is paralysed. The poison itself is therefore a deadly weapon, working together with the gripping action of the fangs.

The fangs are generally sited at the front of the mouth, so that the captured animal is soon killed. However, some species have fangs right at the back

◀ Boa constrictors do not have a poisonous bite but have strangulation down to a fine art, which is just as effective. This opossum is being squeezed to death by an emerald green boa constrictor.

▶ *Like other deep-sea fish, this lamprey of the Melanocetus genus has above its mouth a sort of fishing lure that it waves in front of its prey.*

of the mouth. This is the case in the Montpellier snake, which is found in Europe and whose poison is highly dangerous to animals, but not to be feared by humans. Even if the snake were to bite us, on the hand or the ankle for instance, the fangs, positioned too far back, would not be able to reach as far as the wound.

A fatal embrace

Non-poisonous snakes have a weapon which is as terrifying as venom: strangulation. Boa constrictors and pythons are both very good at this. Once they have seized their prey in their jaws, these large snakes coil themselves round it and strangle it by slowly constricting their muscles. The animal is crushed and, unable to breathe, it dies.

But death is not as quick by this method as it is by poison. The snake can feel its victim's heart beating and lets go only when it has stopped, even with large prey. An anaconda, for instance, can measure more than 10 metres long and weigh 500 kilograms, so it is not surprising that some of its prey are very large, even as large as jaguars and caymans.

Dramas under the sea

In the depths of the sea, certain fish, like the angler fish, actually use a fishing lure to catch their supper. They possess an appendage near their mouths that projects in front like a sort of thread with a small fleshy growth on the end. When a small fish comes by, the angler fish, hiding on the bottom, begins to waggle its 'hook' like a fisherman casting a line. If a fish makes as if to catch the bait, the angler fish casts the thread backwards and opens its mouth wide. The prey is sucked in by its killer and has no way out.

Angler fish that live at great depths (known as abyssal species) where there is no light also possess a fishing lure, but the end is bioluminescent and shines like a little lamp. What sophistication!

Capture by lasso

Not to be outdone, invertebrates also show some imagination in catching food. Tropical spiders of the genus *Dinopis* attach themselves to their web by their hind legs and hold between their front legs a sort of net that they cast when their prey passes by.

HUNTING WITH A HARPOON

In the sea, even primitive animals may have fearsome traps. Cones are tropical-water gastropods. These superb shellfish have an organ called a radula, a sort of rasp which ends in a barbed harpoon. This is linked to a gland producing a poison said to be as strong as curare. As the prey comes past (a worm, snail or fish), the cone uses its breathing siphon to throw its harpoon onto the victim, killing it instantly.

▲ *The bolas spider holds a net, at the end of which is a sticky liquid that will trap a fly or any other insect.*

*▲ The praying
mantis lies in wait
for its prey. This
one is seen eating a
live grasshopper.*

*▼ The spider of
the Dinopis genus
spreads out its
web, spins it round
like a lasso, catches
a fly and wraps it
up before eating it.*

Other spiders, like
the American bolas
spider, of the genus
Mastophora, use a gossamer
thread like a lasso: on the end
it has a ball of sticky liquid
which traps imprudent midges.
Most spiders that spin webs use
them simply as a passive trap. Some
remain hidden nearby with one foot
touching it, and at the slightest quiver
rush to the insect caught in it.

With a fork or a flickering light

The praying mantis, with a wiry body that is
usually green, blends in perfectly with the
vegetation. This enables it to lie in wait for prey.
Insects cannot see it or its terrible front legs, which
are folded up against the abdomen when at rest.
The spiny femur and tibia form pincers that pin
down the unfortunate prey and hold it while the
mantis eats it alive with its large mandibles. It is
curtains for the cricket that thought the mantis
was just a blade of grass.

Insects have developed even more terrible
weapons, like the tail pincers of the earwig, for
example. Harmless to humans, these pincers serve
to hold down the prey earwigs have caught with
their front legs, while they eat it. In fact these
insects use their pincers rather like a fork.

The larvae of dragonflies, also great meat-eaters,
unfold a mouthpiece, called a mask, which
stretches as far as two centimetres in front of them.
In this way, without having to move,
they can capture little insects
that doubtless thought they
were safe flying past at a
short distance from the
motionless larvae.

In the tropics, glow-
worms and fireflies
are everywhere. Each
species has its own
flashing frequency,
which allows males
and females of the same
species to meet during
the breeding season.
Among certain carnivorous
fireflies, the female imitates the
flashes of other species to make them
come to her so that she can eat them.

The funnel of death

The antlion (also known as a doodlebug) is an
insect like a dragonfly, and is essentially
carnivorous. Its larvae use a terrible trap in order
to catch their prey. They dig a funnel-shaped hole
in dry sandy soil, up to five centimetres in
diameter, in which they hide, their fearsome jaws
just visible at the surface. When an ant happens
to come along on this apparently harmless ground,
it suddenly finds itself slipping on the loose sand
and dropping down to the
bottom of the funnel where
the jaws of death are waiting.

There are several species of
antlion. The larvae of some do
not make a funnel-shaped
hole but bury themselves just
under the surface or bits of
debris, leaving only their
head showing. When prey
approaches, they jump out and
seize it in their enormous jaws.
Antlions are known as
doodlebugs after the spiral
trails they make in the sand.

*▲ The end of an
ant! It has just
been caught by an
antlion buried in
the sand.*

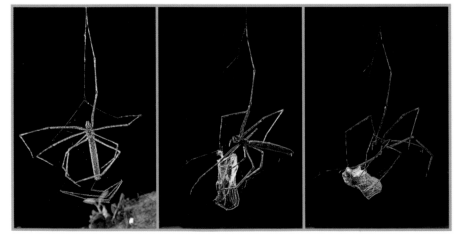

Green ogres

Some plants live in soils that are so poor in nitrogenous matter that they have developed very eccentric behaviour in order to survive – they feed on small animals.

In the tropical forest, an ant strolls along a thick vine, tightly wound round a tree trunk. It comes to a sort of lidded pitcher which exudes a sweet, sugary smell, walks onto the edge of it to investigate, and suddenly slides down a waxy wall to find itself helplessly stuck at the bottom inside a dark well.

The well is in fact the inside of a pitcher plant, *Nepenthes*, which will now produce enzymes to digest this prey. The precious nitrogenous material provided by the ant will be absorbed through the walls of the pitcher. The pitcher is a very special kind of leaf, able by means of secretions from its nectar glands to change into a very attractive trap, and then by the secretion of enzymes to become a sort of stomach to digest its prey.

Proteins for survival

Carnivorous plants exist throughout the world, but they are found particularly in the tropics. They always grow in soils that are very poor in nitrogen: in the forks of trees, or in peat bogs, granitic soils or fresh water. Their survival, therefore, depends on their ability to trap small animals and then digest them. Their methods of capture and the devices they use are very varied: pitchers in *Nepenthes*, bladder traps in bladderwort (*Utricularia*), glue traps in the sundew (*Drosera*) and butterwort, and a spring trap in the Venus flytrap (*Dionaea*). The largest *Nepenthes* in tropical forests are able to trap small frogs and mice, but their usual diet consists essentially of insects, spiders, worms and crustaceans.

It was Darwin, at the end of the 19th century, who succeeded in explaining this astonishing phenomenon. The plant began by defending itself, especially against insects, by secreting a sticky gum. The bodies trapped in this way piled up round their plant predator and enriched the soil, which then fed the voracious host better and better, through its roots. Micro-organisms subsequently decomposed the animals at the spot where they had been caught. The substances obtained in this way could be assimilated directly through the walls of the plant. The plant then secreted its own enzymes, giving itself an autonomous 'digestive system'. At the same time, its traps were improved and nectaries developed to secrete nectar intended as bait for the victims. Environments very poor in organic matter could thus be taken over by these plants with their peculiar metabolism.

▶ *Greed can be a deadly sin. To capture its prey, the* Nepenthes *plant has evolved a pitcher with a delicious scent of sugar, a fatal trap for insects attracted by the smell.*

▶ *The leaves of the sundew (*Drosera*) are covered with sticky hairs that enable it to capture small prey. It invented 'flypaper' long before we did.*

▲ *The leaves of the purple* Sarracenia *mimic a flower. The purple-streaked veins guide insects towards the nectar at the bottom of the pitcher, but hairs pointing downwards prevent the insects from going up again. Digestion of the insect can now begin.*

Stout defences

The best way of defending oneself against an attacker is to put up a form of protection that is able to neutralise the foe's weapons. From mammals to mites, a deterrent possessed by many animals is a hard shell, a kind of 'armour plate'.

What do ladybirds, scarabs, May beetles and weevils have in common? All these bugs are insects of the order Coleoptera. This name, invented by the Greek philosopher Aristotle, means literally 'wing in sheath', for the front wings of the beetles have evolved into rigid cases, which are called elytra.

These insects therefore share the common factor that they are born with a form of armour plate. And it has worked for them, because out of a total of the 1 600 000 known living species, including bacteria, plants and animals, some 350 000 are beetles.

Moreover, the largest insects in the world are beetles, like the African Goliath beetle (which is 12 centimetres long) or the Amazon titan (15 centimetres and over), which frequently weigh 100 grams.

Armour, heavy protection

All insects and their cousins the crustaceans, millipedes and spiders, are endowed with an external skeleton, called an exoskeleton. So they do not have bones and an internal spinal column like we do, but their outer skin is, to varying degrees, hard.

Beetles have developed this feature further, going so far as to cover themselves with a veritable suit of armour. It consists of rigid but articulated plates, which allow some movement.

On the other hand, the weight of this cuirass limits flexibility and their movements are correspondingly slow. But beetles are less vulnerable to predators than other species. Their recipe for success has in fact been used by many other animals.

▲ *The shell of the giant clam, a mollusc which lives in the warm waters of the Indian and Pacific oceans, became unexpectedly popular in European churches where it was used as a font.*

Shell shelters

Molluscs have a soft body. This is a good adaptation to an aquatic environment because the water supports the soft tissue. But this anatomical particularity also makes a predator's job too easy. Their vulnerability forced molluscs to develop a shell as a means of survival (although some, like the slug, have since lost their shell). Water contains calcium, and by a chemical reaction, some living things are able to precipitate this as calcium carbonate, the principal substance of shell. Corals, for example, develop their skeleton and their reefs in this way. But most molluscs build a refuge.

There are two main types of mollusc shell: one forms a cone, often rolled round upon itself, like that of snails, or the flattened cone of the abalone. The other has a bivalve shell, closed by strong muscles that leave no room for anything to squeeze in, like mussels, oysters and clams.

THE CHINK IN THE ARMOUR

The jewel beetles of the family Buprestidae have rich metallic colours and a very hard skeleton. However, they have a weakness: their joints, which allow movements. In the 19th century, the entomologist Jean Henri Fabre showed that these beetles are often caught by a wasp, the digger wasp, which stings them at the point where the thorax joins the abdomen and injects a poison that paralyses them. They then end their life in a burrow, being slowly eaten by the larvae of their attacker.

Protective plating

Vertebrates have found another solution to the protection problem: assembling bony plates. The trunk-fish of tropical seas, tortoises, turtles and also armadillos make use of this defence. In tortoises, the plates are fixed together and form a closed container. But as the head and feet must still be free in order for the animal to move about and take in food, this container has holes in it, which makes the protection incomplete. Moreover, the whole thing is very heavy for these animals to carry which is one of the reasons they move so slowly.

Armadillos have adopted a system of articulated plates. A fixed shield of plates covers the head, another the hind quarters. Between the two is an area of several bands of smaller plates, which is articulated. Only their abdomen, on the underside and covered only in hair, is unprotected, which obliges them to roll up into a ball to protect this vulnerable area. When they have adopted this position, there is nowhere their predators can bite them.

Spiteful spikes

In their choice of protective covering, some animals have preferred to adopt the offensive. Spikes obviously protect their owner, but they also hurt the would-be aggressor. Among invertebrates, sea urchins have this armoury, and among fish, the porcupine fishes, of the family Diodontidae.

Scorpion fish have hollow spines connected to a poison gland, which leave anyone walking on them with a painful memory. Hedgehogs (which are insect-eaters) and porcupines (which are rodents) in Africa and America have put on a suit of prickles. The former are content to roll up into a ball at the approach of danger to protect their underbelly, whereas certain African porcupines are distinctly more vindictive. When attacked, they do not run away but rather run backwards towards their attacker, which thus runs a strong risk of being left with long quills stuck in its body, whereas the porcupine is unhurt.

▲ The Goliath beetle, a giant of the African forest, is a close relative of the rosechafer found in many gardens, which is common but far smaller.

▶ Most species of armadillo are distinguished by the number of articulated bands between the shield at the front of the body and the one at the back. The armadillo shown here has eleven.

◀ ▲ The pangolin, the African cousin of the American armadillo, similarly rolls up into a ball to offer its attacker nothing but the barrier of its bony scales.

◀ The quills of this South African porcupine point towards the back. So the animal attacks by running backwards.

Ingenious predators

Any species that is not vegetarian hunts to a greater or lesser extent. In order that all may find food, the animal world has evolved many strategies, each appropriate to the creature's needs and capabilities.

A lioness prowling through the long grass of the Serengeti, in Tanzania; an eagle circling in the sky above the Rocky Mountains in North America; a heron as still as a statue beside a marsh in the Camargue, in France; a pack of wolves running in the taiga of Siberia: all these are everyday scenes in the wild. And the actors, however different they are, share the same preoccupation: finding meat to eat. To satisfy this need, the insect-eaters, meat-eaters and fish-eaters all have recourse to the same method, whatever technique they use: hunting.

Patience and stealth

Lying in wait for prey does not demand much energy but does require a good deal of patience. This strategy, common to many predators, needs a capacity for standing still, a relatively neutral coat or plumage to blend into the environment, a powerful spring to leap onto the prey at the right moment, and excellent sight to be able to see the slightest movement that betrays the presence of food.

The first step in the art of stealth is knowing how to hide. The large cats with their spotted or striped fur blend in wonderfully with the vegetation, especially in forests or tall grass, where the light and dark parts of their coats imitate the light and shady areas. Concealed in a crack in an old wall, the jumping spider looks like a small pebble. If a midge flies near, it jumps out to capture its prey and inject its fatal poison.

In quite a different habitat, the moray eel lurks in a hole not far off the shore, with only its head and sharp teeth protruding. When a fish arrives within reach of its jaws it springs onto it with its mouth wide open.

The art of staying still

It is similarly an exceptional ability to freeze in position and then propel its long neck in a flash down to a little fish swimming near the surface that makes the heron's fishing strategy so successful. This wader certainly does not blend into the background, but by standing so still it manages to make its prey forget it is there.

Perched at the end of a branch, the spotted flycatcher, a small European passerine, waits just as patiently for an insect to come near. When it sees a victim, it suddenly opens its wings, takes it in mid-air, and then goes back to its observation post to finish off its snack and wait for more.

▲ *Perfectly still, the heron waits for a fish to come near. Then it shoots its head forward to catch the fish in its sharp beak.*

▲ *Like the heron, the collared snake's best ally is patience. This frog is paying the price.*

The kodiaks, large brown bears that live in Alaska, in the United States, stand on a rock in the middle of a river at salmon-spawning time, where they wait for a salmon to leap out of the water and then catch it skilfully in their mouth.

Chasing fast food can be exhausting

For some animals, hunting means running, above all. The performance of the cheetah, the fastest of the big cats and able to hunt its prey at a speed of 110 kilometres an hour, is amazing. But it is also exhausting for the cheetah. And unless it catches its prey in the few seconds after it takes off, the cheetah runs a strong risk of going home hungry. Its phenomenal sprint in fact uses up so much energy that it cannot sustain it for very long.

Speed is easier to maintain in the sky than on the ground. The peregrine falcon, for example, can capture birds, especially pigeons, in mid-air after making a dive-bomb as fast as 170 kilometres an hour. Its tactics consist of rising to a height that allows it to find its prey without alerting it, and then swooping down on it at high speed to take it by surprise.

In the underwater world, fish such as barracuda, tuna and marlin are also excellent hunters. Their sudden spurts of speed happen so quickly that the little fish on which they feed have no chance of escape. All these species count on their speed to surprise their prey in a chase that most of the time will be as short as it is fatal.

▼ *The hunting strategy of lionesses is a model of its kind: they find their prey (here a wildebeest) and together cut it off from its herd, then worry it to tire it out before they kill it.*

Sharing is obligatory

When several animals hunt together, it considerably increases the chance of success. But there is always a drawback: the spoils have to be shared. Lionesses know all about this (see pp. 318-319). Before they can enjoy the fruit of their labours, they have to let the male lions eat first. The young will be served last.

Hunting in a group is also practised by members of the dog family like the African wild dogs (*Lycaon pictus*), which are famous for their endurance and perseverance. These animals go hunting after a long ritual of barking, licking and signs of submission, in order to reinforce the cohesion of the pack. Their strategy consists of spreading out in a phalanx, then approaching a herd. When the

▲*The osprey is a fish-eating raptor that hunts over water. Its acute sight is a wonderful advantage.*

THE FOX IS A PRAGMATIC HUNTER

The fox is opportunistic in the hunting of its prey. It can chase amazingly fast after a wild rabbit, or can wait patiently for a mole to poke its nose obligingly out of its tunnel. It is also quite eclectic in its choice of food, being quite ready to feed on berries when driven by hunger. This adaptability is the most successful strategy for survival.

herd begins to run, they isolate one of its number, preferably a young or weak animal. They hunt it down until it is exhausted, in a relay race which is based on speed, some of them taking a short cut each time their target changes direction. Zebra and wildebeest are their favourite prey. But unlike the lions, African wild dogs allow their young to eat first.

Hyenas, which hunt at night, first work together to scatter the herd in order to isolate an old or sick animal. They will then chase the animal, snapping at it to exhaust it, before finishing it off. They then have to eat their meal as fast as possible if they do not want to see their catch taken by the lions. The old ones start to eat first and then move over for the young.

The lonely hunter

If you do not want to share your food, it is better to hunt alone. This is the chosen method of the tiger. This animal counts on the element of surprise and then on its weight to fell its victim and kill it by strangling it in its jaws. A similar strategy is used by the leopard seal, a large, agile seal which lives in Antarctica. This fearsome predator waits for penguins to come to the surface after a long dive and then seizes one of them. If it is a small penguin, the seal swallows it whole.

The stoat, which lives in northern Europe, and the forests of Asia and North America, runs through the countryside sometimes at speeds of up to 30 kilometres an hour in search of small rodents, birds, insects and carrion. It has an excellent sense of smell and tracks down its prey relentlessly, when it will pounce on it and kill it with a deep bite at the base of the skull. It will calm its own hunger first, and if the hunting is good, it will take home food for its young.

Snakes also hunt alone and each only for itself, because there are no young mouths to feed. Their poison enables them to immobilise prey much larger than themselves. The cobra attacks by throwing its head forward to inject its victim with neurotoxins, chemical substances that affect the nervous system and lead to paralysis and then death by suffocation.

▼ *The waves made by a whale fishing (below, a humpback whale) often attract seagulls, because the turbulence brings fish and microplankton to the surface.*

▲ *The leopard seal is a peerless hunter which feeds on penguins and young seals.*

PASSIVE DEFENCE IN PLANTS

Thorns and other devices

▲ *Cleaning service. The thorns on this acacia have juicy swellings at their base, which attract ants, welcome guests that keep the tree free of fungus and from infestation by other insects.*

In order to defend themselves from browsers, it is in the interest of plants to bear arms, either as thorns, or helpful guests – or to resort to camouflage.

The stony regions of South Africa and Namibia appear to be lifeless, given over only to the mineral world. But suddenly, splendid yellow flowers seemingly placed on top of a pile of stones catch the traveller's eye. These flowers belong to a *Lithops*, a plant which looks like a pair of stones, until it is betrayed by its bright flowers. In this arid environment these plants have modified their leaves to the point where they have become stemless and rounded, in order to retain water better. Most of them are pale grey, pale green or sand-coloured and very thick, and look like a pebble, which enables them to blend in as much as possible with the stones around them and avoid the attention of foraging animals.

▼ *A stone plant. Because of its leaves, which look like stones, the* Lithops *plant blends in perfectly with its background – except when it is in flower.*

Hiding their true nature

Camouflage, however, is exceptional among plants. As plants form the background in most natural scenery, it is – on the contrary – animals that imitate them in order not to be seen. But in the mineral surroundings of desert regions, a plant disguised as a stone has found a good way to protect itself.

The groundnut, or peanut, has found an even more original way to camouflage its fruits: it buries them. Once the flower has been fertilised it dies and the stem bearing it bends over so that the fruit, the peanut, becomes buried in the ground. Safe in the shell, between two and four seeds will develop. The peanut is very rich in oil, and is one of the main agricultural resources of hot regions of the Earth.

▲ *Ostrich policy. After fertilisation, the flowers of the groundnut, bearing the future peanuts, bury themselves and develop their fruit underground.*

Ant bodyguards

Many plants, from the desert cactus to garden roses, are armed with thorns, which are highly effective in warding off the curious, whether animal or human. In Africa, farmers accordingly use thorny branches as hedges round the fields where they keep animals.

Some plants add to this arsenal of prickles, or choose in preference to them, the jaws of ants. These plants have special places in their leaves or stems for these insects, and even go so far as to secrete a sweet nectar to feed them. In exchange, the ants look after their hosts like the most attentive of gardeners. They remove fungus spores, destroy or chase away insects that threaten to eat the leaves or suck out the sap, and by their presence alone dissuade large mammals from grazing too close. Mercenaries are seldom more effective.

The education
of young predators

Hunting cannot be improvised. As in humans, experience is of prime importance in predators, but education also has a part to play. This is why, at least in certain species, the adults teach their young hunting techniques. This apprenticeship, achieved through play, observing and copying, may also require exercises to be repeated over a long period. It is always a crucial time for the young predator who is gradually entering the adult world. Its survival will depend on the skill and mastery it acquires at this stage. Once left to fend for itself, the young carnivore will have no choice but to win or die.

▼ LEARNING THROUGH PLAY

While they are still suckling, lion cubs often play with an adult lion's paws or tail. Crouching on the ground, they will suddenly jump onto a resting adult, as though it were prey. It is obviously through play that the young lions acquire hunting techniques. Kittens are doing the same thing when they jump after a ball or a cork.

▲ FRIENDLY FIGHTS

When young leopards tussle, their claws are retracted and their teeth, although sharp, do not grip their partner's neck very hard. By practising catching, holding, struggling and overcoming, they learn how to use their paws and teeth with skill and speed. Before beginning these mock fights under the eye of an adult the big cats are friendly to each other, and even at the height of the mock battle they show no aggression.

▲ Mock attacks

The art of fighting is a necessary prelude to mastering the art of hunting among members of both the cat and dog families, especially wolves and foxes. These European red fox cubs have adopted a strange posture: forefeet folded together, hindquarters raised. The fight that follows is not a pretence, but the teeth and claws of these cubs are nonetheless not used violently. Each of them is testing, and improving, its agility and speed as a future predator.

▼ Finishing off prey

When a cheetah hunts down a gazelle, she does not finish the animal off. She leaves this job to her young. The female cheetah captures small prey and while it is still living puts it down in front of her offspring (below). If they let it escape, the female immediately seizes it and brings it back until her pupils succeed at their task. This initial contact with prey is the beginning of a long learning process. Months or even years will go by before a young adult succeeds in catching its first victim.

▲ Diving in the big bath

In the waters off the west coast of North America, sea lions bring live fish to their young. The little pupils are now practising on prey already weakened by being caught. They still have to learn to dive in pursuit of a recalcitrant fish and to acquire sufficient agility in the water to finally catch it.

THE REFLEX DIVE

The crested grebe, a fish-eating bird, dives to find its prey. From a very young age, while they are still being fed by their parents, the chicks do likewise, and of course they surface with empty beaks. But the reflex has been acquired. Their apprenticeship will consist of catching prey of increasingly larger size as they themselves grow bigger, so that by the time they are old enough to feed themselves, they will have acquired the technique of the adults.

▼ LOOK AND LEARN

Young lapwings leave the nest as soon as they hatch: the species is nidifugous. As the parents (shown left) do not feed their young, the nestlings have no option but to follow their parents to learn how to feed. Although they have an innate tendency to peck, they have to learn to find places where there are hidden larvae and insects and how to reach them, all by watching their elders.

The young pied oyster-catcher who is lucky enough for once to be fed by its parents, will also, but a little later, imitate what they do in order to catch the earthworms on which this species feeds. In both these cases, learning takes place through copying. Time and experience will do the rest.

▼ EXERCISES IN THE AIR

The peregrine falcon makes sudden swoops in order to catch birds on the wing. The great predator is capable of bursts of dizzying speed, but its complete mastery of the sky is the result of long training. The young bird first catches easy prey, like insects, before taking on birds. The help of its parents in this is of capital importance. The adult captures a bird and lets it drop so that the young one can learn to take it in mid-air. If the young falcon does not manage it, the adult will have no difficulty catching it a second time. Then it will again position itself above the apprentice and drop it once more.

▲ PASSING ON KNOWLEDGE

There are not many species that make use of tools to help them achieve their ends. Chimpanzees, however, use wooden sticks to catch ants inside an anthill, and throw stones at colobus monkeys to kill them, after which they will eat them.

Scientists today are examining the transmission of knowledge in the animal world from one generation to the next, because using a tool has nothing innate about it. Parents must show their young how to do it. The young benefit from a long apprenticeship at their parents' side before they acquire the same skill in the strategies of surrounding and chasing. Some primatologists think that this is actually a form of culture.

▼ THE DANGERS OF LACKING EXPERIENCE

Penguins feed on fish that is brought to them by their parents. But as soon as the young are old enough to be independent, they must suddenly manage alone. Feeding behaviour in this case is not the result of the process of learning from parents. This is why a high mortality rate has sometimes been observed in this species during the first weeks after they leave the nest, as the young birds are not able to feed themselves properly, especially if food is scarce. The only advantage is that when they leave their parents the young penguins are so plump – they are sometimes larger than their parents – that they can live for some time on these reserves of fat.

Deceiving the enemy

Many species of animals make skilful use of their colour and shape to melt into the background and pass unnoticed by predators and prey, sometimes to such an extent that they seem to be in disguise.

The butterfly fish was on its way to graze on a coral polyp, without a care in the world. A few centimetres above the bottom, it swam in front of a large pebble, which swallowed it. In a few thousandths of a second, a stonefish, perfectly camouflaged, had opened its mouth and closed it over the butterfly fish.

The stonefish is a master of homochromy, or blending in with the surrounding colour. Because of its relatively flat, round shape, and its dull-coloured, ragged skin, it can easily be taken for a seaweed-covered rock. Sitting motionless on the bottom, it is almost invisible.

Both on land and in the sea, many animals practise this sort of camouflage, in which their uniform colouring exactly matches that of their habitat. Among them are nightjars, brown-coloured nocturnal birds, which merge into the tree trunks where they rest, and arctic hares, which turn white in winter so that they can easily hide in the snow.

Invisible from above and below

Other animal species have a double homochromy: their back is dark, their front light. Viewed from above, they blend in with the ground, but viewed from below, it is difficult to see them against the bright sky.

Most predatory fish of the open seas, like swordfish and sharks, have a silvery grey back and a white underside. It is only when you are close to them, and when these fish are seen in profile, that the two colours are visible at once. In general, the plumage of birds is also less flamboyant on the back than it is on the front. Seen from afar, they look flat. This deceives their predators for a while, which turn their attention to what looks like more substantial prey.

In contrast, most land animals and fish make use of livery with complicated patterns as camouflage. The stripes of the zebra, the spots of the cheetah and the chevrons of the emperor

◀ The tiger's stripes mingle with the patches of light and dark of the undergrowth. The big cat can therefore move to within a jump's length of its prey and take it by surprise.

▲ This chameleon from the Namib Desert, in southern Africa, has taken on the colour of the rock in the background. This perfect camouflage combined with its total immobility deceives raptors on the lookout for the slightest movement.

angelfish have only one purpose: to break up the appearance of the animal's shape, to separate it into several parts in order to deceive predators or prey. The tiger succeeds in melting into the background, and remains truly invisible because its stripes are seen as part of the undergrowth through which it prowls.

Costume changes

Chameleons, octopuses, flat fish and shrimps are all homochromous. But they are also able at any time to adapt their colour to the colour, shapes and even textures of their surroundings. This ability resides essentially in cells called chromatophores, which are sacs that contain pigments. When the chromatophores are not excited by a nervous or hormonal stimulus, the pigments remain in the centre of each cell, which then has a light colour. When this is not the case, the pigments spread more or less throughout the cell, and the colour becomes darker.

The speed at which a chromatophore passes from one state to the next varies according to the species. The speed at which information governing the change of camouflage travels from the brain (or from the ganglions acting in its stead in the invertebrates) also varies. In many species, this change is governed by both the hormonal and the nervous systems. This applies to flat fish like turbot and soles. A few species like the cephalopods (squid and octopus), however, have only a nerve command, which is much faster than a hormone command because it carries its information at the speed of an electric current. They need only a few fractions of a second to camouflage themselves. By way of comparison, it takes several seconds for a chameleon, with its camouflage governed just by hormones, to change its colouring.

Animal or vegetable?

The stick insects and leaf insects of tropical regions, which are known as phasmids, look like little twigs. Among vegetation they are all the more difficult to see because they are almost completely still during the day. Not for nothing are they called phasmids, which comes from the Latin word 'phasma', which means 'phantom'. These insects come to life and move about at night, but by then their predators are asleep.

Of all nocturnal moths, those of the family Notodontidae show the most remarkable capacity for concealment. Settled in a tree or on a carpet of dead leaves and twigs, the buff-tip moth, for instance, assumes the shape of a small broken branch, with its wings folded against its body making it appear rounded.

▲ The line of the mouth is the only visible part of this completely motionless stonefish. The fish moves only rarely, by leaning on its pectoral fins, which are situated on either side of its head and edged in yellow.

ALL KINDS OF BLUFF

The trunk-fish (right) can, when stressed, suddenly blow up and appear far bigger than it really is. On land, the frilled lizard from Australia, of the family Agamidae, or 'dragons', can suddenly make a large fold of skin fan out round its head and at once look enormous and monstrous, especially as it opens its mouth wide at the same time. The bloody-nosed beetle of Europe, and certain African praying mantises, actually make a red liquid that looks convincingly like blood ooze from their mouths or mouthpieces. These tactics seem to deter predators.

▶ *A twig or a shrivelled leaf? Even though it is big, this Asian stick insect is invisible among the vegetation on which it feeds.*

▶ *Seen from below, the wings of the brimstone butterfly look like leaves. The resemblance is even greater when the butterfly beats its wings, because they then look like leaves fluttering in the breeze.*

▶ *By grouping together, these bugs form a sort of growth like a flower, their colours adding to the illusion.*

▶ *Even disguised as a twig, this stick insect has to guard against predators that are not taken in by it. The only solution is to remain perfectly still.*

Of all reptiles, chameleons are the masters of concealment. Crocodiles and caymans can mimic perfectly a log of wood floating on the surface of a backwater. They let themselves drift slowly along to get near enough to prey coming to the water's edge to drink, and then seize them with their formidable jaws. But young crocodiles also use their log-like appearance to escape the notice of certain predatory birds.

In Africa, some species of bugs stick themselves to a branch and look just like the thorns of a rosebush. Among treehoppers of the sub-order Homoptera, found mostly in tropical regions, the adults arrange themselves in single file along a branch to imitate thorns while the young settle underneath, on the side facing down, and look like a sort of rough bark. Apart from their method of disguise, they are protected further from predators by the presence of ants attracted to the honeydew that they secrete.

The art of being a leaf

The brimstone butterfly (*Gonepteryx rhamni*) is common in Europe. In flight, its yellow colour makes it easy to see. But when it settles, the inner side of its folded wings is the same colour as a young leaf. Hanging from a leaf, the brimstone butterfly goes completely unnoticed.

Another champion at being invisible is the South American green leaf mantis (*Choeradodis rhombicollis*). This praying mantis lives on tree trunks next to a small climbing plant with round leaves that it very closely resembles. It is active at night, but in the daytime it clings to the leaves to blend in with them.

Vertebrates also wear disguises. The horned frog of south-east Asia lives on the ground among dead leaves. A brownish colour, it has fine ridges on its back that imitate the veins of a dry leaf, and appendages on its sides over its eyes that look like the tips of leaves. Below these appendages, two large dark spots imitate the shadows and thus conceal the large eyes of the frog.

This astonishing resemblance to an element in the surroundings or, even better, to another species, is called homotypy. Insects like syrphids, a sort of fly, have assumed the costume of wasps or bees and imitate their behaviour. In this disguise they do not fear meeting creatures wary of their sting, such as human beings.

Escaping from predators

Animals use many strategies to repel their attackers: playing dead, forming a group, squirting offensive substances or simply taking flight.

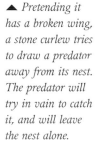

▼ *Perched at the top of a thorn bush, this meerkat (a small African mongoose) has the job of keeping a lookout over the territory of its group. If the sentinel sounds the alarm, the group will all flee to their burrow.*

An opossum looking for food ferrets about near a river in Arizona, in the United States. Suddenly, a coyote jumps out of a thicket. The marsupial sees it, but it is already too late to run away. There is no point either in getting into an unequal fight or jumping into the water. As a last resort, it has recourse to a trick. It lies down on its side with its mouth open, and remains motionless. The coyote approaches, shakes it by one of its front legs, gives it a push, turns it over and sniffs it. Soon losing interest in this dead prey, it turns round and goes off, while the opossum comes back to life and hastily makes its escape.

A well-timed nap

The behaviour of domestic fowls resembles that of the opossum, but it is not spontaneous. It is a surprising fact that when they are laid on their backs with their feet in the air, they will not move. This characteristic is also true of certain wild birds, such as pigeons, lapwings and oyster-catchers. Once turned over, these creatures have a paralysing reflex that immobilises them. But it wears off quickly if you clap your hands or make a noise. This curious sort of paralysis resembling hypnosis is nevertheless a defence mechanism against a predator that might accidentally throw a bird over onto its back. The predator might then assume the bird to be dead, and since many prefer to eat live prey, the bird would have a strong chance of being disdainfully left alone.

Run for life

For some species, running away is the only defence they have. But it is not necessarily a question of just running aimlessly. When the marmots of the Alps or the meerkats of southern Africa see a predator, they run off to their respective burrows, like rabbits and prairie dogs do. But, unlike the latter, they have an elaborate system of lookouts which give them advance warning. Among marmots, a sentinel stands guard and whistles at the slightest sign of danger so that its fellows can take shelter. Among meerkats, a highly gregarious species, the watch system is also sophisticated and relies on sentinels stationed throughout their territory to warn the group with sound.

Umbrella birds

The ability to sound the alarm has meant that certain birds play the part of protector unintentionally. These species, called umbrella birds, do not derive any benefit themselves from the service they provide to other species. Strong defenders of their own territory, they afford a

▲ *Pretending it has a broken wing, a stone curlew tries to draw a predator away from its nest. The predator will try in vain to catch it, and will leave the nest alone.*

▲ *This alpine marmot on the watch warns its fellows with a loud whistle as soon as it sees a threat of danger.*

ATTACK AS A LAST RESORT

It may happen that all the defensive strategies deployed are useless and the predator refuses to give up. When it is cornered, an animal that cannot flee will play its last card and attack. Whether it is a warthog attacking hyenas or a buffalo going for lionesses (right), the fight does not usually go in the weaker animal's favour. Exhausted, urged on purely by its instinct for survival, an animal that knows it is beaten will throw its last energy into one final effort. On this occasion it will be particularly aggressive and dangerous, even to humans. The majority of attacks by animals on people are in fact desperate reflex actions in self-defence.

certain degree of safety to other more timid birds. This is the case with the golden plover of the Arctic tundra, which sounds the alarm as soon as a predator enters its territory. Inadvertently it also warns the dunlin, a rather shy bird that often nests near the golden plover. In temperate Europe, numerous and noisy colonies of black-headed gulls often attract other quieter species, like the black-necked grebe or the gadwall. The latter, a duck, takes up residence in the very middle of a gull colony, because it finds living there the best protection against predators.

Esprit de corps

Instinct often causes animals to group together, seeking safety in numbers. This is, however, a passive coalition. It is intended to inspire fear in the attackers, and in fact often perturbs them.

Shoals of sardines - thousands of them - move like a single fish. When a shoal of these fish sees a Cape fur seal approaching, it swims deeper to regain a safe distance. The shoal may even divide into two, and perhaps surround the seal. In the middle of a shoal sardines cannot be caught: the fish moving in close ranks seem to form a single 'superfish' which reacts like a single sardine and, some say, adopts the behaviour of one. Furthermore, it is easier for them to swim in a group, for each of them has the advantage of the slight depression left by the movement of its neighbour. The only chance the seal has to make a catch is therefore by attacking repeatedly. In the long run, a sardine will fall behind the general movement and sign its death warrant.

Tailing a shark

The seal is a hunter which, in its turn, is hunted. In South Africa seals are the prey of choice for the great white shark. To protect themselves, they move about in small, very mobile groups, which

▼ *These musk oxen have gathered together in tortoise formation against an attack by a pack of wolves. The young are kept safe in the middle of the group.*

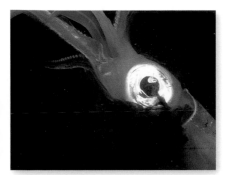

▲ *To cover its escape, the squid (top) flees in a cloud of dark liquid produced from its sac. The electric eel (above) prefers electrocution. On occasion, its attacker is paralysed and ends up in the eel's stomach.*

swim in the wake of the predator. Not being able to see behind itself, the shark ceaselessly tries to turn round, but to do this quickly costs it considerable effort because of its size, and eventually it tires. But if a seal's reactions are too slow it runs a great risk of sharing the fate of the sardines on which it feeds.

No generation gap

Zebra, like wildebeest, form large herds whose function is to protect the young. Usually, the offspring gambol along close to their mothers, separated from predators by rows of adults. But in the event of an attack, it is every animal for itself.

There are, however, a few exceptions to this rule, like the case of the musk ox. This ruminant from northern Europe lives in small groups in summer and in huge herds in winter. When an aggressive wolf is about, the adult musk oxen form a very tight circle in order to protect the younger ones. It is then unusual for the wolf to attack, especially as a rather large ox may charge it in a final attempt to ward it off.

Chemical weapons

Certain prey have another quite radical means of protection: arousing distaste in their attacker. The polecat in particular is excellent at this technique.

Glands situated on each side of its anus manufacture and emit a musk of which the pestilential smell is enough to deter the most motivated predator. The same deterrent is used with a greater or lesser degree of success by other members of the Mustelidae family, such as the weasel, mink, badger and – most famously – the skunk.

The bombardier beetle is also very successful at repelling predators. This insect ejects from the anus a very hot and extremely volatile liquid that leaves searing memories in its attackers.

Ants defend themselves in the same way, by squirting formic acid, an extremely corrosive substance, on the animals annoying them. The technique is used, too, by poisonous species of animals like scorpion fish, centipedes and some sea urchins, which either inject their toxins into their prey or their predators, or vigorously squirt them at the unwanted visitors.

Some fish also use their weapon like a shield. The electric eel, a real live wire, discharges its electricity in an attempt to chase away an animal that comes too close.

The cephalopods – octopuses, squid and cuttlefish – are known for sending out a cloud of dark ink behind them in order to make their escape from a predator unseen.

▲ *Ants squirt formic acid on their attackers. This is a very effective solution to keep birds away.*

▲ *Everyone take cover! To protect itself, this skunk is about to squirt an extremely foul-smelling liquid.*

Deadly toxins

*In animals, poison and venom are terrifying weapons. But many plants,
subjected to the insatiable appetite of animals, also produce toxic substances.*

The triton, a predatory marine mollusc, is five times larger than the cone shell, but this does not deter the cone, which looks as though it will get squashed flat but in fact will knock out the bigger mollusc. A few centimetres away from its attacker, the cone puts out its proboscis, at the end of which is a dart. In the sting is venom, a whitish liquid containing tiny proteins called peptides. Injected into the triton, these peptides block the receptors of its muscle cell membranes. The nerve current can now no longer pass between neurones and

muscles. The triton is in shock. Four more venomous stings and it is paralysed for a short time. If it does not soon flee, it will eventually die from the effects of being stung too often. The cone has won: the large mollusc that was troubling it eventually takes another route.

Fatal colours

In the world of the coral reefs, where many species live side by side, living space is limited and food quite scarce. The competition is tough. For those without the means to escape rapidly or to attack in self-defence, there are chemical deterrents. Thus, certain coral species have become poisonous, and their bright colouring actually acts as a warning: the intruder is made aware of the risks it runs. In tropical waters, an animal that does not flee and has showy colours is probably poisonous.

This is the case with the cone, a magnificent little gastropod whose shell is highly prized by shell collectors as is its flesh by many echinoderms. And sea urchins and starfish number toxic members among their family in most oceans of the world.

▲ *Sea urchins sting by means of a sort of pincer equipped with sense buds and teeth by which venom is injected. These pincers, called pedicellariae, serve also to clean the surface of the testa, the sea urchins' protective shell.*

▶ *The striped cone has got out its harpoon, more than 1 cm long. It is the end of the road for its victim, here a small mollusc.*

▲ *From left to right: the fire worm bears irritating bristles; the scorpion fish (Pterois) has very long and sharp poisonous dorsal spines; the chimaera stabs with a single spike that sticks out in front of its dorsal fin; the little weever fish has poisonous spines.*

In the tropics the spiny sea urchins with long prickles, and those of the *Toxopneustes* genus with their shining colours, are known for being highly poisonous. You have only to touch the fire worm to feel a sharp pain, but its low toxicity causes only skin irritations in humans. In the group of cnidarians (jellyfish, corals, gorgons and sea anemones) only jellyfish are really poisonous, and the magnificent box jellyfish is deadly to man.

The venom of the tropics

There are poisonous species in all groups of aquatic and land animals, including mammals and birds. The 'venom factor' is nevertheless particularly high within the tropical regions. And as tropical animals are not great travellers, there are many of them hunting in biomes where prey is very abundant, like the rain forests and the steppes.

Their toxicity sometimes depends on the temperature or the season. Most of these species are predatory. They inject their prey with poison to make them easy to capture or spray venom over the animal to make it easier to digest. The venom of some species, like that of the cobra, has a property that destroys tissue or blinds an animal so that it cannot escape.

Let sleeping stones lie

Stingrays, eagle rays, chimaeras, silurids, moray eels, weever fish, puffer fish, porcupine fish and scorpion fish are all poisonous. They are all sedentary fish that live in hiding. They sting mainly to defend themselves, since they are quite passive fish, and only rarely sting in order to catch their prey. The worst of them all, for humans as well as for almost all the animals that might attack it, is the stonefish. If it feels threatened, the stonefish suddenly unfolds its dorsal fin, on the front of which the first three spines have been modified to inject incredibly toxic venom. Even sharks are wary of it. Although the stonefish's sting is not usually fatal in humans, it is extremely painful.

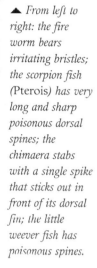

▲ *The black widow, one of the few spiders whose bite is lethal to human beings.*

Sworn enemies of man

Scorpions are invertebrates related to spiders, which sting to kill their prey or in self-defence, but do not always inject venom. Their venom contains a powerful neurotoxin that can cause serious complications in humans. Of 1 400 species, about 30, those of the Buthidae family, are really dangerous.

It is different with snakes: there are five million cases of snakebite a year and 200 000 of them are fatal, particularly on agricultural land. And 550 species out of about 2 400 represent a danger to people, among them cobras, mambas, kraits and coral snakes, belonging to the Elapidae family, and the vipers and rattlesnakes, of the Viperidae family. By far the most to be feared are the forest cobra and the green mamba. Their venom blocks the action of the brain, spinal marrow, muscles and heart. It also causes haemorrhages, necrosis, hypotension and generalised tissue damage.

Of the 40 000 species of spider on this planet, a good many are poisonous, but only about 10 or so are dangerous to humans. Among these are only one mygala, the *Atrax* in Australia and the notorious black widows.

The other poisonous land animals are all insects of the order Hymenoptera (adult females of ants, wasps, bees and

▲ *Raised upright, with its hood extended, the Samar cobra shows its attacker that it is ready to strike.*

▲ *The scolopendra is a fairly primitive relative of insects. It stings its prey in the head, using a pair of sharp appendages called forcipules. Its poison, similar to that of scorpions and spiders, is not very dangerous to human beings.*

hornets, among others), the duck-billed platypus, a mammal found in Australia and Tasmania, the beaded lizard (*Heloderma*), centipedes (or myriapods), shrews and almost all amphibians.

On the other hand, blood-sucking insects (mosquitoes, bugs, horseflies, fleas, lice and ticks) are not poisonous, although they can transmit viral or bacterial diseases.

Plant deterrents

The fire bug, the black and red bug so often found in gardens, is easily bred and is used today as a guinea pig for scientific research. In the 1930s an American laboratory attempted to breed these beetles and failed, because the larvae never reached the adult stage. But in solving the mystery the researchers discovered an unknown toxic substance.

It was noticed that the presence in the cage of newspaper made out of balsam fir stopped the larvae from moulting. This tree, which grows only in the north of the United States, secretes a substance closely related to the juvenile hormone in these insects, which condemned them to remain forever young and never to reproduce.

Poisons manufactured by plants are mostly intended to be a deterrent to herbivorous animals rather than to kill them. Only domestic animals are sometimes poisoned. Wild animals soon leave toxic plants alone because they find the taste unpleasant. However, if their range is restricted and they have no choice but to eat the plants they are affected (*see pp. 244-245*).

The defence of plants, however, is never total. For example the irritating substance in nettles so painful to mammals has no effect on the many caterpillars that eat the leaves. And the substance secreted by the balsam fir affects insects only.

Some insects have even found a way to use plant toxins to their advantage. The monarch caterpillar, a North American butterfly, lives on milkweeds, or asclepiads, a highly toxic plant. But the caterpillar is not affected by it at all. On the contrary, it recycles the plant poison, which serves to protect it from birds, to which the poison is toxic, and who leave the caterpillar in peace.

▲ *Juvabione, a substance manufactured by the balsam fir (above), has the same effect on the larvae of bugs (top) as the juvenile hormone that they secrete and thus prevents their metamorphosis into adults.*

THEN, VENOM, TO THY WORK

Venom is a complex liquid that certainly contains toxic substances, but also has enzymes, hormones, ions, proteins and antibiotics. These molecules, normally not dangerous, reach such concentrations in the venom that they become harmful. The venom can either be transmitted just by passive contact, or, more usually, by means of an inoculating device, with a shape sometimes reminiscent of an enema syringe, sometimes a plunger syringe, with an instrument on the end to inject the prey: a needle, a harpoon point or simply a jaw-like structure. On the right is a hornet's sting.

WHEN THE ELEMENTS RUN RIOT

Nature's towering rage

Rollers off the Californian coast, here at Santa Cruz.

Cosmic collisions

Since the beginning of the formation of the Solar System, thousands of stray celestial bodies have crashed into the planets they have met on their trajectory and made huge craters on their surface.

▲ *The asteroid Ida. Even small, irregularly shaped bodies in the Solar System have immense powers of destruction.*

▲ *If an asteroid were to hit Earth it would be disastrous. The power of such an impact would be equal to that of several thousand atomic bombs.*

In July 1994, observers actually witnessed a space disaster as it happened: fragments of the Shoemaker-Levy comet struck Jupiter. The impact released a large amount of energy: fragments measuring one kilometre across hit the planet at 60 kilometres a second, making colossal craters of almost 10 000 kilometres in diameter. Nowadays, phenomena as big as this are rare. But thousands of pieces great and small are still flying through space and no planet can be sure that it will be safe from them.

Birthmarks

Collisions were very numerous during the early times of the Solar System's existence, when very large bodies were moving through space. The collisions can be explained by phenomena linked to the creation of the planets. When they were made, many planetesimals (the basic and smallest elements of planets) rushed towards the centre of attraction represented by the Sun, colliding with many objects as they hurtled through space. All the planets, all their satellites and all the asteroids must have suffered these attacks from outside and still bear the scars. The enormous impact craters on the Moon give an idea of the huge amounts of energy released when these craters were formed.

Bombardment of the Earth

In the past, Earth was also violently struck by objects sometimes more than one kilometre long. The quantities of matter ejected into the atmosphere were such that the amount of sunlight and consequently the climate were affected.

An event of this nature might explain the disappearance of a large number of species (including the dinosaurs) at the end of the Cretaceous period, 65 million years ago. Little trace remains of these collisions. The craters have disappeared because of the unceasing geological activity on Earth. Of the signs that may still be seen, those dating back the furthest have been found in Quebec, Canada, at Clearwater Lake, where two impact craters measuring 32 kilometres and 27 kilometres in diameter are thought to have been caused by a meteorite that entered the atmosphere 300 million years ago.

Exploding shooting stars

It is thought today that more than 100 000 tonnes of matter enter our atmosphere each year at speeds of between 10 and 70 kilometres a second. Consisting of small specks or larger pieces, this shower of dust is partly due to the comets attracted by the Sun. On their orbit they sometimes cross the path of the Earth. Pieces of matter enter the atmosphere, which causes them to slow down, which makes them heat up. The resulting energy shows as a trail of light: these are the shooting stars that we can see flashing across the night sky.

Most of these bodies are destroyed as they enter the atmosphere, but it may very occasionally happen

◀ *Meteor Crater, Arizona, in the United States, is exceptionally well preserved because it is recent and lies in a dry region where there is not much erosion.*

▶ *On Miranda, a satellite of Uranus, a collision was the cause of this escarpment nearly 20 km high (1).*

▶ *Although it is protected by a dense atmosphere, the planet Venus has also suffered from strikes by celestial bodies.*

BOWLS AND BASINS

The study of craters, impact sites and the type of soil can yield valuable information about the history of the formation of the planets and their structure, and also about the nature and origin of the objects that struck them. Craters more than a few kilometres in diameter take the form of a shallow bowl, measuring 10 to 20 percent of their diameter in depth. The impact in this case caused matter to be ejected and created a rim all round the edge. Craters of a diameter of between 20 and 150 kilometres reveal a much greater impact. They have a central peak, resulting from a rebound, and steep inner walls where landslides occur, gradually building up concentric terraces. In craters larger than 200 kilometres wide, which take the form of a basin, the energy released on impact was such that many successive rebounds may be seen, forming a ring inside the crater.

that fragments measuring several tens of metres in diameter crash into the Earth. The frequency of this event is inversely proportional to the size of the object. Some 40 000 years ago, a metallic meteorite measuring 25 metres across, which corresponds to a mass of 65 000 tonnes, gouged out a crater measuring 1.2 kilometres in diameter and 150 metres deep in Arizona: the famous Meteor Crater. On June 30, 1908, a comet nucleus measuring 40 metres and weighing 30 000 tonnes exploded a few kilometres above Tunguska in central Siberia. No crater is visible, but over a 40 kilometre area trees were uprooted and thrown some distance in the opposite direction from the point of explosion.

One day, perhaps… ?

If a massive object were to hit Earth, it would cause considerable damage. Besides the direct destruction caused at the place of impact and by the associated shock wave, there would be thousands of tonnes of matter scattered in the upper atmosphere which would spread all round the globe. It would be so dark that plant life, deprived of solar energy, would die. Human beings, at the top of the food chain, would die out in their thousands. But such an event occurs rarely: it is estimated that very large celestial bodies only hit our planet three times in a million years, and that only one of these would strike a land mass.

▲ *In 1609, Galileo was the first to observe the Moon through an astronomical telescope and map the craters.*

▼ *The craters of the Moon. Today, space probes allow close observation of the signs of these colossal collisions.*

When mountains burst

The group of volcanoes that ring the Pacific Ocean is extremely dangerous. Their lava has characteristics that make their eruptions particularly explosive and devastating.

On June 9, 1991, Pinatubo, the volcano in the Philippines 100 kilometres from the capital Manila, began to erupt. On June 15, a fearful explosion rocked the mountain. The top of the volcano, which was 1 745 metres high, was blown to dust and lost 300 metres in height at one go. Debris was projected 40 kilometres high into the air. Whereas some volcanoes, like those of Hawaii and Réunion, are only slightly explosive, those in the 'Ring of Fire' of the Pacific are potential bombs.

Gases trapped in magma

If magma manages to foil the mechanisms likely to block it and make it crystallise at depth, it rises to the surface loaded with gases. Above a certain depth, these gases begin to form bubbles (a process known as vesiculation). The richer in gases the magma, the deeper the vesicles are able to form. Because of their very low density, these bubbles tend to rise faster than the magma. They become bigger when they meet and join, and because of decompression, rise more and more quickly.

But the growth and ascension of the bubbles may be checked by the viscosity of the magma. If it is liquid and poor in gases, as it is below the surface of Hawaii, the bubbles form not far down and rapidly reach the surface. On the other hand, if the magma is viscous and rich in gases, they form deeper down and remain trapped in the column of magma. In the end, the pressure they exert there exceeds the resistance of the medium. The whole column then undergoes an explosive rupture that surges to the surface and manifests as an eruption that is often catastrophic.

The gases and debris expelled by such an eruption, called a pyroclastic eruption, reach ejection speeds of several hundreds of metres a second and are projected several tens of kilometres high. The phenomenon is so violent that part of the volcanic structure may even be reduced to dust. The jet is first propelled vertically, making a column (called a Plinian column), then it spreads to form a bulbous, turbulent mass.

Pyroclastic flow: fallback type

Less than half a minute later, the bulbous head and the outermost part of the column begin to fall away, even while the inner part is still rising. A mixture of solid fragments still losing their gas and suspended in superheated gas falls back and flows down the sides of the volcano. This is how the notorious nuées ardentes are formed. Their travelling speed

▲ *The island of Montserrat, in the British West Indies. A nuée ardente flows down the slopes of Mount Soufrière during the eruption of 1995.*

▲ *A Plinian column, a vertical jet of gas and ash, billows from the crater of Mount St Helen's, in the United States, during the 1980 eruption.*

A PREDICTED DISASTER

Mount St Helen's, in Washington State, in the United States, had been dormant for 150 years. During the two months preceding the cataclysmic eruption of May 18, 1980, a large number of earthquakes had indicated a renewal of activity. At the same time, the northern slopes were progressively rising, because of the action of a lava dome swelling a few hundred metres below. When this dome grew bigger than 150 metres, the imbalance was such that the whole flank collapsed at once, causing a gigantic landslide. This landslide suddenly relieved the pressure within the volcanic edifice and thousands of cubic metres of water flowed in. Water coming into contact with the lava dome vaporised and caused a series of lateral explosions which expelled a mixture of rock pieces, ash and steam that destroyed neighbouring forests. The magma gases were then given off in explosions, which generated a wave of pyroclastic flows or nuées ardentes. The result of this catastrophic chain of events was that the northern flank of Mount St Helen's was completely removed and the top 350 metres was amputated.

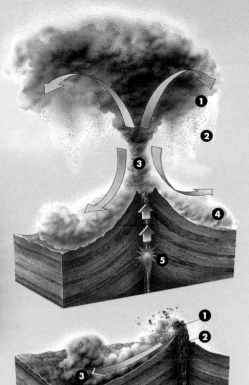

◀ **Fallback type**
1. Expanding bulbous shape
2. Ash shower
3. Explosive column
4. Fallback pyroclastic flow
5. Explosion of the magma column

◀ **Pyroclastic flow of the second type**
1. Explosion of part of the needle because of the pressure of gas
2. Needle of lava
3. Nuée ardente

THE TWO TYPES OF NUÉE ARDENTE

speed of the plume is often more than 50 metres a second. The shock and heat wave that precedes it is extremely destructive.

Pyroclastic flow type two

In some eruptions, the lava is so viscous that it is practically solid when it arrives at the surface, even though it has retained its gas content. Under the internal pressure of the magma, an extrusion in the form of a dome or a needle builds up at the site of the volcanic vent. In this scenario, the gases stored in the column of magma, often under a pressure of several thousand atmospheres, always eventually find their way to the base of the extrusion. Then they escape, blasting away part of the extrusion.

The explosion is directed horizontally and the resulting pyroclastic flow is far more destructive than the fallback flow. Its temperature is almost that of the lava (around 800°C) and it flows down the slopes at speeds that can reach 200 metres a second.

On May 8, 1902, in Martinique, a nuée ardente of this type flowed into the town of Saint-Pierre, having covered in just over a minute the nine-kilometre distance between the town and the volcano Mount Pelée. In a few seconds, the town was wiped out.

(between five and 40 metres a second) and their temperature (reaching several hundred degrees) make them formidable, especially as it is often difficult to foresee where they will go.

As the nuées ardentes progress, gases escape from them, taking with them the finest particles and constituting, above the nuées ardentes, a turbulent ash cloud, or plume, which can reach several hundred or even several thousand metres in height.

This plume has nothing to control it. Able to move uphill or cross ridges, it is totally unpredictable. And even if it does not contain much solid debris, the temperature of the gases within it may still reach several hundred degrees and the

▲ *Plymouth, 1997. The capital of Montserrat was devastated by a nuée ardente. Today, only a few ruins remain of the town.*

295

Killer quakes

The damage done by earthquakes does not depend only on the strength of the quake. Other elements increase the devastation. As there are no short-term methods of predicting an earthquake, damage prevention is more indispensable than ever in high-risk regions.

On January 13, 2001, a large earthquake (measuring nearly 7.9) destroyed part of El Salvador, in central America. A week later, on January 26 at 8.46 local time, another earthquake, measuring 7.7 brought grief to the state of Gujarat, in north-west India, on a vast plain in front of the Himalayan foothills. Several towns and villages were destroyed, and the death toll was thought to be nearly 100 000 in a region that had already suffered a catastrophe of this magnitude in 1819. Although the causes of these earthquakes are known (in this case it was a question of the convergence of the lithospheric plates of India and the Eurasian continent) it is still difficult to predict them and to control their effects.

When the ground turns liquid

Several factors may aggravate the effects of an earthquake. It will obviously cause a number of deaths in populated regions or where the construction of buildings is not up to standard. But the greater the size and duration of the seismic waves, the greater the damage will be.

The effect of the seismic waves depends on the depth of the epicentre. If it is not very deep, about ten kilometres or less, the waves are virtually not attenuated at all over the short distance they travel.

The nature of the substratum and of the soil near the surface is also of great importance. Earthquakes can be magnified in terrain made of alluvium or recent deposits, which are relatively unconsolidated and often waterlogged. These are likely to turn completely liquid and lose all cohesion. In India, this was what happened in Gujarat, when the ground collapsed and volcanoes of sand and mud appeared. The terrain is then unable to support buildings, as was the case in the Kobe earthquake, in Japan, on January 19, 1995. This phenomenon occurs frequently in coastal zones. On December 28, 1908,

▼ *The Japanese town of Kobe after the violent earthquake of January 1995.*

THE EFFECTS OF AN EARTHQUAKE

▲
1. A surface fracture following movement of the fault
2. Landslide
3. Flood (a dammed lake due to a landslide)
4. Rockfall
5. Surface fracture
6. Tidal wave (tsunami)
7. Liquefaction and subsidence (alluvial valley)
8. Liquefaction and subsidence of the shore

along the straits of Messina in Italy, several kilometres of sandy beaches and surrounding areas became liquid and collapsed into the sea, together with roads and houses.

The earth goes berserk

Regions affected by earthquakes suffer other upheavals quite as destructive as the collapse of the ground. In clay areas, large-scale landslides start. They may dam up valleys and create reservoirs of water. More than 200 lakes were formed in this way during the 1783 earthquake in Calabria, Italy.

When the rocks contain water, they spread in muddy flows. In El Salvador, 300 dwellings were covered and destroyed by mud flows only 10 kilometres away from the capital.

Earthquakes also cause rockfalls, particularly from high ground. The pieces of falling rock are sometimes several tens of cubic metres in size, and cause considerable damage, as was the case in the village of Braulins in the north of Italy in 1976.

If an earthquake is shallow and very violent, the movement of the fault propagates to the surface of the ground (*see pp. 182-183*). It then causes surface fractures that shift the terrain above them vertically or horizontally, together with all the buildings along their sides. Anything that has been able to withstand the shaking is cut in two by the subsequent fracture.

Prevention is better than prediction

If these catastrophic scenarios are to be avoided, danger zones must be identified and attempts made to lessen the risks. Today, the most vulnerable regions are well known. But in respect of regions where major earthquakes occur only infrequently (once every several thousand years), the past has to be studied.

Written evidence allows identification of the sites of old destructive earthquakes, which may then be compared with more recent events. Moreover, by studying the line of seismic faults or suspected faults by means of carbon dating of the different layers, it is possible to know how often the periods of activity recur.

Predictions in the long and medium term (from some tens of thousands of years to a few centuries) are possible today. But short-term predictions are much more complex, because they are a matter of determining the place, time and magnitude of an earthquake simultaneously. Animals are sometimes sensitive to certain signs. In order to detect as they do the phenomena that herald earthquakes, all the danger zones would have to have equipment in place that is as advanced as that available in California, which has one of Earth's major faults running through it. However, unlike California, many danger zones are in poverty-stricken areas.

In the absence of methods that are 100 percent trustworthy, we must count primarily on damage limitation. This takes various forms, from making people aware of the action to take in the event of an earthquake, to the organisation of emergency services, from the siting of dwellings to building techniques and composition.

▲ *Traditional dwellings did not withstand the 2001 earthquake in Gujarat, India, which was estimated to have caused more than 100 000 deaths.*

MEASURING THE DAMAGE

Several scales for measuring the intensity of an earthquake have been proposed since that of Mercali in 1902. They are based on observation or eyewitness accounts after a catastrophe. Almost all are graded from I to XII in Roman numerals.

The first degrees (from I to V)	• Effects felt by people • Movement of some objects (doors rattling, bells ringing, clocks stopping, etc.)
Degrees VI and upward	• Damage to buildings (from cracks to virtually total destruction of all types of building) • Topographical changes in the ground (cracks, landslides, liquefaction, etc.)

An earthquake's intensity decreases as you move away from the epicentre, situated on the surface directly above the focus. The scales given above are not to be confused with the Richter scale, which measures the magnitude – the energy – of earthquakes.

The heavies of the sea

Tidal waves, enormous waves caused by gigantic underwater earthquakes, are the most powerful of the natural cataclysms on earth.

The days of August 26 and 27, in the year 1883, were apocalyptic. The little Indonesian volcanic island of Krakatoa, lying between Java and Sumatra, was destroyed in the blast of a terrible explosion. Ash was ejected up to 80 kilometres high in the air. As the volcano collapsed, it set off a shock wave that caused a series of enormous waves along the nearby coastline, the highest of which reached 36 metres. Nearly 36 000 people lost their lives in this tidal wave.

The Japanese call a tidal wave a tsunami: 'tsu' meaning 'harbour' and 'nami' meaning 'sea', or 'the large wave in the harbour'. In fact, it was in Japan – at Awa in 1703 – that the most devastating tsunami in history occurred, thought to have killed 100 000 people. In comparison, consider the current estimate of deaths by tsunamis: in half a century, tidal waves along the Pacific coast have caused the death of 50 000 people.

Huge undersea earthquakes

Tsunamis are monsters from the deep. At depths of thousands of metres, the ocean floor is periodically shaken by enormous convulsions, like the eruption of undersea volcanoes or landslides. The worst of these disturbances in the earth are underwater earthquakes, occurring in regions of particular unrest where continental and oceanic tectonic plates converge. The oceanic plates, which are denser, tend to slide beneath the continental

◀ A tsunami breaks onto Sissano (north coast of New Guinea, Papua New Guinea) on July 22, 1998.

◀ A tsunami in the Pacific: a tidal wave devastates the coast of the island of Oahu, in Hawaii.

plates, causing a sudden rising of the ocean floor over immense distances. On May 22, 1960, 20 000 square kilometres of sea floor suddenly subsided off Concepción, in Chile. The enormous tsunami it caused swept away everything on both shores of the Pacific.

The margin of the Pacific, with its 'Ring of Fire', the perimeter of highly active volcanism, is the most common birthplace of this sort of activity. The principal tsunamis originate in one of the three most highly unstable regions, geologically speaking, on earth: the Chile-Peru oceanic trench, the Alaska-Aleutian Islands trench and the Kamchatka-Kourile-Japan trench. In spite of being a zone where tectonic plates meet, the Atlantic is much more stable.

At the speed of a jet plane

The initial quake at the origin of a tsunami occurs less than 50 kilometres beneath the ocean floor and has a magnitude of at least 6.5 on the Richter scale. The movement of the plates moves an enormous mass of water. But the shock wave created at the bottom of the sea, with a fantastic amount of energy, hardly shows on the surface of the water: there is just a trail of little waves not even one metre high, undetectable in any sea not dead calm.

The characteristics of this wave, on the other hand, are measurable. The distance between two wave fronts is between 100 and 200 kilometres, or about a thousand times greater than between two crests of normal waves. And its speed bears no relation to the speed of normal surface waves, which travel maybe 30 kilometres an hour; it reaches 900 kilometres an hour.

Walls of water 100 metres high

The tsunami, an astonishingly fast creature of the deep, rises from the abyss only as it nears the coast, where the wave triggered by the underwater earthquake comes up against the ocean floor. The friction, together with the more shallow water, causes monstrous waves to emerge suddenly, and rise to a height of between five and 10 metres and sometimes a lot more, up to 100 metres. The waves then follow each other at intervals of from a quarter of an hour to half an hour, continuing for about two or three hours in the case of the most violent tidal waves. The vast amount of energy contained in the wave enables it to travel, without noticeably diminishing, over distances of several thousand kilometres. The tsunami caused by the 1960 Chile earthquake reached the coast of Japan, 17 000 kilometres away, in less than a day.

▲ *In 1755, one of the most terrible tsunamis Europe has ever known reached the coast of Portugal. The pictures above show a simulation of its propagation from the time it was triggered by the earthquake. From left to right: one minute, 15 minutes, 30 minutes and four hours after the earthquake.*

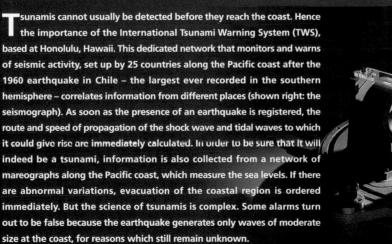

A VITAL ALARM SYSTEM

Tsunamis cannot usually be detected before they reach the coast. Hence the importance of the International Tsunami Warning System (TWS), based at Honolulu, Hawaii. This dedicated network that monitors and warns of seismic activity, set up by 25 countries along the Pacific coast after the 1960 earthquake in Chile – the largest ever recorded in the southern hemisphere – correlates information from different places (shown right: the seismograph). As soon as the presence of an earthquake is registered, the route and speed of propagation of the shock wave and tidal waves to which it could give rise are immediately calculated. In order to be sure that it will indeed be a tsunami, information is also collected from a network of mareographs along the Pacific coast, which measure the sea levels. If there are abnormal variations, evacuation of the coastal region is ordered immediately. But the science of tsunamis is complex. Some alarms turn out to be false because the earthquake generates only waves of moderate size at the coast, for reasons which still remain unknown.

Climatic cycles and abnormalities

In temperate latitudes, our life has been regulated for thousands of years by the four seasons. Apart from variations relating to latitude and longitude, they are quite distinct. At least, that was the impression that modern humans used to have. But since the end of the 20th century, we hear more and more about the changing climate and global warming, and people are worried, not unreasonably, about the role of the 'greenhouse' gases on the future of our environment. Nevertheless, the climate we know has not always been the same. It has been marked in the past by periods of warming and cooling, to the extent that the temperate regions on Earth have at times resembled the tropics and at times been covered in ice, or at least have strongly resembled tundra or northern boreal forest.

▲ **SOME OF THE TOOLS**
On the scale of humanity, it is difficult to identify climatic variations. On the scale of the planet, it is even less possible to be precise. There are, however, a few clues which indicate the state of the climate and the environment a long time before human beings appeared. One example is the air bubbles that have been found trapped in polar ice, discovered by borehole drilling down to almost 3 000 metres at the North Pole, which has yielded information about the climate on Earth 150 000 or even 400 000 years ago. Above, ice core samples in cold storage. Another example is the form and thickness of growth-rings of fossilised tree trunks like the conifers of the *Fitzroya* genus in South America that lived 50 000 years ago.

▶ **STUDYING CLIMATE: A CHALLENGE**
Climate may be defined as the series of atmospheric states above a place, in their usual sequence. People, who see only its consequences, gather data like the figures for atmospheric pressure, wind, temperature and humidity, on a scale of a few tens of years. The climate of a region is the product of the complex interaction between land, air and sea, even if the sea is a long way away. Study of it is empirical: climatologists make observations rather than forecasts, particularly as the role of the driving force behind Earth's climates, the ocean, which sets the air masses in motion and initiates the water cycle, is not fully understood. Development models for the main regional climates (tropical, temperate, arid, cold, polar, mountain) are therefore shaky, although less so for local climates and microclimates, as these are easier to follow. Shown right, a mountain valley in the Vorarlberg, Austria, which would have a microclimate that is easier to predict.

▲ GLACIAL AND INTERGLACIAL AGES OF THE QUATERNARY PERIOD

Human beings appeared on Earth in the Quaternary period. During this period (from 1.8 million years ago until today), our planet has undergone four ice ages interspersed with warmer phases, or interglacial periods. Ice ages, which generally last between 100 000 and 500 000 years, are due in particular to changes in the inclination of the Earth's rotational axis. Ice then covers the temperate regions and at the same time the sea recedes, that is to say the coastline advances because of a drop in sea levels. The interglacial periods are shorter, and characterised by a milder and wetter climate, with a rise in sea levels (the phenomenon called transgression) and increased forestation on land. Shown above, a glacial lake in Iceland.

▶ POST-GLACIAL WARMING

The warming (a rise in temperature and increase in rainfall) that followed the last ice age, between 10 000 and 6 000 years ago, is indicated by an increase in forestation at the expense of the steppe, which until then had covered a large part of Eurasia and Central America. Cold climate species found refuge in the arctic zone or in high mountain regions. Others, like the woolly rhinoceros and the mammoth, did not survive. Around the Mediterranean, where the evolution of climate has been studied at length, warming caused open environments like pasture, boscage and shores, to develop into more forested areas. About 8 000 years ago, a period of climatic equilibrium began, during which dense forest grew up, where oak trees in particular predominated. People formed small agricultural communities. About 6 000 years ago, deforestation was already noticeable. On the right is scrubland, or garrigue, in the Luberon region of France.

▲ NATURAL FLUCTUATIONS

Climate is the average of the natural fluctuations, going from the coldest to the hottest, the driest to the wettest, and so forth. When one of these fluctuations dominates, and finds expression, for example, in the appearance of flowers in the middle of the desert, we have the impression that the climate is disturbed. In reality, these so-called abnormal periods are only extreme fluctuations. The climate can only be considered to have changed when these extremes occur more and more frequently and when there is a lasting change in the way of life of living species because biomes have been modified, when the flora and fauna have been replaced by species typical of another climate and when the basic parameters of the climate (temperature, atmospheric pressure and precipitation) show distinctly different averages. But only repeated observations over a very long period would enable us to tell.

301

▲ WHAT HISTORY SHOWS US

The three great periods of economic growth in pre-industrialised Europe from AD 1000, essentially came about because of the development on the continent of a warmer climate, favourable to agriculture. Between 1000 and 1300, average temperatures everywhere rose by one or two degrees Celsius and humidity went down. The yield of wheat in some places reached levels approaching those in France at the beginning of the 20th century. People were better fed, lived longer and infant mortality rates fell. Between 1300 and 1480, cereal yields fell and the population decreased – a crisis due partly to the climate, which became wetter in the north-west of the continent. Between the end of the 16th century and the middle of the 19th, the European climate went through numerous ups and downs and some extreme conditions. The years from 1692 to 1694 are an example, when famine in France caused almost two million deaths. Reproduced above is a painting by Daniel Turner showing the Old London Bridge in the winter of 1795-96.

▼ DROUGHT IN THE SAHEL

Between 1968 and 1988, the Sahel, which lies between the Sahara and the tropical regions of Africa, suffered years of drought with almost no precipitation. Standing water and plant cover, which once consisted of bushes and acacias, disappeared and the Sahara desert took over, advancing to the south. At the same time, in Guinea the Sahelian zone expanded into what was once forest (shown below is the oasis of Tichit, in Mauritania). These periods of drought caused dramatic drops in agricultural production, which was reduced even further by plagues of locusts. A certain number of species of migratory birds, like the bank swallow, the whitethroat and the sedge warbler, which had flown from Europe to spend the winter in these bushy areas, saw their numbers dwindle. Few of them returned to nest in Europe.

▼ THE GREENHOUSE EFFECT

The uppermost layers of the atmosphere reflect a large amount of solar radiation back into space. The rest is absorbed by ozone, water vapour and carbon dioxide (CO_2) and also by the Earth's surface. Part of this energy is recovered in the form of infrared radiation, in the form of heat that is absorbed by the water vapour in the clouds, carbon dioxide and methane. And the atmosphere grows warmer. This is the greenhouse effect, without which the average temperature of the Earth would be well below zero. Since 1850, the amount of CO_2 has increased by a little more than 25 percent because of the burning of oil and coal. Amounts of methane have also greatly risen as a consequence of the explosion in numbers of cattle (whose flatulence releases methane) and of the extension of rice fields in south-east Asia (below, the island of Bali, Indonesia). After the harvest, flooding the paddy fields causes intense fermentation, which produces methane. The greenhouse effect has therefore increased. Does this suffice to explain the warming of the atmosphere?

Average global temperatures have risen by about 1.5°C since the beginning of the Industrial Age. Seasonal variations are becoming less marked. But over the last 10 years or so, specialists everywhere have been recording extremes of frequency and intensity (drought, torrential rain, cyclones, etc.), in each type of climate, and in each of the seasons. Knowing that it takes thousands of years for the course or average temperature of a mass of air or a sea current to be altered, is this not also a question of natural climatic fluctuation? And what is the precise role of the sea? Does it absorb more carbon dioxide than it gives off? In any case, human activity may modify microclimates, which are very sensitive to differences in the environment. A factory chimney is enough to form clouds, and towns are on average warmer by two or three degrees Celsius than their surrounding suburbs. Below, the grounds of the Chateau of Versailles, in France, devastated by the exceptional storm of December 1999.

▲ THE CLIMATE OF TOMORROW

It is difficult to give a precise forecast of the impact of global warming. An increase in average world temperatures would cause melting of the icecaps, a rise in sea levels (by between nine and 88 centimetres) and a change in the circulation of sea currents. In western Europe, for example, the increase in the amount of fresh water in the Atlantic would cause the salt water to 'dive' deeper, including the Gulf Stream, which would no longer warm the European continent. Europe might then find itself with a climate resembling that of Montreal in winter. More generally, a change in the course of certain sea currents (*see pp. 148-150*) would have an obvious effect on climate all over the Earth, notably by contributing to an increase in the number of cyclones in temperate regions. Shown above is a satellite image of a forest fire in Indonesia.

A NATURAL CLIMATIC DISTURBANCE: EL NIÑO

The eastern region of the Pacific (**1**) is under a high-pressure system. It is the opposite in the western region, above Indonesia (**2**). The trade winds (**3**) therefore blow towards the west, which creates, on the surface of the Pacific, a warm sea current (**4**) flowing in the same direction. The effect of this current is that there is more water in the west of the ocean and less in the east. A reverse current, and one which is cold (**5**) therefore forms below the first, to make up the water lacking in the east. Furthermore, as the layer of warm water is shallow along the coast of Peru and Chile, cold water rises (upwelling, **6**),which is rich in nutrients and feeds the fish populations on which these countries depend. Within a few years, this fine arrangement can turn inside out: this is El Niño. The Indonesian depression moves towards the centre of the Pacific (**7**) at the same time as the high South-American pressure areas weaken. Since the trade winds are not so strong, they are blown back by the westerlies (**8**), which make the warm surface current flow in the opposite direction (**9**). Thereupon everything is upset: even the monsoon is perturbed over the Indian Ocean, which results in floods in India and droughts in east Africa. Cyclones in French Polynesia, fires in Indonesia and Borneo, sandstorms in Australia, floods in South America: all these events have a common link, which is El Niño. This is, nevertheless, just a natural oscillation within the Pacific system.

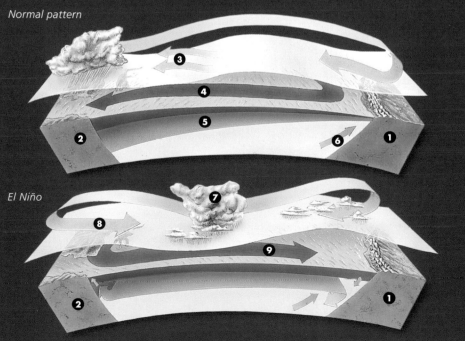

Normal pattern

El Niño

Maelstroms in the air

As a result of differences in temperature between the hot and cold regions on Earth, terrifying, eddying whirlwinds are fomented by the atmosphere.

▼ *Cyclones represent such a danger that the meteorological services put considerable resources into finding, tracking and, especially, forecasting them.*

On December 26, 1999, France was dumbstruck when it awoke. Never in the memory of meteorologists had the country ever suffered such a severe storm. And the nightmare was repeated two days later. Gusts reached 210 kilometres an hour. The final toll was nearly 100 dead and damage costs that reached into the hundreds of millions.

In winter, on average, one storm every 24 hours begins in the North Atlantic. This meteorological feature of temperate latitudes is caused by the combination of a jet-stream – a strong, constant wind

rising at an altitude of about nine to 10 kilometres off Newfoundland, Canada, which may blow towards western Europe at a speed of 400 kilometres an hour – and a depression. Depressions occur all the time at the point where masses of hot, humid air from the tropics meet the masses of cold air from the poles. The cold air, which is denser, slides under the warm air that rises, causing a lowering of pressure on the surface.

Depression and jet-stream, an alliance from hell

Although it takes effect near the surface, the depression begins to interact with the high jet-stream. Following its journey to the west as though on tramlines, it takes in energy from the jet-stream to feed the driving force of the winds that blow round its centre (*see p. 173*) in an anti-clockwise direction. At first moderate, the phenomenon later turns into an enormous wind tunnel. To the south of the depression, its winds join

▲ *The storm that devastated France on December 26, 1999 was exceptionally violent. It blew down millions of trees. Shown here is the forest of Der, in the Champagne region of France.*

forces with the system of westerlies predominant at these latitudes. To the north, on the other hand, they blow against it. The storm reaches its height at the east end of the jet-stream. The depression is at its maximum, the pressure falls (although it rarely goes below 980 hectopascals) and the winds suddenly grow stronger, before the depression rapidly collapses.

This explosive phase usually occurs off the European coasts. But on December 26, 1999, the jet-stream extended abnormally as far as Germany, and the storm reached maturity as it was crossing France. These temperate-latitude storms that last only a few dozen hours contribute, by the thermal exchanges associated with them, to making up the heat deficit between hot and cold regions. They are more violent in the hemisphere experiencing winter, because the work of redistributing energy is greater there than in the other hemisphere, where it is summer.

Is it the fault of global warming?

Were the unusual storms of 1999 a sign of climate disturbance caused by human actions? The increase in the greenhouse effect causes a rise in temperature and an increase in energy imbalances. So there is this extra work for the storms, which should, logically, be more numerous and powerful in order to make the surplus heat circulate between warm and cold points.

There is the same suspicion concerning cyclones, which play an exactly similar role in the climatic system. Scarcely three percent of the energy they release is used to make them function. The rest is all heat that they redistribute in the direction of the poles, via the upper atmosphere.

The energy of an atomic bomb

A cyclone in the Indian Ocean, a baguio in the Philippines, a hurricane in the Caribbean, a typhoon in the Far East and a willy-willy in Australia – these are all different names for the same phenomenon. There are only about 50 of them a year, but since they are multiplying and becoming more violent, there is cause for concern. These climatic monsters are the most powerful atmospheric phenomena. Every second, a cyclone, which covers several thousand square metres, releases an amount of energy equivalent to that of five of the bombs dropped on Hiroshima.

The wind can blow at 320 kilometres an hour, as recorded in the centre of hurricane Allen, in the Gulf of Mexico, in August 1980. They are synonymous with floods: Hyacinth, in 1980, dropped almost six metres of water a square metre on the island of Réunion in 10 days. And the waves they cause are more than 10 metres high, creating catastrophic tidal waves along coastlines.

In the eye of the cyclone

Cyclones are purely a feature of oceans, and tropical. They obtain their fuel from warm waters, and require a sea temperature of more than 26°C at a depth of about 100 metres and over an area more than 15 times greater than that of France.

▲ *Cyclone Fran approaching the coast of Florida in the United States (satellite image). These monsters of the atmosphere coil their cloud spiral over thousands of kilometres.*

▼ *In November 1998, cyclone Mitch destroyed 80 percent of the dwellings in Mangrove Bight in Honduras, central America.*

A VACUUM-CLEANER FROM THE SKY

A tornado is one of the most awe-inspiring atmospheric phenomena. It is a long column less than 400 metres in diameter, round which winds whirl at more than 500 kilometres an hour. In the centre, the pressure drops, so that the tornado acts like a gigantic vacuum-cleaner and throws trees, roofs, animals and cars into the air, and then drops them when it dies down. The spiral, which weaves about erratically, may remain stationary for a moment, and then move at a speed of 70 kilometres an hour over distances greater than 200 kilometres. Tornadoes are common in the centre of the United States, and occur when the masses of hot, wet air from the Gulf of Mexico come into contact with the cold, dry air coming down from the Rocky Mountains. The warm air rises and at the base of very dense cumulonimbus cloud causes the formation of a whirling funnel going down to the ground. Almost 500 tornadoes are formed annually in the United States. They may also form over water, and are then called waterspouts.

▼ Tornadoes sweep away everything in their path. Every year these violent whirlwinds are responsible for considerable damage and it is in North America that they occur most frequently.

If a depression forms at such a place, a formidable chimney gets to work. Every second, it sucks up tonnes of water vapour from the sea to form huge columns of cloud that climb 15 kilometres high. This megapump has only to develop at a latitude above 8° for the Coriolis force (which alters the course of winds and sea currents because of the rotation of the Earth) to start making the winds turn round it, clockwise in the southern hemisphere and anti-clockwise in the northern hemisphere.

Colossal streamers of cloud then wind in a spiral round a narrow central cylinder, which is the eye of the cyclone. This is a strange area, maybe 30 kilometres across, where there is a clear sky and almost no wind, but where pressure falls incredibly low (870 hPa were recorded one day in October 1979 off the Philippines in the eye of cyclone Tip, when the average pressure at sea level was 1015 hPa) and causes huge swells.

Catastrophic tidal waves

At the beginning of a lifetime that may last about three weeks, these spinning tops of wind and water move slowly (at about 30 kilometres an hour), from east to west in both hemispheres. Cyclones follow a parabolic trajectory that curves round to the north and then to the west. Then they reach temperate latitudes.

Because of the lack of warm water, their driving force begins to falter and the cyclone breaks up and becomes an 'ordinary' storm. They receive the same sanction if they approach land, which deprives them of their fuel. This fine theory, however, is sometimes found wanting. The ferocious cyclone Mitch, that ravaged Honduras in 1998, survived several days off-course above central America.

Coastal areas and islands have the most to fear from cyclones. As it reaches shallow water, a cyclonic swell raises waves several metres high, which carry all before them. The Ganges plain, low-lying and overpopulated, has experienced this tragically often. In November 1970, a cyclone killed some 300 000 people in Bangladesh. In November 1999, the Indian state of Orissa saw 15 million people made homeless and tens of thousands of deaths. Devastating cyclones also regularly strike Indian Ocean and South Pacific islands.

The treachery of snow

In a few seconds, an enormous quantity of snow begins to move down the slope, sweeping all before it. This is an avalanche.

▶ *Dogs give invaluable assistance to rescuers and help to find people buried under the snow.*

Very little is needed, maybe just someone skiing past, to make a mountain suddenly move. The spotless covering of snow changes in a few seconds to a fiendish moving carpet that travels down the slope with a sound like thunder.

An avalanche has started. Preceded by a huge rush of wind, the mass of snow grows larger as it sweeps along. It carries away and buries everything in its path – trees, rocks and cabins. In 1998, a group of 32 skiers was killed in the region of Gap in France. These natural disasters cause the death of dozens of people each year.

The culprits: powder snow, wind and a thaw

In mountains, the snow covering is deceptive. Its soft curves and even surface make it look very stable. But it is not. According to the slope and climatic conditions, it can suddenly lose its hold, become detached and an area of several hundred metres can plummet down to the bottom of the valley at a speed that can reach up to 300 kilometres an hour.

Avalanches tend to occur on steep slopes with an angle of between 30 and 45 degrees. On slopes with a less steep angle, the snow does not slide easily, and on steeper slopes, snow falls downhill before it has been able to form a thick cover.

Certain meteorological conditions are liable to set off avalanches. The first of these is heavy falls of powdery snow. This forms a covering on the existing snow layer which is very unstable in spite of its firm, smooth appearance. Powdery snow avalanches tend to be extremely fast and destructive.

The next factor is the wind. By blowing the snow crystals before it, it breaks them up and then binds them together

to form hard plates. It only needs a lack of adhesion to the layer on which they lie and sometimes just a little extra snow to make these plates start to slip downhill.

The third factor is rain or a thaw. As the snow particles become wet, they significantly disturb the balance that holds the blanket of snow in place. Wet snow, which is very dense, moves down the slopes much more slowly than powdery snow does, but it is just as destructive and causes just as much erosion as the other kind.

▲ *An avalanche above the ski resort of Telluride, in southwest Colorado, in the United States. Powder snow can move downhill at a speed of 300 km an hour.*

HOW CAN AVALANCHES BE PREVENTED?

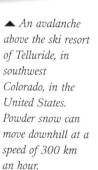

In order to try to prevent avalanches from starting, fences or nets are set up across routes at risk in order to hold back the snow covering. But trees continue to afford the most effective natural protection. Above a certain altitude, walls or tunnels can divert the course of an avalanche and slow it down. When there is certain danger and not enough protection, a preventive measure involving the use of explosives to start an avalanche has to be envisaged (pictured right).

When rivers burst their banks

Rivers in spate or flood are the cause of approximately 40 percent of deaths from natural disasters in the world. It is a tragic toll that could be reduced by greater protection of the areas at risk.

▲ *The Ouvèze in spate at Vaison-la-Romaine, France, in September 1992. Sudden heavy rain causes an increase in the rate of flow of watercourses, all the more suddenly if the ground is on a slope, cleared of vegetation or densely built up.*

In September 1992, torrents of rain fell onto the south-east of France, especially in the region of the Vaucluse. The Ouvèze, a small river flowing down from the Baronnies, was suddenly in spate and its swollen waters started racing furiously, causing the death of 38 campers on the river banks at Vaison-la-Romaine.

Flash floods

A flood of comparable violence occurred at Laingsburg, in South Africa, on January 25, 1981, devastating the town centre in a few hours, taking the lives of 104 people and leaving only 21 houses standing. Floods like these, called flash floods, happen when short but heavy falls of rain do not leave the water time to soak away. Streams suddenly fill and very quickly swell the river they flow into, causing a sudden increase in the rate of flow. These flash floods may occur in mountains, in gorges, on hillsides or on densely built-up slopes. They may appear within just a few hours, as in Laingsburg or Vaison-la-Romaine.

Because they are so strong, flash floods often sweep away soil, trees and any other material in their path. They may destroy bridges and houses built near the minor bed (the normal course) of the river, and cause the death of people who venture onto the banks in heavy rain or who try to cross at a ford.

Slow floods

On plains, rivers in spate cause floods that are called slow. Generally, rain falls for several days in succession, causing the water to rise gradually. Then the river overflows its minor bed and floods vast tracts of land.

The phenomenon can be aggravated by groundwater rising, by the water downstream slowing down because the banks have not been properly maintained, by obstruction caused by devices for crossing the river, and, near the sea, by exceptional tidal movement that prevents water from flowing into the sea.

In China, during the course of the last 3 500 years, almost 1 500 major floods of the Yellow River have been counted, which have caused the death of millions of people. Floods generated by cyclones also regularly devastate Bangladesh. The flood of 1991, for example, caused the death of approximately 150 000 people.

▼ *Dhaka, the capital of Bangladesh, in September 1998. A young woman crosses the flooded streets of the Rampura quarter with the belongings she has managed to save.*

◀ Flooded urban areas in Bangladesh. In this densely populated, poor country, storms and the catastrophic floods they generate all too frequently cause thousands of deaths.

▲ A flooded village in Bangladesh. Only a quarter of the land in the country is more than 3 m above sea level. When there is a cyclone, the surface of the sea rises as a result of low pressure and violent winds. Flooding is bound to occur.

Predictable magnitude

Computer analysis of the flow of rivers enables spates and disastrous floods to be reproduced in theoretical models, which integrate factors like climate, relief and whether there is building or plant cover. We know that deforestation, consolidation of small agricultural plots of land and the construction in urban areas of dwellings, roads and parking areas impermeable to water all decrease the amount of water soaking away and increase runoff.

Moreover, scientists think that global warming will probably lead in certain regions to an increase in the frequency of heavy rain. The gravity of floods or spates liable to occur according to a given average frequency can therefore be estimated, but unfortunately it is impossible to foresee when they will occur.

Preventive measures

The first step in taking preventive measures is to map carefully those areas that carry a high risk of flooding. If these are uninhabited, building on them must be forbidden or severely limited. If not, holding reservoirs should be built upstream, the flow of water in the minor bed should be contained by dykes and the evacuation of this flow downstream must be facilitated. Dedicated warning networks, which rapidly set safety plans in motion, should also be set up. These flood-warning systems take into account the risk of heavy rain, flow and overflow conditions in the rivers and the vulnerability of the areas involved according to population density, infrastructure and property.

Although it is possible to reduce the risk in this way, it must nevertheless be admitted that spates and floods will always remain one of the natural phenomena that cause the most damage to people and their property.

RUNOFF AND INFILTRATION

Rainwater is partially absorbed into the ground: this is infiltration. When the ground is saturated, the water runs over the surface and forms small streams that feed into rivers, which may also receive water coming from melting snow or underground reservoirs.

1. Urban area
2. Heavy precipitation
3. Deforested zone
4. Heavy lateral runoff
5. Low infiltration
6. Flood zone
7. Inflow of groundwater
8. Consolidated agricultural area

When the ground moves

A great variety of land movements change the face of the Earth. The most spectacular of them displace several million cubic metres of rock.

In 1970, as a result of a large earthquake, a mass of rock and ice fell from a summit in the Andes, the Huascarán (in the Cordillera Bianca region of Peru). It then began to slide downwards, taking with it quantities of snow and ripping up debris and pieces of rock from moraines. The whole accumulation fell more than 3 000 metres and travelled almost 15 kilometres, burying the village of Yungay and its inhabitants under a layer 10 or so metres deep.

It is thought that this avalanche of rocky debris mixed with ice and water may have reached a speed of 300 kilometres an hour. It even went up slopes and jumped obstacles, which made people think that it moved along on a cushion of air.

Dangerous slopes

The instability of large masses of rock depends on many factors. Understandably, it is more frequent in steep mountain regions. Old glacial valleys in particular have steep walls that are no longer supported by ice and these are particularly unstable.

On October 9, 1963, several villages were destroyed in the disaster of Vajont, in the Italian Dolomites, one of the most catastrophic land displacements in the 20th century. A whole slab of mountainside – about 300 million square metres of rock – crashed down into a dam, causing a sort of tidal wave that engulfed the valley of the Piave and swept away everything in its path. More than 2 000 people were killed as a result.

The whole chain of the Alps bears scars of these great landslides. Often, calcareous ridges collapse, forming huge piles of broken rocks, like that on the plateau of Assy, opposite the Mont Blanc massif.

Moving clay

Water makes clay rocks heavy and more likely to slide. They can easily begin to move even in places that are not mountainous, because a slight slope is enough. In regions with a tropical or wet climate, where very thick rocks (up to 100 metres) are broken down into clay, earthquakes may trigger potentially catastrophic land movement. Landslides following the El Salvador earthquake in January 2001 engulfed the village of Santa Tecla and killed hundreds of its residents.

When a landslide of clay rocks is limited to a small area and does not go deep, it can be controlled by stabilising work like draining off the water and building retaining structures. Unfortunately, this is not possible along the Normandy coast in the Calvados region, where the sea keeps the process going by digging out clay from the base of the landslide. Between Honfleur and Deauville the landslide of the Graves cirque, which reaches right to the beach, cannot cease of its own accord and the coastline continues to fall into the sea.

▼ *In 1970 in Peru, a densely populated valley in the Cordillera Bianca was completely destroyed by a landslide.*

▲ *In El Salvador, a landslide set off by the earthquake of January 13, 2001 buried some 300 houses and killed several hundred inhabitants of Santa Tecla village.*

THE COEXISTENCE AND ASSOCIATION OF SPECIES

Deals, alliances and solidarity

A clown fish and a sea anemone: two species in a symbiotic relationship.

A fruitful collaboration

To ensure their reproduction, plants must have their pollen, containing their male cells, transferred to other flowers. Many make use of animals as conveyors by producing deliciously sweet nectar for them.

Imagine we are in tropical Asia. A small nectar-drinking bird flutters around balls of mistletoe, which has explosive flowers. In passing, it pecks the brightly coloured flower bud. Under the pressure the bud bursts, ejecting pollen that covers the bird's feathers while it plunges its beak into the corolla to drink the juice. Without this intrusion, the flower would not have been able to open and would have faded, sterile. Avoiding this fate was well worth the sacrifice of a few mouthfuls of nectar.

The exchange of friendly services

Hummingbirds in the Americas, honey-creepers in the West Indies, sunbirds in Africa and Asia, honey-eaters in Australia, flower-peckers in south-east Asia, and white-eyes in New Zealand are some of the species of nectar-drinking birds, especially numerous in tropical regions. In the corollas of flowers they find their favourite foods: small insects and nectar.

As the birds fly from one flower to the next, they carry and deposit pollen and thus unwittingly take

◀ *Nectar-drinking birds like this sunbird from Senegal are particularly attracted to red- and orange-coloured flowers. They have a sufficiently long beak to reach the bottom of a narrow corolla.*

▲ *This bat's long snout and tongue enable it to lick the nectar from flowers and eat the sweet flesh of fruit.*

▼ *From left to right: a bumblebee has filled the pollen-baskets on its hind legs with the pollen it will carry away; the hummingbird moth gathers pollen from flowers while it hovers above them; the long, thin proboscis of this sulphur butterfly enables it to feed from almost any flower; the rose beetle eats the stamens of the flowers as its body is too smooth to retain much pollen.*

part in the plant's reproduction process. But they are not always indispensable (except to the explosive mistletoe) because insects and the wind also play the part of pollinators.

The close collaboration between birds and flowering plants goes back a long way. The most primitive species of many plant families are fertilised by birds, and these birds, or their ancestors, were already pollinating the first flowers to have evolved. Then bats, and particularly insects, in their turn served as efficient intermediaries. In temperate regions, insects are the only animals that guarantee the transfer of the precious pollen, except in North America where several species of hummingbird act as pollinators.

Harmonious partnerships

Flowers that attract birds are large, often red or orange in colour and have a deep corolla. They even have tissues strong enough to withstand pecks that are often far from gentle. Each plant is adapted to its pollinator. So flowers pollinated by bats or moths are strongly perfumed, whereas those that attract insects are generally smaller and contain a colour we cannot see: ultraviolet.

The partners in this arrangement have also evolved to be as efficient as possible: there is no chance that their tongues will be too short to reach the nectar.

Perfumes of the night

In the tropical night, little bats fly daintily to the flowers, attracted not by the colours but by the heady scent of the corollas, and cling to them with their claws, plunge their noses into the mass of stamens and lick off the nectar. Their atrophied jaws, their long noses with excrescences, and their long, rough, hairy tongues to collect the pollen or nectar show how well adapted they are to this particular diet. The pollen also collects on the fur of their backs or heads.

The survival of the African baobab tree, the American calabash tree, the Asian kapok tree and the Mexican candle cactus depends on these small flower-loving bats.

A job for experts

Birds and bats do not take particular care of their food source, which is evident in the damaged or torn flowers they leave behind after their visits. This same lack of consideration is also seen after the departure of certain unspecialised insects like earwigs, bugs or beetles.

But most insects, which represent the majority of pollinators, are well adapted and each has a speciality. Flies and small wasps have a fondness for tiny flowers grouped in a capitulum (as in the daisy family) or umbels (like the carrot). Bees, with

▲ *The bee orchid adopts the shape and smell of a female solitary bee. To increase the illusion, some areas of the flower reflect ultraviolet rays – invisible to us, but perceived by insects.*

their long tongues, and especially butterflies, with an extending proboscis, prefer to frequent larger flowers, which have a deep corolla.

The interdependence between flower and pollinator may be very close. In 1862, Darwin described an orchid from Madagascar which had a corolla 29 centimetres deep that contained four centimetres of nectar. He deduced from this that there must be an insect living in the same region with a proboscis measuring 25 to 30 centimetres in length. Although it was ridiculed in certain quarters, his theory was verified in 1903 by the capture of a sphinx moth that had a proboscis 25 centimetres long.

The honey bee, which from birth to death eats only nectar and pollen, bears the marks of its dependence on flowers. It has modified mouthpieces that form a tongue to lick up nectar and its feet are provided with brushes and combs to collect the pollen and a basket with a strong spike on which to fix the pollen balls that it makes.

Charms and traps

With its magnificent white flowers with red markings, measuring more than 40 centimetres in diameter, the giant stapelia of southern Africa has a strange beauty. Its perfume is just as unusual. In fact it gives off the sweet smell of decomposing meat, which is why its common name is 'carrion plant'. This plant has chosen to be pollinated by flies, which live off decomposing flesh. So you are not likely to find any at the florist's shop. The rafflesia

of Sumatra, the world's largest flower with a diameter of almost one metre, and arum lilies, have opted for the same strategy. The flies often find themselves trapped in a closed chamber where they become covered in pollen before they eventually discover the way out.

Certain orchids, like the bee orchid (*Ophrys*), deceive insects rather than feed them. By its shape and colour, the flower mimics the female of the visiting bee. But above all, the plant synthesises a scent that imitates the sex pheromone emitted by the female to attract suitors. The males, completely deceived by this female impersonator, come to mate with the flowers. They leave with pollen adhering to their bodies, ready to fertilise the next deceiving flower.

Fruit and vegetables at the mercy of insects

Did you know that without insects, fruit and vegetable stalls would be almost empty? Much research has been done by agronomists on the part insects play. For example, 60 percent of flowers on apple trees visited by bees bear fruit, compared with only five percent on apple trees protected from insects. And these few apples are less well formed than the others.

The pollination of certain closed flowers, like those of the legumes (peas, lucerne and clover), is not possible without insects. The growing of vanilla, a Mexican orchid, only became possible away from its region of origin when it was understood that it had to be hand-pollinated, in the absence of its natural pollinator.

AERIAL ACROBATS

Because each flower contains only a little nectar, animal pollinators are obliged to visit several flowers each mealtime. So why waste time landing and taking off again? Some insects (like the sphinx moth), long-tongued bats, hummingbirds and this swallowtail butterfly, refuel in flight. They are able to hover, and even fly sideways or backwards while feeding. To siphon out the nectar, they unroll a proboscis, put out a long tongue or plunge a long curved beak into the heart of the corolla.

Undesirable guests

Parasitism is widespread, whether it is bacterial, fungal or animal. Differing from predators in the way in which they search for food, parasites rely on mechanisms of an incredible complexity.

It looks like a terrific windfall. The shrimp-like gammarids, little crustaceans of which the lesser black-backed gull is so fond, have come out of hiding. Usually buried in the sand, these creatures have come to the surface and are wriggling about madly. The gull swoops on them. But the bird is going to make itself ill by swallowing them, because the strange behaviour of its prey has been brought about by a parasite. The gammarids have been infected by a flatworm, more exactly a trematode named *Microphallus papillorobustus,* which has lodged in their nervous system. By perturbing the freshwater shrimps, the parasite has succeeded in attracting the attention of its final, or definitive host – the gull – where it will be able to finish its development to sexual maturity.

Easy food

Less violent and brutal than species that hunt, parasites feed and grow by stealth at the expense of another organism, which they allow to remain alive for as long as it is necessary to them. Hunters get food, parasites get lodgings as well. There are three major types: those that live on skin (ticks and leeches); those that lodge in organs connected to the exterior of the body, like the intestines and the urinogenital system (the tapeworm and amoebas); and those that live in organs or tissues not communicating with the outside, like muscles, nerves, ganglions and blood (the plasmodium that causes malaria and the schistosome that causes bilharzia).

Suckers and clones

If you were to compare a parasite with a related non-parasitic species, you would see that the first is on average bigger than the second. This is because there is no competition for food in or on a host, which allows parasites to reach their maximum size. In addition, the attachment organs (hooks, hairs, suction discs) of parasites are well developed, at the expense of sense and motor organs. This is especially true of internal parasites.

Parasites are very fertile, and are also able to reproduce by cell division. Most pass through several hosts before settling down to reproduce sexually in their final host. This is the case with plasmodium (which infects human beings by way of the anopheles mosquito), and the schistosome (a fluke which lives on a freshwater mollusc and infects humans with bilharzia, or schistosomiasis, through contact with the larvae).

▲ *The* Planorbis *pond-snail (top) is one of the hosts of the schistosome, which carries bilharzia. The larva of the parasite changes to a sort of tadpole inside it, and is able to infect humans. Mildew (above) is a fungus that destroys crops, including potatoes.*

▲ *Plasmodium, which carries malaria, develops in the blood cells until they burst.*

◀▼ *Ticks, by biting their hosts, may inject them with agents of infectious diseases.*

Living cheek to cheek

Altruism is unknown in nature. But balanced relationships are sometimes established between two species, each depending on the other for their survival. This is symbiosis, which is much less common than parasitism.

▲ *Symbiosis with a tree is a necessary condition for survival for mycorrhizal fungi like this boletus.*

It is the 'hunting season' in Europe. Special dogs and even pigs are out sniffing the ground, searching for truffles. This fungus is not difficult for the animals to find as it has a strong fragrance. But they find the truffles only under oak trees; there are none anywhere else. There are even many of them under adolescent oak trees, visible to the naked eye as there is no grass under these trees. What makes the truffle live only under an oak tree? Food. And what makes the oak tree allow it to? Food.

The fellowship of fungus and roots

Oak trees and truffles are linked by a sort of feeding contract called symbiosis. The fungus receives most of its energy from the tree and in return provides it with essential minerals (such as phosphorus, nitrogen, etc.).

More precisely, like all fungi called mycorrhiza (which live at the foot of trees and include the boletus and amanita mushrooms), the truffle cannot digest the humus in the soil. Unlike other types of fungi, it is therefore unable to find for itself the organic carbon compounds it needs, like sugars and vitamins. But by inserting its very long mycelial filaments, or roots, into the tree's roots, it finds an abundance of these organic compounds, manufactured by the leaves of the tree by photosynthesis and carried to the roots by the sap.

The tree also benefits from the fungus. The mycelial filaments increase the absorption surface of its roots, which enables them to capture more mineral salts and water.

Furthermore, the fungus is able to produce substances like amino acids, which are useful for the function of the tree.

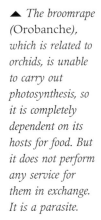

▲ *The broomrape (Orobanche), which is related to orchids, is unable to carry out photosynthesis, so it is completely dependent on its hosts for food. But it does not perform any service for them in exchange. It is a parasite.*

Stealthy profiteers

Many trees live in a symbiotic relationship with fungi, as do other plants. In some, the symbiosis is so efficient that the plants no longer have very much photosynthetic activity. The nutritional contribution of their root fungi is such that the plants barely need their leaves, which have become reduced and are little more than scales. This is true of gentians, heathers and other ericas, and orchids. But do the fungi involved profit from this

◀ *The clown fish and the sea anemone have established a truly symbiotic relationship. It suits both of them, and each would have difficulty living without the other.*

▶ *About 20 species of pearl fish in the Carapidae family live in the digestive tube of invertebrates. The fish (right) has set up house in a sea cucumber. Its relationship is one of parasitism rather than symbiosis, because the host animal does not benefit in any way from this forced association and can do nothing about it.*

association? Apparently not. As is often the case, symbiosis has developed into virtual parasitism to the sole profit of the plant.

The plants of the *Orobanche* genus, known as broomrapes, have become entirely parasitic. Lacking chlorophyll, and without a fungus to help them feed, these plants related to orchids are totally dependent on other plants and attach themselves to their roots to siphon off the sap. There are even those that specialise, attaching themselves to clover, thyme, ivy, sedum, and so on.

Animal symbiosis

Symbiosis also exists among animals. The best-known example is the relationship between the clown fish and sea anemones. The fish live in pairs between the tentacles of the anemone and, protected by mucus against their poison, are safe from predators. Moreover, they feed on the leftovers of the anemone and save it from death by suffocation.

Other fish have opted for an even closer form of protection. Certain pearl fish, like *Caropus apus*, from the time they become adults, live inside sea cucumbers and starfish. Protected against predators, they reproduce there. There is little or no advantage to the invertebrate in this arrangement. Here again, it is really a matter of parasitism rather than symbiosis.

Rarely a fair share

Examples of perfect symbiosis (also called mutualism) are in fact quite unusual. This is because the necessary conditions are not easy to meet. Symbiosis is by definition an association of living organisms in a relationship balanced between profits and losses for each of them. Each must benefit from it, gaining much while losing nothing. Often, symbiosis is vital for the species involved, which could not survive for long without it.

One of the rare examples of a perfect symbiotic relationship, perhaps the only one, is that of lichen, those attractive patches of colour on rocks, bark and even the stones of old buildings. Lichen is an organism that is formed by the association of an alga and a fungus. The union of these two life forms is such that the lichen is considered as a single organism, like any other plant. Lichens live in environments where their fungi and algae could not live on their own. They also manufacture substances that their symbiotic partner cannot. And they do not reproduce sexually.

How have lichens been able to adapt and evolve through the ages? This is still a mystery, as is the origin of this miraculous example of symbiosis.

▲ *Lichen is the quintessential example of symbiosis. The alga and fungus that form lichen are physiologically interdependent. Highly sensitive to pollution, lichen serves as an indicator of air quality.*

HIDDEN INTESTINAL ALLIES

A number of symbiotic relationships exist within organisms. Our intestines, for instance, right from birth absorb bacteria, protozoa and yeasts that, by developing, aid our digestion. This is symbiosis. It is particularly efficient in ruminants: without the bacteria and the protozoa in their stomachs, cattle would be unable to assimilate the indigestible cellulose of the leaves on which they graze. The same applies to termites, which would be unable to eat wood without the flagellate protozoa living in their intestines. Another profitable association is the one in flashlight fish (shown right), which harbour phosphorescent bacteria in pouches under their eyes, an ideal arrangement for attracting prey.

United we stand

For some animal species, hunting or fishing in a group is an efficient way to find food and consequently improve the chances of survival of each individual member.

▲ *White pelicans hunt in groups. They surround the fish (top) and then simultaneously plunge forward to catch them (above).*

Above a large lake in the Danube delta, white pelicans are circling. Suddenly they swoop down and settle on the water. Then they form a semi-circle and begin to fish, with their beaks wide open and wings outspread, in order to give shade so that the reflected light does not get in their eyes. They quickly surround their prey. The fish are trapped. Unable to escape, they will end up in the fearsome pouches of their predators.

Flocking to the scene

The brown pelican hunts in the same way as the white pelican. But it may also dive down, sometimes from a great height, with its neck stretched out like an arrow, to plunge its beak directly into the blue waters of the Pacific.

This technique is also used by the gannet, an Atlantic seabird. When it has found a shoal of fish, it dives down to it and seizes its prey in its beak. This manoeuvre will not escape the notice of one of its fellows, who will hasten to join it. Then another will come, and another…until in a few minutes' time there will sometimes be hundreds of gannets – or pelicans – to be seen busily fishing at the same spot. A gathering as large as this may also attract gulls, skuas and petrels, even dolphins and tunny, which are all fond of little fish. There may even be a fishing boat, as crews are well aware of the reason why such large numbers of animals congregate.

Elaborate strategies on land and sea

Dolphins also hunt in groups, but under the sea. More highly evolved than the seabirds, they communicate among themselves as they go after a shoal of fish by making sounds in the form of clicks and whistles, and by remaining in sight of each other. The sound activity plays its part in spurring on each member of the group.

Orcas (also known as killer whales), which are known to stay in the same group over 20 years or more and so have strong social cohesion, have developed highly sophisticated attacking and surrounding tactics. Working in a group of up to three individuals, these tactics allow them not only to hunt almost always successfully, but also to set their sights on prey larger than themselves, like whales.

The strategy of wolves is to work in a pack to hunt down their prey until it is exhausted, and then to surround it. The animals take turns to lead the hunt. Here again, their howling excites the pack and enables them to remain in contact with each member. Faced with the hunting pack, the victim has little chance of escape. Once more, it might be much larger than its predators, like an old reindeer or a male moose.

Lionesses also work in groups but surround their prey silently. While one or two females hunt the prey, the others approach stealthily and remain in ambush, ready to leap out at the unfortunate wildebeest or unwary zebra should it come their

▼ *Orcas do not only hunt out at sea. They sometimes land on a beach, either singly or in a group, near a colony of seals. Then they suddenly go for an animal that has not moved fast enough away from the water's edge.*

▲ The dance of the vultures round a carcass follows complex hierarchical rules. Dominance is not always linked to age or sex.

way. Hyenas use similar techniques in their nocturnal hunts (*see pp. 274-276*).

The signals of vultures

Vultures are scavengers and so do not kill their prey. They feed on the carcasses of animals that have been killed by other species and await their turn in the order of feeding, or on the bodies of animals that have died from illness or accidents. They devote the hottest hours of the day to gliding in the sky searching for their next meal. When a vulture finds food, it gradually descends lower and lower while flying in large circles and settles on the ground only when it is sure there is no threat of danger.

This manoeuvre attracts other vultures, sometimes from very far away, and they join the first one at the banqueting place. Each bird will

---

MEETING ON THE WEB

American spiders of the genus *Mallos* spin a communal web. When an insect is caught, all the members of the group hasten to the feast. What is remarkable is that there does not seem to be any fighting over the spoils. The spiders do not attack each other but share the prey. It is likely that at these times they emit a 'social' pheromone which prevents them killing each other.

---

have access to the spoils according to its rank in the hierarchy.

As vultures have the advantage of good long-distance sight, it is generally this that enables them to find their food supply. This tried and tested technique allows them to practise the fruitful policy of cooperative feeding.

Abduction by Amazons

Slave-maker ants, also known as Amazon ants, carry out raids on other species, such as aphids. But in addition, they periodically leave their own den and, led by 'guides', make their way to a neighbouring anthill. They put to rout the inhabitants, seize the eggs of their victims and make off with them.

When the captive eggs have passed through the nymph stage, the prisoners will be set to do all the domestic work of the anthill. Amazon ants are in fact well suited to fighting but are unwilling to take care of their offspring. It is this attitude that explains their energetic strategy of recruitment by press gangs.

▲ Like lionesses, wolves hunt in a group. Here they are trying to separate an individual musk ox from the herd.

▲ Slave-maker ants use ants that they have captured to do their domestic work for them.

In the tundra: the lemming cycle

In theory, a food chain functions in the shape of a pyramid. At the top is a super-predator. Below it is a succession of different species each of which preys on the smaller species below it in the chain. In reality, things are sometimes more complicated. In the arctic tundra, for instance, the fluctuation in the number of lemmings (small rodents about 10 centimetres in length that live in communities and are related to voles) undoubtedly has an effect on the predatory species above them in the food chain. But it also affects other animals that are quite unconnected to the lemmings.

▼ 1 - PORTRAIT OF A LEMMING
In winter when snow lies on the ground, the lemmings protect themselves from the cold by taking refuge in dens and burrows as deep as 90 centimetres underground. When spring arrives, they leave their winter quarters to look for food. In summer, led by the males, the animals return to their winter burrows. Lemmings can reproduce incredibly fast throughout the year and reach sexual maturity particularly early. But they have many predators, like the arctic fox, the stoat, and birds like the gyrfalcon, the skua (a sort of seagull) and the snowy owl.

3 - HAPPY PREDATORS

Usually the number of predators (right, a snowy owl) remains modest. But it increases in years when there is a population explosion of lemmings. An abundance of food can only encourage reproduction and allow the adults to raise their young adequately so that mortality rates go down. By the end of the season, the numbers are very high. There is then a close correlation between the number of predators and the number of prey.

2 - GOOD YEARS ARE DANGEROUS TIMES

Every three or four years there is a population explosion among lemmings. From a density of about one animal a hectare there is an increase to almost 200 a hectare. There are many different reasons for this proliferation: a decrease in the number of predators that year; a mild winter, which means more successful breeding; luxuriant vegetation in summer. What is stranger, is that every 30 to 40 years this density reaches a peak of about 2 000 individuals a hectare. Competition for space and food becomes tough, and the animals become aggressive. Obliged to move away, they undertake often lengthy journeys that may lead masses of them to the edge of water (shown in the drawing on the left), lakes, rivers or even the sea, and they will die trying to make the crossing. Consequently people speak of mass suicides because whole colonies of lemmings die in the course of these migrations. Their numbers then fall to very low levels.

4 - THE WADERS ARE SAFE

Small waders nest in the tundra: plovers, sandpipers, godwits, turnstones and geese of the *Branta* genus. In normal times, the young fall prey to predators. At a time of population explosion among the lemmings, though, the foxes and falcons leave the waders alone and instead hunt the small rodents, which are easier to catch. The birds can breed in peace and raise their young successfully, which results in a rise in their numbers.

5 - COMPLETING THE CIRCLE

In the spring following a mass exodus of lemmings. the predators (shown below, arctic foxes) have difficulty finding enough food and their ranks diminish. So they fall back on the birds that return to nest in the Arctic after spending the winter in southern and western Europe. The chicks are highly sought after, and few of the young survive to become adults. The numbers go down until the lemmings are once more abundant.

Joint ownership

Related species may well live together in the same habitat, as long as they make different use of it.

A scientist studying the insects of the tropical forest in Peru has recently discovered no fewer than 650 species of beetle living on the same tree. Studies conducted on other cases of cohabitation make it possible to suggest that each of these 650 species has found its own ecological niche, so that a different diet, and different times and places for meals allow each of the beetles to live on good terms with the others.

Allocation of living space

The trees in the South American rain forest give shelter to many other inhabitants, like tree frogs. Some of them live at the foot of trees in the leaf mould, others in the branches, others among the leaves, and others on the trunks. Some are also found at the top of the trees, in the canopy, and never go down to the ground. The numerous insects on which the frogs feed have also chosen to live in separate parts of the trees.

This division of territory occurs because the large trees of the rain forest experience variations in microclimate between their base and their canopy. They accordingly offer different micro-habitats, to which different animals adapt without competing with each other.

▼ *This Amazonian tree frog lives at the top of large trees and comes down only to lay its eggs. This means that it does not compete with the species of frogs that live on the lower levels.*

The coexistence of related species did not happen without some sort of special adaptation that allowed each species to exploit part of a habitat and an ecosystem to the maximum, but without rivalry. And when they live side by side it is essential, of course, that they do not go after the same prey.

Adapting for peaceful coexistence

The Galapagos Islands are the home of many animals, among them finches – actually a species of passerine – that were studied by Darwin and enabled him to develop his theory of evolution.

These species are morphologically very similar to each other and frequent similar environments. But over the course of time, their beaks evolved and adopted various different forms that allowed them to reach different food sources. This evolutionary process meant that species were not eliminated because of competition, but on the contrary were allowed to spread in quite an exceptional manner. In this limited environment, a single founding species has given rise to 14 others according to the process known as adaptive radiation.

The same thing has occurred among the Hawaiian honey-creepers of the Drepanididae family, which are endemic to the Hawaiian islands. According to their species and the shape of their beaks, which can be short and pointed, short and curved, or long and curved, they eat nectar, grain, fruits or insects, which allows them to remain on good terms in the same environment.

By the same process, bats that are exclusively insectivorous are thought by some scientists to have become nocturnal to avoid competing with other animals like swallows and swifts, which also eat insects. This theory is supported by the fact that in the Azores, a group of islands in the Atlantic Ocean where bats have no competition, they are both numerous and diurnal.

▲ *Although very similar in shape, each species of 'finch' in Darwin's Galapagos Islands has developed a particular beak shape. Each therefore feeds in a well-defined area. This reduces competition, which in island conditions might be considerable.*

Farming communities

Some complex ant communities have developed astonishing methods of animal farming and crop production in order to feed themselves.

The rosebush is full of aphids. One dose of insecticide and the gardener is rid of them. But in protecting the plants the gardener may have destroyed a whole farm.

Aphids excrete a sweet honeydew that ants love. This substance is made of the partially digested sap the aphids suck out of the plants. So to protect this source of honeydew the ants carefully farm herds of aphids, killing any predator that dares to venture near their 'livestock'. And just like good farmers, the ants regularly 'milk' their herds to extract the precious liquid. They have only to stroke the abdomen of the aphid with their antennae for the aphid to exude globules of the sugary substance from its anus.

▲ *This red ant is 'milking' an aphid. It extracts a sweet liquid substance of which it is very fond.*

Honeypot ants

Among the honeypot ants of North America and Australia a certain number of workers live permanently suspended and motionless in the nest. Since the honeydew on which this species lives is secreted by the aphids during a short period only, the workers act as larders for the colony and store the honeydew in their abdomens for their companions to eat when food becomes scarce. They end up looking like swollen drops of honey. The community will not forget to feed their helpful wet nurses in return.

Fungus growers

In South America, leafcutter ants of the genus *Atta* are actually capable of making fungus gardens, and control all the stages of its cultivation.

These ants spend their days cutting up the leaves of a tree or other plant. They may defoliate a whole plantation in this way. Then other ants in the colony chew these leaves very finely with their mandibles and spread the masticated substance on the ground to enrich the humus. On this spongy layer they will lay pieces of a microscopic fungus, which will grow on the leaf mixture until the fungus crop covers the selected area. When harvest time comes, the leafcutter ants feed on this fungus together with the sap of the chewed leaves.

▲ *These large hanging balloons are actually worker honeypot ants. Their abdomens are swollen with a sweet substance that will be food for their fellow ants, which will feed these worker ants in exchange.*

▼ *Leafcutter ants carry bits of leaf back to their anthill. They will use this material to grow the microscopic fungus on which they feed.*

Safety in numbers

The massing together of individuals of one or more species serves, above all, to provide better protection against predators. It also makes moving about and looking for food easier.

▲ For birds (above, a penguin colony), one of the reasons for breeding colonies is the reduction of an individual's exposure to predators. As part of a crowd, each individual has statistically less chance of falling victim to the claws or talons of predators.

There is a smell of guano in the air. A ceaseless babble of sound indicates the direction of the source and grows louder as you approach it. From the top of a hill you come upon it suddenly: an immense colony of penguins extending over several hectares. Paradoxically, a large concentration of animals like this gives an impression of vulnerability, as a meteorological accident could wipe out the whole colony in an instant. Yet the penguins gathered on this sub-Antarctic island in the Indian Ocean have good reasons for living so closely packed together.

Breeding in a group is safer

When animals congregate in order to reproduce, as in the case of the penguins, they all have to contribute to the protection of each of the members of the community, particularly against predators. But the protection they provide is passive. None of them has the responsibility of defending its fellows. It is just that there will always be more pairs of eyes on the lookout for danger and more birds ready to give the alarm and face the aggressor.

A species which breeds alone will have to devote a considerable proportion of its time guarding its nest and territory, and will have less time for searching for food and bringing up the young. In a group, however, the time allocated to keeping watch is shared among all the adult animals.

Feeding in a group is easier

Moreover, a colony is a communication centre. When the parent birds return with food for their young, they inform their companions of where supplies are most plentiful. This behaviour has recently been identified in the kittiwake, which nests on cliffs on either side of the Atlantic Ocean. This contact with information is perhaps one of the main advantages seabirds have when they live in colonies, and is probably also true of birds that have colonies on land, such as seagulls, pink flamingoes and rooks.

The same also applies to many gregarious mammals, like rabbits, meerkats and prairie dogs, and also seals and sealions, which form groups in a similar way for better defence against predators. Lions, baboons and deer too congregate in family groups to find food.

◄ Cave-dwelling bats congregate for hibernation and also for the protection of the group during the breeding season.

▼ These fur seals are going fishing. They remain close to each other, which allows them to stay in constant voice contact and to sound the alert if danger threatens.

Group rules

Among seals, harems kept by dominant males form colonies for protection against enemies (mainly orcas and leopard seals), but each lives as it likes and feeds when it wishes. On land, the females have only one rule: they must stay within the limits determined by the male.

Among bats, on the other hand, the colony observes strict rules for feeding times. At nightfall, all the individuals living in the same cave go out to hunt. They will all return at dawn.

Estuary birds (such as ducks and small waders) following the rhythm of the tides, scatter over the mudflats at low tide in search of food. It is every bird for itself. But as soon as the tide is in, they all regroup on patches of dry land remaining out of the water to clean themselves and sleep in these roosting places. As a flock, they are better able to deter predators.

Gregarious, up to a point

Between animals that live in huge communities (like weaverbirds and African passerines) and those that live alone, which means on their own territory where they do not allow any of their kind (like bears), animals adopt many intermediate kinds of behaviour.

There are also semi-communal species, animals which live in more or less loose colonies, which means that they are never very far away from each other but not on top of each other either. Each has its territory, but the proximity of individuals of the same species gives an added degree of safety. Among birds, small waders like lapwings, stilts and avocets are good examples, as is the larger snow goose. Among mammals, this way of life is found among rats and ground squirrels, among others.

Group travel

There is certainly nothing like a compact group when it comes to escaping from a predator. And group travel for migration has further advantages. Inexperienced individuals can learn routes known to the adults, and it is probably easier for several animals together to find food sources than it is for one animal alone.

So it is with whales and many migratory birds which migrate as a flock and stay together through the winter, whereas in the breeding season each tends to remain in its own place. This applies to swans, swallows and raptors like eagles and kites. Outside the breeding season individuals show less aggressive behaviour. From a physiological point of view, the gonads decrease in size. From a nurturing point of view eggs no longer need to be defended. Defending the territory is no longer of prime importance.

▲ *The red-billed quelea (above, far left) is said to be the most common bird in the world. Flocks, especially around watering places in the African savanna, can number several million birds. The black-winged stilt (above centre) and the guillemot (above right) are totally dependent on favourable environmental conditions to nest, and congregate by the thousand.*

THE DANGERS OF DENSE POPULATIONS

Beyond a certain point, the group effect that protects individuals gives way to a mass effect that threatens their lives. The presence of too great a number of animals may lead to a high death rate, a decrease in fertility among females and a slower growth rate in the young. Numbers are regulated until a lower population density allows the gregariousness that favours the survival of the species to be resumed. This phenomenon is observed among locusts and to a lesser extent among lemmings.

Animal alliances

Why not form a partnership with another species for the purposes of defence, health care or feeding? When the association is advantageous to both parties it is called 'mutualism'.

In the woodland and savanna of eastern South Africa, amid the general concert of birdsong, an expert ear can detect a sharp whistle sounding from time to time. A honeyguide is making its way among the leafy branches. This bird, related to woodpeckers, is very fond of honey and it has found a large wild bees' nest. But it is quite unable to deal with it. Is this a problem? Not really, because while it is making its way up above, whistling, below it there is a mammal on the ground following the sound.

The animal following the honeyguide is a ratel, or honey badger, related to the European badger. It too likes honey, and it is following the honeyguide in the hope that it will reveal the location of the bees' nest. When the bird reaches the spot, it gives a special cry. It is time for the honey badger to act. With its powerful claws it scrapes out the combs full of honey. When the badger has finished, it will be the honeyguide's turn to help itself, as the badger always leaves it some of the spoils. In South Africa, people have

also learned to follow the honeyguide. Traditional hunters, like the San, also leave an offering for the honeyguide.

Not everyone believes this charming story. Some scientists think that this behaviour is too complex not to be learned. But the honeyguide is a parasitic bird raised by parents of different species, which show no sign of this behaviour. More prosaically, other scientists suggest that the honeyguide only takes advantage of the leftovers of the ratel. But this hypothesis has not been verified either.

In the anthill, food is available around the clock

Ants produce a lot of waste matter in their nests. To get rid of it, they have recourse to 'cleaning ladies'. They allow into their nest a variety of small creatures, such as mites, spiders and collembolans or springtails, which feed on the rubbish left by the ants and find, besides good food, excellent protection in the anthill, so they settle there permanently.

Other ants open their doors to special caterpillars, those of the spring azure, a small blue butterfly. These caterpillars also receive board and

▲ According to a good story that not everyone believes, the ratel or honey badger, fond of honey and bee larvae, has a fruitful partnership with the honeyguide, a bird which is thought to lead it to bees' nests.

▼ The caterpillar of the Adonis blue butterfly is a real milch cow for ants, which enjoy the aromatic substances it secretes.

lodging, and in return they produce drops of aromatic substances of which the ants are very fond.

In the tropics, in Guyana, certain ants allow bees to coexist with them. In this case the detritus (such as dead larvae) discarded by the bees is appreciated by the ants as a source of food.

In Malaysia a bargain is made between the herding ant *Dolichoderus cuspidatus* and a mealybug of the genus *Malaicoccus*. The ants, which are nomadic, regularly take the mealybugs out of their nest to 'pastures', or juicy young branches, and back again. They live only on the honeydew produced by the mealybugs, and would die without it. In return, the mealybugs, which are not very mobile, need the ants to carry them about and to protect them against predators. Without the ants, they would soon die out.

▲ *The greater honeyguide. This inhabitant of tropical woodland is especially common in Africa, and to a lesser extent in Asia. It is the only bird able to digest beeswax.*

Cleaning service

Pilot fish of the *Naucrates* genus accompany sharks as they move about. For a long time it was thought that these fish guided the sharks towards possible prey. But in fact it has been found that they feed on the remains of the sharks' meals and at the same time keep them free of parasites and even tidy up their excrement.

The remora, or shark-sucker, has a sucking plate on the top of its head and usually lives attached by means of this to a shark (although sometimes it is fastened to a whale, turtle or dolphin). The remora acts as a hygienist, by regularly cleaning its host of the parasites lodged in its skin. In return it gets a meal, and also close protection. The shark is certainly not short of attendants. Wrasse wander round inside its open mouth and clean out all sorts of parasites.

In coral reef environments inhabited by dozens of species, other wrasse are hard at work. In six hours, as many as 300 cleaned fish have been recorded. Certain shrimps of the *Paraclimenes* and *Stenopus* genera also feed on the parasites of fish. They attract their 'clients' by waving their antennae at them from the bottom of a crack in the middle of the reef.

Guide fish for the blind

Shrimps of the *Alpheus* genus are blind. But the fish that is called Luther's goby, which is a marine bottom-dweller, is not. And the two animals both need to find food and lodging. Since the shrimp can dig, it might as well invite the goby into the hole it has dug. In exchange, the goby accompanies the shrimp when it goes out and guides it to food supplies. Outside the hole, the shrimp remains in constant contact with the goby by means of its antennae, which touch the fins of the fish. Lodging in exchange for food under these circumstances is a reasonable bargain.

The pest control experts

In the African savanna it is always hot and there are plenty of ticks, horseflies and other kinds of flies looking for a moist food supply. This they find in the form of the blood of large mammals like buffaloes, giraffes, elephants, hippopotami and rhinoceroses, which are perpetually bothered by

▲ *When it comes to cleaning a shark, remoras, wrasse and pilot fish get to work vigorously.*

▼ *This goby is guiding a shrimp to a food source. Gobies are essentially bottom-dwellers in temperate or warm seas.*

these bloodsuckers.. Fortunately, they have a friend in the bird commonly called an oxpecker (*Buphagus africanus*).

This passerine, which is the size of a large starling, specialises in feeding on the parasites that it finds in abundance on the skin of large animals. You will see small groups of oxpeckers stalking about on the back of a giraffe or an elephant, or even a cow, thoroughly scouring it all over in search of food, while the animal just leaves the birds to their task without showing the slightest sign of annoyance. Their guests sometimes feel so at ease that they go to sleep or even mate on the host animal.

In the Galapagos Islands, certain Darwin finches, too, act as hygienists for the marine iguanas, picking their skin clean of bothersome parasites.

Food providers

The large mammals in the temperate regions of Eurasia and America also need helpers to keep them free of irritating insects. Starlings and even magpies are sometimes seen perched on the backs of cows or sheep, looking for parasites to eat.

But this food can be obtained in a less energetic fashion. Grazing cattle disturb insects in the grass and periodically cause clouds of them to fly up into the air. This is a windfall for birds like cattle egrets or little wagtails, which in this way find an abundant and easily caught supply of food between the legs of horses and cows.

◀ With the utmost placidity, this rhinoceros allows oxpeckers to tickle its nostrils. This is because they free it from horseflies and other parasites.

THE HIDDEN TALENTS OF THE RAT

The rat is surely the most adept commensal creature, the species that has best been able to live in close association with humans. For centuries it has lived close to us and at our expense. Even if the rat was for a long time the carrier of serious diseases, and even if we wage constant war against it, it nevertheless plays an important role in the disposal of our waste. According to scientists, if there were no rats in the large cities in the world, the subsoil would be saturated with detritus and rubbish of all sorts.

NATURE'S RECORD-HOLDERS

The champions of the natural world

*A leafcutter ant carries a
piece of a stem to its nest.*

Colossi with feet of clay

Being large and tall is not without its drawbacks in the natural world. A large size, often found in conjunction with a complex organism, makes it impossible to adapt quickly to a sudden change in the environment.

On the Pacific coast of North America the sequoias, which can reach 110 metres in height, are the tallest trees in the world. Their longevity is also extraordinary. Many specimens are still going strong at more than a thousand years of age, and some may be as much as 3 000 years old. However, these trees, which appeared on Earth 65 million years ago, are today dying out, because of the deforestation that started a long time ago and the more recent pollution, among other things.

◀ The giant sequoia, which grows naturally only in California, grows quickly when young. But the wood is soft and is not used commercially.

Extreme size in the plant world

As far as records for plants go, the rafflesia wins the cup for the largest flower in the world. Its red corolla can measure 90 centimetres in diameter and weigh up to seven kilograms. This plant from south-east Asian rain forests is a parasite on species related to the wild grape. It is characterised by the fact that when in bloom it gives off the sweet, putrid smell of rotting meat.

▼ The rafflesia, the largest flower in the world, has fallen victim to its reputation for possessing medicinal properties, and is rare today.

▶ *The manta ray is endowed with pectoral fins that give it a 7 m wingspan, and it can weigh up to 3 tonnes. Every day, its huge mouth sucks in hundreds of kilograms of plankton.*

▲ *This bullfrog weighs at least a kilogram. Native to North America, it was introduced into the Gironde region of France in 1980, and is so happy there that it is a threat to the local amphibians.*

Local inhabitants thought that it had medicinal properties. These have never been demonstrated, but the belief contributed to its becoming rare in the wild.

The giant cactus, the saguaro, grows in Arizona and Mexico. It can easily reach a height of five or six metres, but one measuring 17 metres has been recorded. Its candelabra-shaped branches are a typical feature of deserts in the United States.

Small-scale giants

When it comes to animals, giants are found even among usually small species. Some Goliath frogs weigh almost a kilogram, and one was captured in Guinea that tipped the scales at 3.6 kilograms.

In the insect world, among scarabs, the Goliath beetle can measure 11 centimetres and weigh between 70 and 100 grams. But the record for size is held by a rhinoceros beetle, *Dynastes hercules*, an inhabitant of central America which measures up to 17 centimetres. The giant stick insect of the Malay Peninsula, *Pharnacia serratipes*, is the longest insect in the world, with a length of 55 centimetres.

190 tonnes in the sea

The animal kingdom acquired giants with the advent of the dinosaurs, but the largest animal ever to have existed is the blue whale, measuring between 24 and 27 metres in length and weighing between 100 and 150 tonnes. However, a specimen with a length of 33.5 metres and weighing about 150 tonnes has been identified.

What is the use of being so big? Its size intimidates its predators, the large sharks and particularly orcas. And an enviable advantage of being so big is that it can keep its body temperature warm, as there is less heat loss per square metre of skin surface.

In water, Archimedes' upthrust counteracts weight and allows the blue whale's skeleton to grow safely up to a certain physical limit. Aquatic invertebrates, with no internal skeleton, can grow much larger: 50 metres long, including the tentacles, in the case of some giant squid.

▼ *This giant Australian spider, belonging to the family of wolf spiders or Lycosidae, is characterised by its large size, and also by its three pairs of eyes.*

▲ Spider crabs of the Macrocheria *genus may have a spread of 3 m from the tip of one leg to the tip of another. These mud-dwellers live in groups along the Japanese coast, where they feed on organic waste.*

Giants that eat dwarfs

Sharks are also frontrunners in the giant stakes. The largest of them all is the whale shark, at between 13 and 18 metres long. This species feeds on plankton.

The organisms that make up plankton are only a few millimetres in size and not very nutritious, so any sea animal has to swallow considerable quantities of them each day.

The ancestor of the whale shark was large, and its descendants have been able to develop a stout and capacious digestive system that allows proper digestion. Whale sharks, which are filter feeders, have become enormous so that they can profit better from the modest plankton, and eat considerable quantities of it all the time.

The maximum size of predatory species, however, is much less than that of filter feeders. They spend so much energy hunting and feeding that they would wear themselves out if they were bigger, even if they were feeding continuously. This is the reason why the white shark and the tiger shark, tireless ocean predators, are only very rarely more than 7.5 metres long. This is large enough, all the same!

Growing big to eat leaves

On land, where gravity acts against large size and weight, animals are less huge. Over a certain limit, they would give way under their own weight. The African elephant is incontestably the largest land-based animal. It can reach four metres in height and weigh five tonnes.

The ungulates were the lords of the wild when their ancestors appeared. Ungulates were able to survive only by eating the toughest plants left by the other species. But hard leaves and fibrous stems are difficult to digest. The digestive system needs a long time to soften them enough to free the few nutritive elements they contain. Natural selection therefore encouraged successive species of elephant to grow bigger in order to accommodate longer and longer intestines and to possess strong grinding jaws.

What about the dinosaurs? The biggest of the herbivorous dinosaurs weighed up to 80 tonnes. Such a mass made them virtually invulnerable, but more importantly, in the hot period when they lived, it enabled them to regulate their body temperature. Natural selection therefore acted in their favour, as small reptiles were not warm-blooded, and mammals were too small to be rivals.

Dinosaurs ruled the animal world until a cataclysm occurred with a speed and a magnitude that did not give these large animals time to adapt.

HUNTED GIANTS

When they live in environments where they are too large to be threatened by predators, nature's giants cannot adapt quickly enough to the danger represented by humans. The mammoths disappeared after the last ice age, but they were already threatened by hunters. Of course there was more food if one brought down a mammoth rather than a roebuck. For the same reasons, of the members of the *Sirenia* genus, which includes dugongs and manatees, it was the largest that was the first to die out. This was the rhytina or Steller's sea cow, which was seven metres long. Of the birds, the moa of New Zealand and the elephant bird of Madagascar, each a sort of enormous ostrich measuring up to three metres tall, fell victim to indiscriminate hunting (right, the egg of an elephant bird). They died out, respectively, at the end and at the beginning of the 17th century. The dodos from the southern islands of the Indian Ocean died out too because of their large size, which was almost that of a turkey, their inability to fly and the fact that they lived in an island environment.

Uncommon strength

Being a Hercules does not necessarily mean that you have to be a giant. In the wild, small creatures are often capable of incredible feats.

▲ *The strength of insects (above, a rhinoceros beetle) is due to the hard surface and robustness of their carapace.*

The plantain is an astonishing thing. This modest-looking plant is actually an arrant opportunist, capable of dwelling in the heart of a city and growing anywhere at all, even on the asphalt of the pavements or bursting through tarred surfaces to poke its stem in the air. What power is it able to draw on to accomplish this? It is thought that the water pressure in its stems and roots, which slither into the smallest gap, is responsible for this achievement.

▲ *It is water pressure in the stem of the plantain that is thought to give it its strength and enable it to grow almost anywhere.*

Major powers

Far away from towns, in the tropical forest, live the strangler figs of the *Ficus* genus, to which other types of fig tree belong. Stranglers are large canopy trees, growing up to 45 metres in height, but they are able to grow only when they are attached to another tree.

The seed of the strangler fig germinates on the trunk or branch of another tree, and as the fig grows its roots eventually reach the ground. But strangler figs twine round their support and grow so vigorously that in the end they strangle even the largest trunks. The host tree finally dies from the pressure of this invasive guest. It is poetic justice that the strangler fig itself will also die without its precious support.

The indisputable champion among weightlifters is the rhinoceros beetle, which can carry about 850 times its own weight on its back. To achieve the same feat, a person weighing 75 kilograms would have to lift almost 64 tonnes. This is actually a customary achievement for many insects, made possible by their hard carapace, which gives them great mechanical resistance, and by their economical use of energy, which they reserve for carrying out demanding work.

In a strong position

Of nature's strongmen, the white rhinoceros and the African elephant are in the top rank. No obstacle can stop them for very long. Charges by a rhino are famous for being rare and for their extreme violence. This two-tonne 'truck', which can charge with a speed of almost 40 kilometres an hour, can knock down or shatter any obstacle less solid than a large baobab tree.

The African elephant, which can weigh as much as 6.5 tonnes, is famous for tending its territory. It uproots trees not only for food (it needs to eat about 140 kilograms of food each day), but also when it thinks they are too big or that there are too many for it to be able to move about freely. A herd of elephants can consequently make paths through wooded areas and gouge out waterholes in thick clay merely by travelling back and forth.

▼ *In tropical forests, ground space is limited, to the point where many of the trees are parasites on others. One example is the strangler fig.*

Small but tough

Although adaptation to environmental conditions has led to a reduction in size in certain animal species, the smallest of them often play a fundamental role in maintaining the balance of ecosystems.

When you walk along the paths in any great forest are you aware of the carnage you are creating? You may be a gentle walker, but with each step you are squashing, or at least upsetting, a miniature ecosystem.

In the midst of the dead leaves that are strewn over the ground there are flora and fauna that are invisible to the eye, but they fulfil an essential function. There are earthworms, for instance, which swallow earth, digest the organic particles, and actively participate in the decomposition of the humus. Over one square kilometre of forest, the biomass (the mass of living organisms) of large mammals is 220 kilograms, that of mice 500 kilograms and that of earthworms 100 tonnes.

▼ Like all shrews, the pygmy white-toothed shrew is a voracious mammal. Eating insects and sometimes other animals, it hunts day and night among bushes and in the burrows of small mammals. It uses its long whiskers as a radar system.

Small worlds

Very small animals live in large numbers in environments where prey is small, the living area limited (because even the smallest ecological niche must be occupied) and, paradoxically, the danger is great: the smaller one is, the easier it is to hide for safety. To be dwarf-size is an advantage in these environments.

This is the case in all biomes where organic matter is degraded, like the leaf mould on the forest floor, or the space between grains of sand. Waste-eating animals like mites and worms feed here on microscopic particles. Since each animal degrades the organic matter on which it feeds and makes it even more microscopic, the size of species within a biotope decreases the lower one goes down the food chain.

Other waste-eaters, like crabs, live between the tentacles of a sea anemone or among coral polyps and feed on the leftovers of their hosts and the sediment that might cause them discomfort. A multitude of minute animals also adhere to underwater plants, filtering the water on these promontories. This particularly applies to invertebrates like the hydrozoan jellyfish and the bryozoans, or moss animals, that encrust underwater surfaces.

Dwarfs among shrews

There are also dwarf-sized vertebrates. On land, the smallest is a shrew, the pygmy white-toothed shrew.

▲ A springtail, or collembola, feeds on young moss. Springtails may be the oldest insects in the world. They are wingless, do not undergo metamorphosis, and spend their lives on the surface of the ground. Their size varies between 0.5 and 0.55 cm.

▶ *The freshwater hydra is the simplest of the cnidarians, animals with tentacles. A few millimetres in height, this efficient predator can reproduce from pieces of itself.*

▶ *The adult harvest mite measures 1 mm in length. It is a particularly tenacious parasite that bites the skin of mammals in order to feed on their blood.*

This insect-eater lives in southern Europe. It is voracious, with a long, hairy snout, is 3.5 centimetres long, not counting the tail, and weighs between 1.5 and 2 grams.

It is difficult to imagine that a mammal's skeleton, heart and head can weigh so little. Yet the pygmy shrew forages and hunts just as well as the other types of shrew. But it no doubt fails to live as long as the others do because its metabolism is faster. Because of the high ratio between its surface area and its weight, the pygmy shrew loses a lot of body heat, and it constantly has to replace lost energy. So its heart beats more quickly. At more than a hundred or so beats a minute, the shrew's heart tires more rapidly than the heart of an elephant, which beats only 20 times during the same period.

The pygmy shrew has a counterpart in the air: a bat from Thailand called Kitti's hog-nosed bat, which weighs about the same as the shrew. It was discovered in 1973, and is very rare.

All families have their dwarfs

Examples of dwarf animals can be found in all animal families. Here are just some of the more noteworthy examples. A snake in the West Indies, the worm snake, or Lesser Antillean threadsnake (*Leptotyphlops bilineatus*), is only 10 centimetres long, which makes it the smallest snake in the world.

The Cuban frog is no larger than 1.2 centimetres, and a spider, the patu digua of Samoa, scarcely reaches 0.4 millimetres.

Similarly, the bee hummingbird, found in Cuba, is the smallest bird in the world. It measures less than six centimetres and weighs 1.6 grams. Its Lilliputian size is the result of competition between species. Faced with a struggle for survival against other hummingbirds with which it shares an environment – all hummingbirds eat nectar from flowers – the bee hummingbird grew gradually smaller, its only chance of survival being for it to become a dwarf. Its minute beak allows it to reach the nectar of tiny flowers that other species of hummingbird cannot reach. When it becomes smaller, an animal can exploit certain environmental resources abandoned by its neighbours or beyond their reach.

▲ *The bee hummingbird can gather nectar from the smallest flowers in tropical America. To do this it has to hover: its wings beat 78 times a minute and function like the blades of a helicopter.*

THE REVENGE OF THE MAMMALS

It was no doubt their size that saved the first mammals. At the time when the dinosaurs reigned supreme on Earth, mammals were chiefly nocturnal and small in size. Being easily able to hide from predators, which were probably not very interested in them anyway, as they were so small, they adapted more quickly to the hostile environment that preceded the disappearance of the dinosaurs and many other animals, 65 million years ago. So they survived and evolved to become as we know them today.

Tough enough for any test

Certain animal and plant species, which withstand changes in their environment particularly well, are able to establish themselves in any habitat.

▼ *A champion adapter, the rat has followed man all over the world, even though he wages constant war against it.*

Besides the crew and the provisions on board the three-masted vessel that sailed from Bordeaux, France, on November 23, 1746 for the islands in the South Pacific, there were also a few goats, two cows, hens, pigs, five cats and the same number of dogs. At least, that was the official list, for in the hold there were stowaways: black rats. Some months later, those rats were going to disembark on the enchanted shores of a Polynesian island and settle there peacefully.

At first sight harmless, this unintentional introduction was to cause a great disturbance in the ecological balance in this part of the world, because the rats proved to be redoubtable rivals of the native species.

◀ *At the end of the 19th century, starlings that had stowed away on boats first set foot on the soil of the New World. Today these birds are found all over North America and other parts of the world.*

Man's best friend

The two species of rat (the black rat, *Rattus rattus*, and the brown rat, *Rattus norvegicus*) have followed human beings in almost all their wanderings over the Earth, usually preferring to do this by stealth, as they have never had a good reputation. After all, the rat is the carrier of bubonic plague, typhoid and rabies. It was the cause of the huge outbreaks of plague that ravaged Europe between the 14th and 18th centuries. The Black Death alone, which spread from central Asia in about 1330, is thought to have killed almost 70 million people. The rat carries fleas, which infect humans with the disease when the fleas bite them.

Since then, the rat has continued to keep people company. It has left the countryside to live in towns, making its home below ground level, in cellars, drains and tunnels, where it manages to find food almost unnoticed. Completely omnivorous, an excellent

▲ *The few eggs laid by this bluebottle are enough to create a new fly population. In this way, carried by the various forms of human transport, flies have been able to colonise the whole of the Earth.*

swimmer, remarkably fertile and rapidly reaching maturity, it possesses all the advantages to be a winner at adaptation. The best proof is that it has learnt to coexist with humans – 3 000 years ago in the case of the black rat and only 300 years ago for the brown rat – whatever the circumstances.

Today the rat participates in all kinds of medical experiments. Without it, the great progress in treatment of the last few decades doubtless would not have been possible.

Born colonialists

The fly is also highly proficient at adapting. In the cabin of a large Airbus taking off from New York for Dubai are two unlisted passengers – flies. On arrival, they at once fly off, and moreover, if they are male and female, they will soon make sure they have offspring. It is because of this exceptional vigour that flies are found today throughout the world. In the heart of the Sahara or the Gobi Desert, far from anywhere, a fly will come to visit a motorist driving with the windows down and will travel in the vehicle for a few dozen kilometres. This is the way to colonise the whole planet with little effort.

Champions at survival

The waterbears of the phylum Tardigrada are perhaps the most extraordinary of all animals. They are minute (less than one millimetre long) and live in salt or fresh water. They look like a cross between annelid worms (like the earthworm) and arthropods (such as insects, arachnids and crustaceans). At the threat of danger, they first pull in their telescopic feet, then make themselves a second skin, or cuticle, which will totally enclose them. When these precautions are complete, they allow themselves to dry out. Then they are able to withstand temperatures ranging between -272°C and 150°C and also pure alcohol, ether and radioactivity, the only animals in the world able to do so.

Shut away like this and looking as though they are reduced to dust, these minute creatures are blown away by the wind. About 100 years later, they will awake in Greenland, the Himalayas ... or in a plant collection.

▲ *Invisible to the naked eye (they measure between 0.1 and 1 mm), the waterbears are gifted for adaptation. They can tolerate cold, heat, alcohol and radioactivity.*

DEADLY AND INVASIVE SEAWEED

Originating in the Caribbean, the alga *Caulerpa taxifolia* stopped over in several aquariums in Europe, where it acquired an incredible ability to adapt, before it ended up in the Monaco aquarium. In 1984, a few strands were thrown into the Mediterranean Sea. Contrary to all expectations, they survived in the cold water, for *Caulerpa taxifolia* is a 'mutant' which grows faster than other algae and tolerates the most polluted environments. Furthermore, it feeds both like an alga and a fungus, and it is toxic to most Mediterranean herbivores so it does not risk being eaten. Finally, it has learned not to fear the cold winters. Consequently this alga can take root anywhere and rapidly colonise the seabed where it is poor in nutrients. Today it has invaded 2 000 hectares between Genoa and Nice.

Fastest, highest, farthest…

Some animals travel a long way for a long time without sparing themselves. Others are amazing sprinters but sometimes tire quickly.

▲ *With its elastic skeleton, large thoracic cavity, slender legs and efficient heart and lungs, the cheetah is the world's sprinting champion reaching a speed of 110 km an hour.*

Before the fascinated eyes of tourists, two magnificent athletes engage in an impressive race of death. The 'athletes' are a cheetah, sprinting at 110 kilometres an hour, and an impala, not only a swift runner but also able to jump amazingly high. As the big cat's teeth are about to clamp on the antelope's thigh the cheetah slows down with fatigue, and the antelope escapes.

Speed, quick reactions and endurance are major advantages for survival in the wild. This is true for the prey trying to escape from predators and for the predators intent on filling their stomachs.

High-flyers and aerial speedsters

In the sky, acrobatics go hand in hand with speed. Some bird species are expert performers. The peregrine falcon and the saker falcon, for instance, are powerful speedsters with pointed wings for greater velocity. These wings enable them to hunt down other birds and catch them in mid-flight after dives at a speed of more than 150 kilometres an hour. Some record-holders even reach a top speed of 300 kilometres an hour. It is easy to see that these feats do not leave much hope for unfortunate birds like pigeons.

Vultures and eagles have another advantage. Their extraordinary respiratory adaptability allows them to endure extreme atmospheric conditions. They often soar at a height of between 3 000 and 4 000 metres above summits and sometimes even much higher.

But the champion of the skies, the perfect aerial athlete, is surely the swift. Its tapering body, sickle-shaped wings and short legs all contribute to its highly efficient aerodynamics. And this small bird spends the whole of its life on the wing (it covers almost a million kilometres a year, many of them while it migrates), sleeps and mates in the air, and lands in temperate regions only in order to nest. In the European winter, it spends weeks and months in flight in the equatorial regions of Africa. But it also excels in speed. During the breeding season amazing hunting-races take place, in the course of which the swift may execute dives of 200 kilometres an hour.

The Olympic Games of the animal world

If there were Olympic games that included all animal species, human beings would come last in all the races. Our best weightlifters, for

◀ *When frightened, the springbok, an antelope of southern Africa, can reach a speed of 90 km an hour and can perform leaps 15 m long while running.*

◀ *When alarmed, the young springbok whistles and then moves off in a bounding trot that changes to a full gallop, which gives it a 90 percent chance of escape.*

▶ *Rüppell's vultures are high-flying athletes. One bird, moving effortlessly because of a very fast air current, was encountered by an aircraft at an altitude of more than 11 000 m.*

instance, can lift about three times their own weight, the world record in 1998 being 205 kilograms. A male gorilla with an average weight of 300 kilograms, would in theory be able to lift 1.6 tonnes, as its muscle power is eight times as much as ours. This feat pales into insignificance compared with that of the ant, which can lift 50 times its own weight. On a human scale, this would mean that our champions could carry three tonnes at arm's length.

Pole-vaulting allows human athletes to jump as high as possible relying on their own strength. In this sport, the champions go over the bar at six

metres, or about three and a half times their own height. As a comparison, the klipspringer, an African buck twice the size of a hare, can jump nine metres high, that is, 15 times its height at the withers, to escape from a predator. For an equivalent jump, a person would have to jump 28 metres high.

We would also bring up the rear as far as pure speed is concerned. The greatest human sportsmen run a hundred metres in about 9.85 seconds, and reach a top speed of 36 kilometres an hour. This is a poor performance compared with the 72 kilometres an hour of the ostrich and the 110 kilometres an hour of the cheetah.

◀ *The powerful legs of the ostrich function like connecting rods on either side of a body that remains surprisingly still when the bird is running. Consequently the bird makes a series of long leaps more than running.*

▲ *No vertebrate goes deeper under the water than the sperm whale. One was detected by sonar at almost 2 300 m down. This is quite a feat, as the whale holds its breath while diving, like all mammals.*

There is no hope of victory for our swimmers either. Swimming 100 metres freestyle in about 48 seconds, our top athletes move through the water at 7.5 kilometres an hour. This means they would trail behind even the slowest fish and far behind the gentoo penguin (which swims at 35 kilometres an hour), dolphins (at 50 kilometres an hour) and the greatest sprinter in the aquatic world, the marlin, an undisputed gold medallist at 110 kilometres an hour.

A few leagues under the sea

With weights to assist in the descent and an inflatable parachute as an aid to return to the surface again, a good diver can go down to more than 100 metres under the water. Without this equipment the diver could go down as far as 60 metres just using fins and would be able to last without breathing for six minutes. It would be a very worthy performance, comparable to that of the sea-otter, but the difference is that the sea otter dives down to more than 90 metres every day without needing any preparation, either physical or mental.

The emperor penguin that lives in Antarctica has no need to practise either. Although it is a bird, it regularly dives to a depth of 265 metres to catch fish. A return trip takes on average about 15 minutes.

This result is not as impressive as the feats of the humpback whale (which dives to a depth of 355 metres and can survive 20 minutes of apnoea) or of the bottlenose dolphin (diving down to 540 metres but in just five to seven minutes).

At these depths there is not much competition, but even so, Weddell's seal, perhaps the most formidable diving machine in the animal world, can spend nearly an hour hunting at a depth of 600 metres.

The northern elephant seal descends 900 metres, but for less time (45 minutes). Its skills, however, are totally overshadowed by those of the sperm whale, which measures 18 metres in length and feeds mainly on giant squid. As these squid live in the abyssal regions of the ocean, the sperm whale has to go to find them at a depth of between 1 200 and 2 250 metres, where the pressure is 226 times that at the surface of the water. This obviously does not disturb the whale, as it can remain at these depths for nearly 90 minutes without breathing.

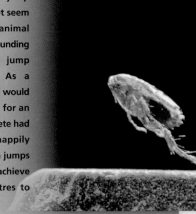

WINNERS OF THE LONG JUMP

The latest known record for a flea jump is 33 centimetres. This might not seem very spectacular, but for an animal measuring 1.5 millimetres it is an astounding feat nevertheless, because this jump represents 220 times its size. As a comparison, a person 1.76 metres tall would have to make a leap of 385 metres for an equivalent result. If this human athlete had to compete with a jerboa, which happily clears 4.5 metres, and a frog, which jumps 3 metres, he or she would have to achieve leaps of 80 metres and 17.5 metres to compare with these opponents.

Glossary

A

Acarians Very small arthropods, including ticks and mites. Some are free-living, like dust mites, and some parasitic, like ticks.

Accretion In astronomy, the aggregation of particles of matter due to gravitational force.

Acrosome A structure containing enzymes found at the end of a sperm, which facilitates the penetration of the ovule at the time of fertilisation.

Actin A protein found in animal cells, active in muscular contraction.

Actinista A class of fish of which the only non-fossil representative is the coelacanth.

Adaptive radiation A type of evolutionary adaptation which, from a common ancestor, leads to a great diversity of species each exploiting a different ecological niche.

Alkaloid A substance of plant origin with molecules containing at least one nitrogen atom and having a specific physiological action.

Alluvium Material (sand, mud and gravel) deposited by a watercourse on the bottom or sides of its bed.

Altocumulus A white or grey medium-altitude cloud made up of large white flecks or appearing as a sheet.

Altostratus A medium-altitude cloud forming a grey covering layer.

Anticyclone An outflow of air from an area of high atmospheric pressure.

Antiparticle A particle with the same mass as a given particle, but with an opposite electrical charge.

Aquifer A porous or permeable land formation containing water.

Aragonite A sort of rock made of crystallised calcium carbonate.

Archeobacteria A group of bacteria probably of very ancient origin with characteristics different from those of other, more modern bacteria.

Arenite Coarse sand resulting from the weathering of crystalline rocks like granite, gneiss and diorite.

Arthropods From the Greek 'arthron', meaning 'joint', and 'podos', meaning 'foot'. A phylum, or division, of invertebrate animals with articulated feet, represented by insects, spiders and mites, scorpions, millipedes and crustaceans.

Atmosphere A layer of gases surrounding the Earth and other celestial bodies (stars and planets).

Aven (French) A chasm in a calcareous plateau often communicating with underground passages.

B

Bactericide A substance that kills bacteria.

Biomass The mass of the total number of living things present at a given moment within a certain area.

Biome A sizeable ecological community.

Biosphere The total number of ecosystems in the sea and on the land.

Black hole A region in space where the gravitational force is so strong that light cannot escape.

Boson Name given to any particle that obeys a law of atomic physics called Bose-Einstein's law. Gluons, for example, are bosons.

Bryozoan From the Greek 'brion', meaning 'moss', and 'zoion', meaning 'animal'. Small aquatic invertebrates called moss animals that live in large communities.

C

Calcareous Describes rocks that contain calcium carbonate.

Calcite A type of rock made from crystallised calcium carbonate, of which stalagmites and stalactites are formed.

Calcium carbonate A mineral that makes up calcareous rocks, found mainly in three forms: aragonite, calcite and dolomite.

Caldera A large, roughly circular crater with vertical sides, formed by the collapse of the central part of a volcano.

Canopy The upper storey of tropical forests including the tops of trees.

Capitulum Group of small flowers clustered on the bulge of a stem (the receptacle), often looking like a single flower. An example is the flowers of the daisy family.

Carboniferous period The last geological period of the Palaeozoic era, from 360 to 286 million years ago.

Cavernicolous Used to describe an organism that lives in caves. By extension, also used of birds that nest in a cavity.

Cenozoic era A geological era that encompasses the present, and extends from 65 million years ago to today. It includes the Tertiary and Quaternary periods.

Centrosome A small cellular structure that plays a part in cell division.

Chiroptera The name of the order to which bats belong.

Chlorophyll The green-coloured plant pigment that captures the sunlight necessary for photosynthesis.

Chromatophore A pigment cell on the skin of some animals like chameleons and cuttlefish, which enables the skin to vary in colour.

Chromosome An element of a cell that is made of a DNA molecule coiled round specialised proteins.

Chromosphere An atmospheric layer round a star, such as our Sun, between the photosphere and the corona.

Cirrocumulus A high-altitude cloud forming lines of small white flecks.

Cirrostratus A high-altitude cloud forming light, whitish layers.

Cirrus Very high-altitude clouds scattering the sky with white streaks.

Class A taxonomic group into which a phylum is divided and which itself is divided into orders. The class Mammalia, for example, contains the order Carnivora.

Cnidarians A group of small, aquatic invertebrates, mostly marine, belonging to the phylum Cnidaria. Examples are corals, sea anemones, gorgonians and jellyfish.

Coleoptera From the Greek 'koleon', meaning 'case', and 'pteron', meaning 'wing'. An order of insects possessing a pair of hard, front wings called elytra, which form a protective case for the flight wings when at rest. An example is the scarab beetle.

Commensalism An association between two different species, one of which feeds on the remains of the other's meals, such as the relationship that exists between humans and rats.

Convection current A current that forms in an air mass because of differences in temperature between two regions.

Convective layer An area of the Sun situated below the photosphere.

Coriolis force A force due to the Earth's rotation, which diverts the trajectory of the winds and sea currents to the left in the southern hemisphere and to the right in the northern hemisphere.

Corona In astrophysics, the very thin outer layer of a star's atmosphere.

Cosmology The study of the Universe, its beginning and evolution.

Craton A very ancient, thick and stable part of the Earth's crust.

Cretaceous period The last geological period of the Mesozoic era, from 146 to 65 million years ago.

Crossopterygians A sub-class of mostly fossil fish with fleshy fins on a stalk like a limb, and comprising the Actinista and Rhipidista. The only living representative is the coelacanth.

Cumulonimbus A large, dark, puffy cloud moving vertically and producing rain- and hail-storms.

Cumulus A white cloud, level at the bottom and billowing at the top.

Cuticle Among arthropods, this is the rigid, impermeable outer layer of the animals, made of chitin and proteins.

D

Decapod A member of the Decapoda, an order of crustaceans in which the thorax bears five pairs of functional feet. Examples are shrimps and crabs.

Delta The outlet of a watercourse into the sea or a lake, where sediment has accumulated.

Dendrochronology A dating method based on the study of the growth rings of trees.

Depression In meteorology, an atmospheric disturbance corresponding to a low-pressure area.

Diapause In invertebrates, a period when the development of an egg, embryo, larva, etc. stops.

Dinoflagellates Also known as peridians, these are a phylum of aquatic, unicellular organisms that have two mobile filaments, the flagellae. These organisms form part of plankton.

Diptera From the Greek 'dipteros', meaning 'having two wings'. An order of insects with a single pair of wings. Examples are flies and mosquitoes.

DNA (Deoxyribonucleic acid) A molecule supporting genetic information made up of units called nucleotides, which are the 'letters' of the genetic code.

Dorsal ridge A range of peaks under the sea extending along the areas where oceanic plates diverge. It has a trench running along its axis, out of which basalt lava flows.

E-F

Echolocation The orientation method of some animals (like whales and bats), based on the emission of ultrasound and echoes of it that bounce off obstacles or prey.

Ecosystem The total community of living beings (biocenosis) considered together with the environment in which they live (biotope).

Elapids Members of the Elapidae, a family of venomous snakes from tropical regions. Examples are the cobra and mamba.

Electron An elementary particle with a negative electrical charge.

Elytra The pair of tough forewings of Coleoptera (beetles) and Orthoptera (grasshoppers and crickets) that, when the animal is at rest, cover the membranous wings used for flight.

Enzyme A protein with the function of hastening chemical reactions in living organisms.

Epicentre The point on the surface of the ground situated vertically above the focus of an earthquake.

Estuary The mouth of a river affected by tidal movements.

Exoskeleton The rigid, external skeleton of many invertebrate animals. An example is the shell of a crab.

Facula A hot, bright spot on the photosphere, the visible surface of a star.

Family A taxonomic grouping into which an order is divided and which itself is divided into genera. The family Felidae, for example, contains the genus *Felix*, which includes the cats.

Fault A break in the Earth's crust with the separated parts displaced in relation to each other.

Fungicide A substance that kills fungi.

G

Gabbro A crystalline rock of the same composition as basalt, which has slowly crystallised deep underground.

Gastropod A member of the Gastropoda, a class of aquatic or land molluscs that move along on a muscular foot (the word means 'stomach foot') and generally have a spiral shell. An example is the snail.

Gene A portion of DNA carrying genetic information corresponding to the synthesis of a protein, and a determinant in physical characteristics.

Genus The taxonomic grouping into which a family is divided and which itself is divided into species. For example, the seabird, the tern, belongs to the genus *Sterna*, of which there are several different species.

Geode A spherical, hollow rock lined with crystals on the inside surface.

Geodesy The study of the shape and dimensions of the Earth.

Geophysics A discipline of physics that studies the Earth.

Geothermal energy Energy produced by the heat of the Earth.

Geyser A hot water spring that spouts at more or less regular intervals.

Gluon An elementary particle responsible for interactions between quarks.

Gneiss A rock with a foliated structure containing alternating layers of dark mica and lighter feldspar and quartz.

Gonad A male or female organ that manufactures sex cells or gametes. An example is a testicle.

Granulometry Measurement of the size of the different grains or particles composing sediment (sand) and the relative proportions of each.

Greenhouse effect Restitution by the atmosphere, in the form of heat, of part of the energy from the Earth's radiation.

H

Hercynian Used to describe a geological fold formed during the Palaeozoic era.

Herd A group of cattle, or wild ruminants, such as elephants, zebras or wildebeest.

Homochromous (adj.), homochromy (n.). Protective colouring, enabling an animal to blend in with its surroundings.

Homoptera From the Greek 'homos', meaning 'same', and 'pteron', meaning 'wing'. An order of insects possessing two pairs of identical wings and feeding on the sap of plants. An example is the aphid.

Homothermal, homothermic, homothermous Warm-blooded; used to describe animals that maintain a constant body temperature, whatever the external temperature may be. Examples are mammals and birds.

Honeydew A sweet, liquid substance produced by aphids from the sap of plants of which they are parasites, and frequently farmed by ants.

Hormone A substance manufactured by a gland which, carried in the bloodstream to a target organ, modifies its action.

Hydrogeology The part of geology that deals with underground water.

Hydrographic Relating to the study of seas and watercourses.

Hydrolysis The decomposition of a molecule by water.

Hydrozoan A member of the order Hydrozoa of small, mainly marine, colony-forming invertebrates. An example is the hydra.

Hymenoptera From the Greek 'hymen', meaning 'membrane', and 'pteron', meaning 'wing'. An order of insects with two pairs of membranous wings. Examples are wasps and bees.

I-J-K

Infrared A light ray of which the wavelength, situated outside the visible spectrum, is below that of red.

Infrasound A sound with a frequency too low to be audible to the human ear.

Interstellar space Space between the stars in a galaxy.

Ionisation The transformation of neutral particles into ions (electrically charged particles).

Ionosphere The level of the Earth's atmosphere between about 60 and 1 000 kilometres above the surface, where there is a concentration of electrons formed from the solar radiation entering the atmosphere.

Isotopic dating Determining the age of a geological layer by studying the radioactivity of certain elements like carbon 14, which decreases over time.

Jet stream A regular, high-altitude air current that can blow at up to 400 kilometres an hour.

Jurassic period A geological period of the Mesozoic era, from 208 to 146 million years ago.

Kimberlite A magmatic rock that may contain diamonds.

L

Lapilli Very small volcanic projections with a diameter less than six centimetres.

Leguminosae A superfamily of plants with fruit in the form of pods. An example is the pea.

Lepidoptera The order to which butterflies and moths belong.

Light-year The distance travelled by light in a year, which is 9.46 million million kilometres.

Lithosphere From the Greek 'lithos', meaning 'stone'. The rigid outer layer of the Earth, made of the Earth's crust and the upper part of the mantle.

Lithospheric plate Also called a tectonic plate. Each of the pieces making up the lithosphere, the surface of the Earth, which constantly move in relation to each other.

M

Macromolecule A large-sized molecule, like that of DNA.

Macroplankton Those organisms among plankton that are visible to the naked eye, like jellyfish.

Magma The mass of liquid rocks in fusion that issues from the Earth's mantle.

Magma chamber A cavity in the Earth's crust underneath volcanoes, in which magma collects and where a volcanic eruption begins.

Magnetic pole The point of intersection of the magnetic axis of the Earth and its surface, slightly different from the geographical pole.

Magnetite Natural iron oxide.

Magnetosphere The area round the Earth subjected to the Earth's magnetic field.

Magnitude The amount of energy released by an earthquake. There are different scales of magnitude (including the Richter scale), according to the type of wave used in the calculation.

Mandibles The mouth appendages in arthropod invertebrates.

Mangrove A tree typical of swamps in tropical regions, often with submerged roots.

Mantle The inner layer of the Earth, between the crust and the core.

Mareograph An instrument used to measure sea levels.

Melon In zoology, a fatty lump in the forehead of whales and dolphins that plays a part in the emission of ultrasounds.

Meristem In botany, the tissue formed by cells able to divide.

Mesosphere A layer of the atmosphere situated between the stratosphere and the thermosphere.

Mesozoic era A geological era, between 245 and 65 million years ago, embracing several periods – the Triassic, Jurassic and Cretaceous.

Metamorphism The transformation at depth of solid rocks, caused by a large increase in temperature and pressure.

Metamorphosis In zoology, the complete modification in shape and structure of certain animals (especially insects and amphibians) that allows them to pass from the larval to the adult stage.

Metatarsal One of the bones in the foot.

Microplankton The part of plankton that is made up of microscopic species (unicellular algae, the larvae of crustaceans and molluscs, etc).

Mollusc A member of the phylum Mollusca, a division of invertebrates comprising animals with a soft body, usually protected by a shell. Examples are snails, mussels and cuttlefish.

Monotreme A very primitive egg-laying mammal of the order of Monotremata, which is confined to Australia, Tasmania and New Guinea. The only examples extant are the echidna and the duck-billed platypus.

Moraine Rocky material carried or deposited by a glacier.

Morphology The form and structure of organisms, including their size, appearance and patterning or colour.

Mustelidae A family of small or medium-sized carnivorous mammals with short legs, possessing anal glands that produce an unpleasant-smelling secretion, such as the skunk.

Mutation In genetics, a modification of the hereditary material of a living being, either spontaneous or artificially produced (for instance by radiation), and transmissible to the next generation.

Mutualism A durable association between two living species, from which each benefits.

Mycelium From the Greek 'mukês', meaning 'fungus.' It refers to the vegetative part of fungi composed of very fine filaments that are usually underground.

Mycorrhiza, mycorrhizal From the Greek 'mukês', meaning 'fungus' and 'rhiza', meaning 'root'. Symbiosis between the mycelium of a fungus and the roots of a plant.

Myriapod From the Greek 'murias', which means 'ten thousand', and 'podos', meaning 'foot'. A member of the group Myriapoda, which includes millipedes.

N

Nebula An interstellar mass of gases and dust that looks like a cloud; with stars, the main component of galaxies.

Necrophagous Describes an animal that feeds on dead organic matter.

Necrosis The death of animal or plant tissue.

Nectary A plant organ that secretes nectar.

Neoteny The persistence of the physical characteristics of the larval stage in an adult animal.

Neurone (also spelled neuron) A cell specialised in the transmission of nerve impulses.

Neutrino An elementary particle without an electrical charge and generally with no mass.

Neutron An elementary particle without an electrical charge that together with protons forms the nuclei of atoms.

Nidicolous Describes the young of bird species that after hatching remain in the nest where they are fed by their parents.

Nidifugous Describes the young of bird species that leave the nest a few hours after hatching.

Nimbostratus A low cloud forming a grey layer and giving lasting precipitations.

Nuée ardente From French, meaning 'a burning cloud', a glowing cloud of gas, dust and rock fragments caused by a volcanic eruption.

Nymph In insects, the immobile form, sometimes protected by a cocoon, assumed by larvae as they undergo the metamorphosis that will change them into adults. In butterflies, the nymph is also called a chrysalis.

O

Ocellus 1) The simple eye of invertebrates. 2) The round mark looking like an eye on the plumage or skin of an animal.

Ommatidium From the Greek, 'ommatos', meaning 'eye'. Refers to each of the elements forming a complex eye in arthropods.

Ophiolite A group of stratified rocks thought to have originated in the lithosphere of the sea but now exposed by upthrust and visible on land.

Order A taxonomic grouping which is a division of a class and which is itself divided into families. For example, the order Carnivora contains the family Felidae, or the cats.

Organelle A specialised part of a cell serving as an organ.

Ozone A gas of which each molecule is made of three oxygen atoms. The ozone occurring naturally in the upper atmosphere absorbs a large part of the ultraviolet radiation from the Sun. In recent decades it has been diminishing in areas, which has allowed through higher levels of ultraviolet light.

P

Parasite An animal or plant that lives on or in another living organism and feeds on it without offering any benefit in exchange.

Parhelion A very bright point of colour appearing beside the Sun, due to the reflection of light on ice crystals.

Parthenogenesis From the Greek 'parthenos', meaning 'virgin', and 'genesis', meaning 'birth'. A form of reproduction whereby an embryo develops from an unfertilised female sex cell (or ovule). It is practised by some fish and aphids.

Passerine A bird belonging to the very large order Passeriformes, which generally includes small songbirds that perch and most living species of birds. Examples are tits and robins.

Patagium A membrane found in some animals (like flying squirrels), which extends along the flanks between the front and rear legs and allows them to glide when leaping from one tree top to another.

Pelagic Describes an organism found in the open sea.

Percolation The infiltration of water into porous soils.

Peridotite A magmatic rock that is found in the upper part of the Earth's mantle.

Permafrost An area of land that is permanently frozen.

Pheromone A chemical substance produced by an animal which is detectable from a long way away by an individual of the same species. It often has a function in attracting mates.

Photogenesis The production of light by some living creatures. It is synonymous with bioluminescence.

Photon An elementary particle carrying light.

Photoreceptor A specialised cell in the retina that captures light and transforms it into a nerve impulse.

Photosphere The visible surface of a star or the Sun, situated below the chromosphere.

Photosynthesis The process by which plants manufacture organic matter using water and carbon dioxide from the air and solar energy.

Phreatic Refers to underground water near the surface which comes from precipitation.

Phylum A taxonomic grouping into which a kingdom is divided and which is itself divided into a subphylum and a class. For example, the phylum Chordata contains the subphylum Vertebrata (the vertebrates) which itself contains the class Mammalia, or mammals.

Phytoplankton A form of plankton made of plant organisms.

Planetesimals Elements that by aggregating end up by forming planets.

Planetoid A planet in the process of being formed.

Plankton Name given to the microscopic animals and plants floating in the sea or fresh water and moving according to the currents.

Plasma In physics, a fluid made of gas, ions and electrons, with properties different from those of natural gases.

Plate tectonics Movements at the surface of the Earth of the plates that make up the lithosphere.

Pollen A powder produced by the stamens of flowers composed of a multitude of grains each of which contains male reproductive cells.

Pollinator An agent responsible for carrying the pollen from one flower to another. The main pollinators are animals that feed on nectar (birds, insects, etc), the wind and water.

Polyp A small marine invertebrate often secreting a rigid calcareous shell,which lives attached to a support, singly or in colonies (such as coral reefs).

Precambrian period Geological period extending from the formation of the Earth (about 4.6 thousand million years ago) to the Palaeozoic era (about 540 million years ago).

Precipitation In meteorology, rain, hail, sleet, snow or dew; the amount of rainfall etc.

Prion An infectious particle related to proteins, an accumulation of which causes brain lesions. Prions are incriminated in mad cow disease and Creutzfeldt-Jakob disease.

Prism A multi-faceted solid with two identical parallel bases (polygons) and whose sides are parallelograms. White light is reflected in a prism, emerging as a rainbow.

Protist A member of the kingdom Protista, a large group of unicellular organisms including bacteria, Protozoa and algae.

Proton A positively charged particle which, together with neutrons, forms the nuclei of atoms.

Protozoa A phylum of unicellular animal organisms.

Pulsar A strongly magnetic, dense celestial source emitting various types of radiation at very regular intervals.

Pupa The nymph of dipterous insects like flies.

Pyroclastic flow A flow of fragments of volcanic rock that moves rapidly downhill during a volcanic eruption.

Q-R

Quark A particle which is part of the composition of hadrons (a family of elementary particles including neutrons).

Remiges The large, stiff feathers in the wings of birds, used for flight.

Richter scale A scale of measurement for earthquakes that goes from 1 to 9.

Rift valley A succession of subsidence trenches extending over a distance of several hundred or several thousand kilometres on land or under the sea.

S

Schistosity The parallel arrangement, or foliation, of certain rocks, like schist.

Sensillum A sense receptor, of taste, smell or touch, in some invertebrates (arthropods).

Sexual dimorphism Differences in shape and colour between the two sexes of the same species, other than the difference between the sex organs themselves.

Shield volcano A broad volcano with a flattened cone formed by very liquid basalt lava flows.

Silt Deposit of fine particles, finer than sand grains, carried by water or wind.

Sleet Precipitation consisting of drops of water frozen on their surface.

Solar wind The flow of particles escaping from the atmosphere of the Sun.

Species A taxonomic grouping into which a genus is divided.

Spermaceti A whitish oily substance contained in a reservoir in the head of the sperm whale.

Stamen The part of a flower that produces pollen.

Stratocumulus A low-altitude grey cloud bringing occasional rain.

Stratovolcano A volcano with a cone formed of pyroclastic rocks alternating with lava flows. It is synonymous with a composite volcano.

Stratus A low, uniform cloud forming a grey layer.

Subduction The movement whereby one lithospheric plate sinks below another.

Subsidence Downward movement of an air mass or part of the Earth's crust.

Supernova A short and very bright explosion which is one of the last stages of very large stars. Because their nuclear force is exhausted, the balance between this force and the gravitational force is disturbed, they become unstable, and an enormous explosion is produced.

Swim bladder The sac full of gas possessed by most fish that enables them to float.

Symbiosis A permanent or even obligatory association between two living organisms, from which each benefits.

T-U

Tarsal In vertebrates, the bones forming the rear part of the foot. In insects, the end of the foot.

Taste bud A group of cells in the tongue that is specialised in the perception of information about taste.

Taxonomy The classification of plants and animals into groups based on certain similarities. These groupings, the names of which have Latin and Greek roots, have a hierarchy and descend from kingdom to phylum, class, order, family, genus and species.

Telluric Adjective meaning of or similar to the Earth, as in 'a telluric planet'.

Telomere Each of the two ends of a chromosome.

Tethys Sea The ocean that, 150 million years ago, separated the two continents on the Earth's surface, Laurasia and Gondwana.

Thermonuclear reaction The interaction of atomic nuclei during which great amounts of energy are produced by the formation of larger nuclei (fusion).

Thermosphere Part of the upper atmosphere extending from about 60 to about 1 000 kilometres and containing electrons from solar radiation. It is synonymous with the ionosphere.

Tibia In insects, one of the leg segments.

Topography The configuration of the landscape features of a region.

Trace element A metallic or non-metallic element present in animal and plant organisms in minute quantities but indispensable to their function.

Travertine Calcareous deposit formed of grey or yellowish concretions in regular layers, often lying on the surface.

Trematodes Members of a class of parasitic flatworms, the Trematoda. An example is the liver fluke.

Troposphere The lowest layer of the atmosphere.

Tsunami Gigantic wave caused by an underwater earthquake that affects the surface of the sea and causes a high tide and coastal flooding.

Tundra A region surrounding the area of the North Pole characterised by bushy and herbaceous vegetation, often dwarf species, and animals that change colour in autumn and spring to blend in with their surroundings.

Ultrasound Sound with a frequency above 20 000 hertz, inaudible to the human ear.

Ultraviolet Light radiation invisible to the human eye. In the spectrum it is situated beyond violet (with a wavelength shorter than that of violet).

Umbel Group of flowers on pedicles of the same length that all start from the same point on the stem. An example is the parsley flower.

Ungulates The name given to herbivorous mammals with hoofs. Examples are the horse, elephant and rhinoceros.

Upwelling Rising of cold water from the depths of the sea to the surface.

V to Z

Vibrissa A long tactile bristle of certain mammals. An example is the whiskers of a cat.

Viperidae Class of venomous snakes like vipers and rattlesnakes.

Viviparous Describes an animal whose egg develops in the uterus and the young are born without a covering, contrary to oviparous, or egg-laying animals.

Waterspout Tornado that forms over water.

Zooplankton A type of plankton consisting of animal organisms.

Index

Index

PHOTOGRAPHIC CREDITS

Abbreviations : l = left, r = right, m = middle, t = top, b = bottom, c = centre.

Cover : background : G. ZIESLER ; mlt : JACANA/Müller ; ml : JACANA/NITS ; bl : STOCK IMAGE ; r : NATURAL EXPOSURES. INC./D. J. Cox.

5t : MINDEN PICTURES/F. Lanting ; 5mt : O. GRÜNEWALD ; 5mb : BIOS/D. Halleux ; 5b : COSMOS/EYE OF SCIENCE/GELDERBLON ; 6t : COSMOS/NPHA/Hellio & Van Ingen ; 6mt : SUNSET/FLPA ; 6mb : O. GRÜNEWALD ; 6b : BIOS/Ph. Henri ; 7t : COSMOS/WILDLIGHT/P. Jarver ; 7mt : O. GRÜNEWALD ; 7m : BIOS/Cavagnaro ; 7mb : JACANA/A. Larivière ; 7b : MINDEN PICTURES/F. Lanting ; 8t : BIOS/G. Lopez ; 8mt : BIOS/O.S.F/P. De Oliveira ; 8m : BIOS/O.S.F/S. Winer ; 8mb : BIOS/H. Hall ; 8b : BIOS/R. Amann ; 9 : MINDEN PICTURES/F. Lanting ; 10l : COSMOS/S.P.L ; 10br : COSMOS/S.P.L/T. & D. Hallas ; 11tl : COSMOS/S.P.L/Royal Observatory Edinburgh ; 11tr : COSMOS/S.P.L/European Southern Observatory ; 12t : COSMOS/S.P.L ; 12b : E. BALL ; 13t : HOA QUI/IMAGES ET VOLCANS/Krafft ; 13bl : Ph. BOURSEILLER ; 13b : HOA QUI/Ph. Bourseiller ; 14t : MAP/A. Guerrier ; 14b : BIOS/F. Bruemmer ; 15t : BIOS/O.S.F/S. Kuribayashi ; 15ml : BIOS/J. J. Alcalay ; 15mtr : JACANA/H. Cong Hoang ; 15mrc : BIOS/W.W.F/M. Rautkari ; 15mb, 16tl : BIOS/O.S.F/S. Kuribayashi ; 17tr : SUNSET/HORIZON VISION ; 17bl : SUNSET/D. Perrine ; 18 (background) : CIEL ET ESPACE/NASA/STBCI ; 19t : COSMOS/S.P.L/D. A. Hardy ; 19b : COSMOS/S.P.L/NASA ; 20mr : BIOS/G. Martin ; 20bl : BIOS/J. C. Munoz ; 20/21tr : PHONE/R. Valter ; 21tr, br : BIOS/M. Gunther ; 21bl : BIOS/Y. Tavernier ; 22t : BIOS/T. Thomas ; 22b : BIOS/O.S.F/R. Kuiter ; 23t : BIOS/M. & C. Denis-Huot ; 23mt : BIOS/M. Harvey ; 23ml : SUNSET/R. Maier ; 23br : PHONE/AUSCAPE/J. M. La Roque ; 24 : BIOS/STILL PICTURES/SHORT/UNEP ; 25tr : PHONE/R. Valter ; 25tl : BIOS/FOTONATURA/M. Harvey ; 25br : BIOS/OKAPIA/R. Philips ; 25bl : BIOS/C. Guihard ; 26tr : BIOS/D. Heuclin ; 26m : BIOS/P. Rismiller ; 26b : BIOS/J. J. Alcalay ; 27 : O. GRÜNEWALD ; 28tl : BIOS/D. Delfino ; 28mb : PHONE/R. Valter ; 28b : BIOS/Klein-Hubert ; 28/29t : PHONE/P. Goetcheluck ; 29mtr : BIOS/J. P. Delobelle ; 29mbr : BIOS/J. M. Prévot ; 30mr : BIOS/B. Marielle ; 30mc : BIOS/D. Heuclin ; 30/31 : BIOS/J. M. Prévot ; 31tr : BIOS/F. Gilson ; 31mr : BIOS/J. Mayet ; 32tl : JACANA/F. Danrigal ; 32bl : JACANA/P. Pilloud ; 32tr : BIOS/D. Bringard ; 32trm : PHONE/P. Goetcheluck ; 32br : JACANA/Y. Gladu ; 33 : J. M. DAUTRIA ; 34tl : BIOS/L. Bengt ; 34mr : SUNSET/G. Lacz ; 34/35tr : HOA QUI/P. Reimbold ; 35bl : BIOS/G. Lopez ; 35tr : JACANA/Ch. Favardin ; 35br : JACANA/H. Dalton ; 36b : BIOS/Klein-Hubert ; 36/37t : BIOS/M. Denis-Huot ; 37tr : BIOS/J. L. Moigne ; 37b : BIOS/G. Lacoumette ; 38tl : RAPHO/J. Pickerell ; 38b : BIOS/M. Rapilliard ; 39 : EURELIOS/L. Bret ; 40tl : COSMOS/S.P.L/SPACE TELESCOPE SCIENCE INSTITUTE ; 40bl : COSMOS/S.P.L/NASA/SPACE TELESCOPE SCIENCE INSTITUTE ; 41t : COSMOS/S.P.L/Hester & Scowen ; 42 : JACANA/W. Wisnieski ; 43tl : MAGNUM/E. Arnold ; 43tr : COSMOS/S.P.L/K. Lounathaa ; 43br : SUNSET/F. Stock ; 44tl : JACANA/T. Walker ; 44tr : C.N.R.I/PHOTOTAKE/Dr D. Kunkel ; 45 : BIOS/D. Halleux ; 46tl : PHONE/Ch. Courteau ; 46bl : BIOS/J. L. Rothan-P. Arnol ; 46mr : PHONE/C. Carre ; 47 : BIOS/FOTONATURA/V. Visniewski ; 48t : BIOS/J. M. Prévot ; 48ml : BIOS/A. Compost ; 48brt : BIOS/P. Pernot ; 48br : BIOS/Klein-Hubert ; 49tr : C. RIVES ; 49b : BIOS/G. Dif ; 50tl : BIOS/O.S.F/R. Packwood ; 50br : BIOS/G. Martin ; 51tl : COSMOS/S.P.L/A. Pasieka ; 51ml : C. RIVES ; 51r : JACANA/S. de Wilde ; 52b : BIOS/O.A.S/Y. Tomokazu ; 53tl : BIOS/G. Dif ; 53tr : HOA QUI/P. Reimbold ; 53m : BIOS/Ch. Decout ; 53br : BIOS/PANDA PHOTO/J. Watt ; 54mt : JACANA/PHOTO RESEARCHERS ; 54tr : SUNSET/A.N.T ; 54tr : JACANA/E. Sipp ; 54b : SUNSET ; 55bl : BIOS/C. Ruoso ; 55r : BIOS/M. & C. Denis-Huot ; 56b : BIOS/M. Roggo ; 56tr : BIOS/SCIENCES PICTURES ; 56/57mrb : BIOS/M. Roggo ; 57tr : BIOS/P. Arnold - S. Kaufman ; 57mr : BIOS/C. Guihard ; 57br : BIOS/G. Blondeau ; 57/58t : BIOS/M. Roggo ; 58tl : BIOS/D. Watts ; 58tr : BIOS/O.S.F/D. Cox ; 58b : BIOS/R. Seitre ; 59t : BIOS/D. Watts ; 59tl : BIOS/R. Seitre ; 59mr, bl : BIOS/Klein-Hubert ; 60l : BIOS/G. Martin ; 60bl : PHONE/R. Valter ; 60br : PHONE/J. Collection ; 61tl : JACANA/P. Lorne ; 61tr : BIOS/O.S.F/R. Kuiter ; 61bl : BIOS/M. Harvey ; 61br : BIOS/D. Watts ; 62t : PHONE/J. M. Labat ; 62b : PHONE/AUSCAPE/G. Robertson ; 63 : COSMOS/EYE OF SCIENCE/GELDERBLON ; 64 (background) : CIEL ET ESPACE/AAO/D. Malin ; 64br : COSMOS/S.P.L/CFH/J. C. Cuilandre ; 65tr : COSMOS/S.P.L/NASA ; 65bl : COSMOS/S.P.L/NASA/SPACE TELESCOPE SCIENCE INSTITUTE ; 65bm : CIEL ET ESPACE/AAO/D. Malin ; 65br : COSMOS ; 67 : COSMOS/S.P.L/NASA ; 68tl : COSMOS/S.P.L/NASA ; 68tm : COSMOS/S.P.L/NASA ; 68mr : CIEL ET ESPACE/USGS ; 68br : COSMOS/S.P.L/NASA ; 69t : NASA/JPL/Collection Photothèque planétaire d'Orsay ; 69b : COSMOS/S.P.L/D. Haroy ; 70t : CIEL ET ESPACE/W. K. Hartmann ; 70b : COSMOS/WOODFIN CAMP/J. Blair ; 70/71m : CIEL ET ESPACE/FOTOSMITH/R. Haag Coll. ; 71tr : COSMOS/S.P.L/NASA ; 71mr : I.S.Williams/The Australian National University, Canberra ; 72mr : CIEL ET ESPACE/NASA/ESA ; 72bl : COSMOS/S.P.L/CERN ; 72/73, 73br : COSMOS/S.P.L/ARSCIMED ; 74/75 : COSMOS/S.P.L/CFH Télescope/J. C. Cuillandre ; 74b : COSMOS/S.P.L/ARSCIMED ; 74b : EURELIOS/Ph. Plailly ; 74b : C.N.R.I ; 75t : COSMOS/NASA/SPACE TELESCOPE SCIENCE INSTITUTE ; 75mt : NASA ; 75t : METEO FRANCE ; 75b : COSMOS/NASA ; 75b : C.N.R.I/Phototake ; 75 : A.K.G PARIS/CAMERAPHOTO ; 76ml : COSMOS/S.P.L/Dr Gopal Murti ; 76br : C.N.R.I/Prof. S. Cinti, University of Ancona ; 76/77 : COSMOS/S.P.L/A. B. Dowsett ; 77 : COSMOS/S.P.L/D. Scharf ; 77 : COSMOS/S.P.L/Dr J. Burgess ; 78tl, tm : C.N.R.I ; 78tr : C.N.R.I/R. M. Tektoff ; 78b : BIOS/F. Bavendam ; 79tl, tm, tr : C.N.R.I ; 79ml : COSMOS/S.P.L/A. & H. Frieder Michler ; 79br : COSMOS/S.P.L/M. Kage ; 79 (background) : PHOTONONSTOP/B. Morandi ; 80 : BIOS/G. Martin ; 81tl : BIOS/M. Gilles ; 81tr : BIOS/J. M. Prévot ; 81bl : BIOS/J. P. Theron ; 82tl : PHONE/C. G. Phone ; 82tr : PHONE/C. C ; 82b : BIOS/O.S.F/Science Gallery ; 83 : COSMOS/NPHA/Hellio & Van Ingen ; 84tl : JACANA/E. A. Soder ; 84bl : BIOS/C. Ruoso ; 84/85bl : PHOTOCEANS. COM/A. Rosenfeld ; 85tr : PHOTOCEANS. COM/R. Rinaldi ; 85mrb : PHOTOCEANS. COM/T. Sarano ; 85ml : BIOS/Y. Lefevre ; 85br : SUNSET/NHPA ; 85/86tl : PHOTOCEANS. COM/A. Nachoum ; 86ml : SUNSET/J. Warden ; 86tl : SUNSET/Brake ; 86tr : SUNSET/T. Leeson ; 86b : BIOS/O.S.F/P. Parks ; 86/87m : BIOS/WILDLIFE PICTURES/B. Odeur ; 87br : BIOS/O.S.F/M. Fogden ; 88tl : BIOS/FOTONATURA/M. Harvey ; 88ml : HOA QUI/ZEFA ; 88bl : BIOS/J. L. Le Moigne ; 88tr : BIOS/FOTONATURA/Montford ; 89tl : BIOS/O.S.F/H. Pooley ; 89ml : BIOS/FOTONATURA/D. Nill ; 89tr : BIOS/M. Harvey ; 90 : BIOS/G. Martin ; 91tr : BIOS/P. Arnold - T. Mangelsen ; 91mr : COSMOS/INTERNATIONAL ST./T. Murphy ; 91br : BIOS/M. &. C. Denis-Huot ; 92tl : JACANA/A. Ducrot ; 92mr : PHONE/J.F. Hellio - N. Van Ingen ; 92b : BIOS/J. Ch. Vincent ; 93t : BIOS/E. Bruemmer ; 93mrt : SUNSET/NPHA ; 93mr : PHONE/F. Gohier ; 93bl : BIOS/T. Moreau ; 94t : BIOS/J. Ph. Delobelle ; 94b : BIOS/R. de la Harpe ; 95t : BIOS/F. Bavendam ; 95ml : BIOS/M. Duquet ; 95mr : PHONE/C. C ; 95br : SUNSET/ANIMALS ANIMALS ; 96t : BIOS/D. Bringard ; 96mr : BIOS/D. Halleux ; 96br : BIOS/Klein-Hubert ; 97tr : BIOS/R. Seitre ; 97mr : BIOS/N. Gasco ; 97/98 (background) : BIOS/G. Lopez ; 98tl : BIOS/G. Lacoumette ; 98tr : BIOS/J.M. Lenoir ; 98b : BIOS/M. Duquet ; 99tr : SUNSET/HOLT STUDIOS ; 99mr : JACANA/R. Konig ; 99ml : BIOS/D. Heuclin ; 99br : RAPHO/E. Philippotin ; 100tl : BIOS/B. Fischer ; 100mr : BIOS/O.S.F/M. Breiter ; 100/101b : BIOS/P. Arnold - D. Blell ; 101t : BIOS/F. Bruemmer ; 101mr : JACANA ; 102tl : O. GRÜNEWALD ; 102/103t : BIOS/O.S.F/H. Hall ; 103t : SUNSET/ANIMALS ANIMALS ; 103mr : PHONE/J. F. Ferrero ; 103bl, b : JACANA/G. Ziesler ; 104t : SUNSET/ALASKA STOCK ; 104b : BIOS/R. Seitre ; 104/105 : COSMOS/J. W. Warden ; 105tr : BIOS/P. Kobeh ; 105ml, 106t, mr : BIOS/P. Arnold - J. L. Amos ; 106bl : BIOS/J. L. & F. Ziegler ; 107 : SUNSET/FLPA ; 108tl : BIOS/P. Kobeh ; 108bl : BIOS/WILDLIFE PICTURES/Henno ; 108mr : BIOS/F. Gilson ; 108br : JACANA/J. P. Saussez ; 109/110 (background) : JACANA/Grinberg ; 109tl, ml : BIOS/H. Ausloos ; 109mr : BIOS/J. L. Rotman ; 109b : BIOS/O. Gautier ; 110tl : BIOS/F. Bavendam ; 110tr : BIOS/Th. Thomas ; 110b : BIOS/PANDA PHOTO/Allan ; 111tr : COSMOS/NHPA/S. Dalton ; 111mtr : BIOS/T. Crocetta ; 111tl : BIOS/G. Martin ; 111btl : BIOS/D. Heuclin ; 111bl : BIOS/B. Pambour ; 112tl : O. GRÜNEWALD ; 112m : BIOS/P. Arnold - J. Kieffer ; 112/113bl : BIOS/N. Dennis ; 113tl : BIOS/J. Alcalay ; 113tc : O. GRÜNEWALD ; 113tr : BIOS/Klein-Hubert ; 113ml, br : BIOS/M. Gunther ; 114tr : BIOS/D. Heuclin ; 114ml : BIOS/Klein-Hubert ; 114b : JACANA/R. Tercafs ; 115t : BIOS/Ch. Meyer ; 115b : BIOS/T. Lafranchis ; 116l : BIOS/M. Nicolotti ; 116m : BIOS/H. Van den Berg ; 116bl : BIOS/E. Boyard ; 116br : BIOS/R. Puillandre ; 117tl : BIOS/M. & C. Denis-Huot ; 117tr : BIOS/O.S.F/D. Cox ; 117tr : BIOS/D. Delfino ; 117bm : SUNSET/G. Lacz ; 117br : SUNSET/R. Maier ; 118tl, tr, brm : BIOS/Klein-Hubert ; 118bl : BIOS/J. L. Zimermann ; 118/119t : BIOS/P. Arnold - T. Mangelsen ; 119tl : BIOS/Klein-Hubert ; 119tr : BIOS/WILDLIFE PICTURES ; 119b : BIOS/H. Ausloos ; 120tl : JACANA/G.I. Bernard ; 120mbr : WILD BIRD FEDERATION TAIWAN/Gr. Chang Show-hua ; 120/121 (background) : BIOS/STILL PICTURES/T. Bangum ; 121tl : BIOS/P. Arnold - T. Schiffman ; 121mbl : BIOS/R. Seitre ; 121tr : JACANA/G. I. Bernard ; 122tl : BIOS/H. Ausloos ; 122tr : JACANA/VISAGE/Varin ; 122bl : BIOS/STILL PICTURES/W.W.W/T. Bangum ; 123tr : JACANA/J. Stevens ; 123mr : BIOS/R. Seitre ; 123bl : JACANA/P. Wild ; 124tl : BIOS/M. Lane ; 124tr : BIOS/J. Roche ; 124bl : BIOS/G. Bortolato ; 124br : BIOS/T. Stoeckle ; 125t : BIOS/Th. Montford ; 125ml : BIOS/G. Lopez ; 125b : BIOS/M. Roggo ; 126l : PHONE/A. Le Toquin ; 126tr : BIOS/E. Barbelette ; 126mr : BIOS/M. Gunther ; 127tl, tr : PHONE/A. Le Toquin ; 127bl : BIOS/P. Guinchard ; 127br : JACANA/Rouxaime ; 128tl : SUNSET/NHPA ; 128tr : BIOS/WILDLIFE PICTURES/Henno ; 128rtb : BIOS/Th. Montford ; 128br : SUNSET/ANIMALS ANIMALS ; 129 : O. GRÜNEWALD ; 130br : HOA QUI/E. Valentin ; 130/131b : COSMOS/S.P.L/T. Craddock ; 131t : BIOS/B. Marcon ; 131mr : PHOTONONSTOP/C. Moirenc ; 131br : HOA QUI/San Bughet ; 132bl : COSMOS/S.P.L/R. de Gugliemo ; 132br : N. Bariand ; 133tl : L. D. BAYLE ; 133tm : COSMOS/S.P.L/R. de Gugliemo ; 133tmr, tr, mb, bl : L. D. BAYLE ; 134tl : BIOS/J. L. Lemoigne ; 134tr : JACANA/J. P. Thomas ; 134b : JACANA/R. Konig ; 135tl : JACANA/D. Legay ; 135tr : JACANA/G. Hofer ; 135bl : HEATHER ANGEL ; 135br : BIOS/G.P.L/S. Harte ; 136 : BIOS/Klein-Hubert ; 137t : O. GRÜNEWALD ; 137bl : BIOS/J. Ph. Delobelle ; 137bm : BIOS/OKAPIA/G. Zimmert ; 137br : HOA QUI/N. Cuvelier ; 138 (background) : COSMOS/S.P.L/D. Parker ; 138bl : COSMOS/TV VIDEO ; 139tr : COSMOS/WOODFIN CAMP ; 139b : COSMOS/S.P.L/J. Mead ; 139/140 : G. DAGLI ORTI ; 140t : O. GRÜNEWALD ; 140tr : SUNSET/ANIMALS ANIMALS ; 140bl : HOA QUI/C. Valentin ; 141ml : PHOTONONSTOP/J. D. Sudres ; 141b : O. GRÜNEWALD ; 142t : HOA QUI/Th. Perrin ; 143tl : PHOTONONSTOP/A. Le Bot ; 143tm : IMAGES/Carlier ; 143tr : IMAGES/J. Nogrady ; 143bl : BIOS/Klein-Hubert ; 143btr : BIOS/D. Bringard ; 143br : RAPHO/R. & S. Michaud ; 144bl : COSMOS/H. T. Kaiser ; 144tr : O. GRÜNEWALD ; 144br : MEDIALP/Fournier ; 145t : BIOS/M. Acquarone ; 145 : J. M. DAUTRIA ; 146tl : JACANA/I. Arndt ; 146mtl : JACANA/W. Layer ; 146btl : BIOS/D. Heuclin ; 146tr : BIOS/J. Dani ; 146b : BIOS/H. Pfletschinger ; 147 : BIOS/Ph. Henri ; 148g : BIOS/C. Ruoso ; 148br : SUNSET/FLPA ; 149tr : BIOS/Th. Thomas ; 149mlt : HOA QUI/J. Hagenmüller ; 149mlb : HOA QUI/D. Repérant ; 150tl : COSMOS/S.P.L/NASA ; 150bl : G. LEVEQUE ; 151t : CORBIS-SYGMA/P. P. Poulain ; 151b : O. GRÜNEWALD ; 152h, mr : COSMOS/S.P.L/NASA ; 153tl : VANDYSTADT/S. Cazenave ; 153r : PÊCHEUR D'IMAGES/Ph. Plisson ; 154tl : HOA QUI/LE MONDE ; 154/155b : ALTITUDE/J. Wark ; 155tl : ALTITUDE/Y. Arthus-Bertrand ; 155tm : PHOTONONSTOP/D. Ball ; 155tr : HOA QUI/C. Sappa ; 156t : ALTITUDE/Y. Arthus-Bertrand ; 156bl : O. GRÜNEWALD ; 156/157b, tr : BIOS/M. Rapilliard ; 157mr : BIOS/M. Gunther ; 157br : BIOS/P. Bertrand ; 158l : HOA QUI/IMAGES & VOLCANS/Krafft ; 158r : PHOTONONSTOP/C. Moirenc ; 159tr : O. GRÜNEWALD ; 159tl : RAPHO/E. Kashi ; 159bl, 160t, 160/161b : O. GRÜNEWALD ; 161t : BIOS/J. L. Zimmermann ; 162t : ALTITUDE/Y. Arthus-Bertrand ; 162/163 : CORBIS-SYGMA/D. Conger ; 163tr : CORBIS-SYGMA/T. Bean ; 163mr : PHOTONONSTOP/A. Even ; 163br : ALTITUDE/M. Gootschalk ; 164tl : MEDIALP/Moiroux ; 164tr : RAPHO/G. Gerster ; 164br : HOA QUI/E. Beracassat ; 165 : COSMOS/WILDLIGHT/P. Jarver ; 166, 167t : COSMOS/S.P.L/NASA ; 167mr : CIEL ET ESPACE/OMP/J. L. Leroy ; 167bl : COSMOS/S.P.L/NOAO ; 168mr : COSMOS/S.P.L/NASA ; 168b : COSMOS/S.P.L/R. Royer ; 169 : BIOS/P. Fagot ; 170b : COSMOS/S.P.L/NASA ; 170/171t : COSMOS/S.P.L/F. Espenak ; 171r : COSMOS/S.P.L/J. Sanford ; 172 (background) : METEO France ; 174 : COSMOS/S.P.L/S. Fraser ; 175t : BIOS/A. Compost ; 175tr : BIOS/L. Thouzeau ; 175br : CORBIS-SYGMA/BETTMANN ; 176t : COSMOS/S.P.L/P. Parviainen ; 176tr : ALTITUDE/Y. Arthus-Bertrand ; 176br : BIOS/R. Seitre ; 176/177 : JACANA/F. Zuardon ; 177tl : METEO FRANCE ; 177mt : COSMOS/S.P.L/G. Post ; 177tr : BIOS/Klein-Hubert ; 177br : BIOS/F. Suchel ; 178 (background) : COSMOS/S.P.L/P. Parviainen ; 178tr : COSMOS/S.P.L/J. Ourst ; 179tr : COSMOS/S.P.L/S. Fraser ; 179b : COSMOS/S.P.L/Dr F. Espenak ; 180tl : B. PELLEQUER ; 180tr : COSMOS/S.P.L/T. Craddock ; 180b : COSMOS/S.P.L/P. Parviainen ; 181 : O. GRÜNEWALD ; 182 : HOA QUI/IMAGES & VOLCANS/Krafft ; 183tr : GAMMA ; 183ml : J. C. BOUSQUET ; 184tl : ALTITUDE/M. Gottschalk ; 184br : ALTITUDE/G. Rossi ; 184/185 : ALTITUDE/M. Sekido ; 185t : J. M. DAUTRIA ; 185b :

The *Amazing Secrets* of Nature
is published by The Reader's Digest Association

First edition
First printing